Scientific Laws, Principles, and Theories

Scientific Laws, Principles, and Theories

A Reference Guide

ROBERT E. KREBS

ILLUSTRATIONS BY RAE DÉJUR

Greenwood Press
Westport, Connecticut • London

Library of Congress Cataloging-in-Publication Data

Krebs, Robert E., 1922–
    Scientific laws, principles, and theories : a reference guide / Robert E. Krebs.
        p. cm.
    Includes bibliographical references and index.
    ISBN 0–313–30957–4 (alk. paper)
    1. Science. 2. Science—History. I. Title.
    Q158.5.K74 2001
    509—dc21   00–023297

British Library Cataloguing in Publication Data is available.

Library of Congress Catalog Card Number: 00–023297
ISBN: 0–313–30957–4

First published in 2001

Greenwood Press, 88 Post Road West, Westport, CT 06881
An imprint of Greenwood Publishing Group, Inc.
www.greenwood.com

Printed in the United States of America

The paper used in this book complies with the
Permanent Paper Standard issued by the National
Information Standards Organization (Z39.48–1984).

10 9 8 7 6 5 4 3 2 1

To Carolyn

My constructive critic, pre-editor, proofreader, supportive wife, and friend.

Contents

Introduction

The development of universal scientific laws, physical principles, viable theories, and testable hypotheses has a long history. Humans are unique in that they can think about and contemplate the world around them, conceive ideas to explain natural events and processes, and then make use of what was learned. Early explanations of natural phenomena were interesting but not very reliable. Not until a few thousand years before the birth of Christ would writing and recording of history provide evidence of how humans related to the events and phenomena of nature. Some of the early Egyptian, Greek, Islamic, and Asian theories, as well as those from other cultures, demonstrated great insight into the structure and functions of animals, plants, matter, meteorology, agriculture, earth, and astronomy. Our ancestors had the curiosity but lacked the means for truly accurate explanations and conclusions about nature.

Included in this volume are a few ancient classical ideas and concepts that were descriptions of nature, and often very inaccurate. This ancient classical philosophical process resulted in many dead ends. Modern physics began when people learned how to explain nature by objectively observing events, asking questions that could be answered by making reliable measurements, using mathematics, and then considering probabilities to make reasonable predictions. This process led to operational facts, which continued to be upgraded and corrected by the self-correcting nature of modern science. However, testing theories and hypotheses in a controlled situation developed late in human history, and thus science as we think of it was slow to advance in ancient times. The development and implementation of scientific processes increased the growth of knowledge, as well as the rate of growth of science from the seventeenth to the nineteenth and twentieth centuries. During this period, science accelerated at an astounding exponential pace, and this growth will most likely continue throughout the twenty first century and beyond, particularly in the biological sciences based on quantum theory. None of our current understanding of the universe, nature, and ourselves would have been possible without men and women using the processes and procedures of science.

This book presents a historical aspect for the important laws, principles, theories, hypotheses, and concepts that reflect this amazing progression of scientific descriptions and explanations of nature. These did not just appear out of thin air. They are related to the period and people who developed these explanations of our universe. The laws, theories, hypotheses, and concepts are listed alphabetically—in most cases, according to the name of the person credited with formulating the theory or concept. Some of these, as well as the names of the scientists who conceived of them, are familiar. Others are less well known. Inventions and discoveries are included only if they contributed to the development or understanding of a particular scientific law. Only laws, principles, and theories related to the physical and biological scientific fields are included.

This book is designed for high school and college students as well as for general readers with an interest in science. A glossary is included, and terms contained in it are highlighted in bold in the text of the book. The Bibliography contains sources for additional information. The index will assist the reader in locating entries associated with particular persons and subjects.

NOTE: Many of the original statements of laws, principles, theories, and so on are highly technical. Also, some of the original statements are lost, are in a foreign language, or include technical jargon. They therefore have been paraphrased and restated to make them more comprehensible. The paraphrased statements are printed in italics for easy identification.

DEFINITIONS

Scientific Physical Laws

Scientific laws are generalized descriptions of how things behave in nature under a variety of circumstances. We live in an ordered universe, made comprehensible by applying the concept of physical laws and using rational and intelligent powers of reasoning. Scientific physical laws describe how things work in the universe. In other words, any phenomenon, event, or action that occurs and behaves in the same manner under the same conditions, and is thus predictable, can be stated as a scientific law. These are sometimes referred to as *fundamental laws, universal laws, basic laws*, or just *scientific* or *physical* laws. At least five characteristics apply to all scientific laws: (1) They can be expressed mathematically. Mathematics is the rule of the scientific enterprise. (2) They are not always exact. They may need future adjustments as more knowledge is gained concerning the natural phenomenon as expressed by the law. (3) The natural system may be complex and contain many pieces, but the law describing the phenomenon is always simple. (4) Most important, scientific laws are universal. (5) By using statistical probabilities, they can be used to predict future physical events.

There are only a few general, fundamental, and overriding postulates for all physical laws and from which all scientific laws are derived. One is the unifying concept of *symmetry*, which is found in physical laws, principles, and theories. The scientific definition of symmetry relates to the position of something regardless of its orientation in space and time. It will always act and react in the same way. If it obeys symmetry, it will behave the same later as well as sooner, no matter when or where it is located in space and time. Symmetry also enables an object to rotate on a fixed axis in any orientation regardless of where it is located in the universe, and it will move at a uniform velocity in a straight line regardless of orientation. Symmetry in our everyday lives is considered two-fold—bilateral (lengthwise) and radial (crosswise). Anything that can be reoriented or changed, keeping its same basic geometry, will be symmetrical. An example is a two-dimensional square. Despite its orientation, it will look the same (unless you view it from the edge of a piece of paper). Some three-dimensional figures and objects when examined can exhibit two kinds of symmetry—both bilateral and radial. For example, a drinking glass can be cut lengthwise (bilateral) or crosswise (radial). However, there is a concept held by cosmologists that perfect symmetry must be broken in order for the universe to have been created. Perfect symmetry is beautiful, but it is like a vacuum—formless. A perfectly symmetrical universe was formless until some event, such as a single particle, was introduced into this perfect vacuum, thus breaking the perfect symmetry, causing the big bang and resulting in the formation of all energy and matter in the universe.

Another fundamental or basic concept for all physical laws is *conservation*. The laws of conservation of matter and energy are related to symmetry because everything is in balance—somewhat like the reflection in a mirror or a process that obtains equilibrium. The same is true for quantum mechanics, which connects the minimum principle with the laws of conservation. There are antiparticles for all particles, and positives ($+$) for negatives ($-$). In other words, they are symmetrical but opposites and are conserved (no loss or gain). The fundamental property of mass is inertia. *Inertia* is the property of an object (mass) that offers resistance to any change in its velocity when a force is applied to the mass. For our earth and solar system, this explanation is adequate, but it is inadequate for the relativity of space and time. Einstein's theories of relativity redefined mass as no longer absolute when considering the vastness of space, as expressed in his famous equation $E = mc^2$, which equates mass and energy as being conserved. (See Einstein for more on this principle.) Other examples of conservation principles are Newton's three laws of motion and the theory of kinetic energy of particles.

Another example of a generalized principle involves *mathematical rules*. In other words, the answer reached will always be the same. Examples are the square of distance related to force and the inverse relationships between events, as in the mathematical explanation of gravity, which states there is a direct relationship between the masses of two objects but an indirect relationship as to the square of the distance between them. This type of mathematical phenomenon is common in generalizing about nature and often leads to the formation of physical constants (e.g., c for the speed of light) derived to maintain complete symmetry and conservation for mathematical calculations of physical events.

Scientific physical laws deal with events that are measurable and thus predictable, at least at the high end of the scale of probability. This means that physical laws are not always exact, but they describe events close enough to be considered consistently predictable, valid, and reliable. Descriptive generalizations, which may not be universal, are usually invented to allow for a more exact description of reality. Scientific laws can be further confirmed or not confirmed by the discovery of new and additional empirical facts. Scientific physical laws do not address human nature and behavior, religion, the social sciences, e.g., history, or the paranormal, which are not measurable in a mathematical sense, predictable, or universal.

Philosophers and scientists categorize scientific laws as either *empirical* or *theoretical*. Empirical laws are based on empirical (viewed) observations of phenomena that can be directly measured and thus are referred to as observables. Theoretical laws, often referred to as abstract laws, deal with unobservable submicroprocesses. Even though these microevents and microprocesses may be too small for our current instruments to measure, they may still be included in physical laws. Little should be made of these distinctions since the two classes of laws do meld to form universal laws of science (see Table 1).

Several systems of measurement of both observable and nonobservable events have been developed. A number of international units of measurement were developed for the Systéme international d'unités (SI) now used by all scientists. This system, and the international metric system, differ from the English and other systems of measurement in that the SI and metric units are easily converted between and among each other since each is based on the decimal system.

An example of a universal law that developed as both an empirical law and a theoretical law is the law of gravity, which describes the force of attraction between two bodies with mass. But the law does not explain what gravity is. It is described as a universal force because it exerts a similar attraction on and between all objects that have mass anywhere in the universe. This force of attraction depends on the relative mass of each object, while the force is inversely proportional to the square of the distances separating these masses. Newton's law of gravity is one of the greatest generalities of all science based on both theory and observable facts.

Several scientific laws, in the narrow Newtonian sense, are deterministic. Causality determines the future state or existence of any system from what we know about the system. In other words, scientific laws are an expression of statistical laws. After several hundred years, it was discovered that Newton's laws did not explain some phenomena of nature. Therefore, Einstein, Planck, Heisenberg, Bohr, Schrödinger, and others developed new laws based on new data to explain

Universal Physical Laws as Conditional statements

(requires necessary conditions)

PHILOSOPHY ⇒——————⇒-----------------------------⇐——⇐SCIENCE

Empirical laws ⇒	Scientific physical laws	⇐Theoretical laws
Directly observable (with eyes)		Unobservable (with eyes)
Direct measurements		Indirect measurements
Uses mathematics		Uses statistical methods
Describes macroevents		Describes microevents
Explains macroprocesses		Explains microprocesses
Generalizations		Abstractions
Uses deductive logic		Uses inductive logic
Develops empirical laws		Adds to empirical laws
Predicts unknown events/facts		Verifies predictions

Table 1: The formation of new scientific physical laws is the result of combining empirical and theoretical laws.

things in a different way. The laws of relativity and quantum theory and mechanics do not really change how we see things on earth, but they do explain much in the micro and macro of the four dimensions of space and time.

To summarize, there are only a few very fundamental physical laws of nature. Most scientific laws can be related to and derived from characteristics such as symmetry (or at least partial symmetry), transition in time (zero time or a delay makes no difference), conservation of energy/matter (there is no total loss or gain), entropy (disorganization), relativity (space/time frame of reference), and quantum theory (unitary Standard Model). Some scientists conclude that the two most important foundations for understanding the nature of the universe are quantum physics and the general theory of relativity, both of which have been established experimentally and will drive scientific research and discoveries of the twenty-first century.

The most important characteristic of a basic scientific law is its statistical predictability of applicability for any place or time in the universe. Physical laws have one other thing in common: they are not exact. Although they describe very complex phenomena, they are very simple statements expressed as universal agreements.

Scientific Constants

Scientific constants, also referred to as *physical constants*, are included in scientific laws in the sense they relate two or more variables and never change their values. They are known also as *fundamental constants* or *universal constants*; as either parameters or fixed values, they never change regardless of where they are used in the universe. The speed of light in unobstructed space (vacuum) is an example of a universal scientific constant.

Scientific Principles

A scientific principle, also referred to as a *physical principle, generalized principle*, or *fundamental principle*, is similar to a scientific law. It is the statement (of principle) that the same physical law(s) and constant(s) of nature will be applicable. An example is Einstein's principle of relativity, also referred to as a law or theory, which is based on non-Euclidean geometry and non-Newtonian physics that states the laws of physics are the same in all (universal) frames of reference (*see* Einstein). Another example is the uncertainty principle related to quantum mechanics used to explain the nature of subatomic particles and energy (*see* Heisenberg). An important distinction is that relativity is explained by abstract theoretical mathematics, while quantum mechanics and indeterminacy are based on statistical probabilities. Both are statements of scientific principles based on physical laws.

Scientific Theories

Scientific theories are a type of model designed as general explanatory statements about the workings of nature. As established explanations of how things work in nature, they are the end point of scientifically gathered evidence about specific events that may incorporate other laws and hypotheses. All theories are derived by humans and as such are linguistic constructions of assumptions made by scientists. They are similar to, but much more than, educated guesses; they result from critical observations, experimentation, logical inferences and creative thinking. They are not the same as what we think of as guesses in everyday life. They are neither undocumented statements nor uninformed opinions. People often come up with "theories" based on notions or assumptions that are accepted by faith. These "nonscientific" theories are without experimental proof and are seldom based on verifiable facts, or measurable evidence. The main test of validity (truth) of an idea, concept, or theory is found in the results of the experiment and, when possible, a controlled experiment. Theories advanced by scientists can be described as predictions based on the scientist's knowledge of a probable occurrence within a given set of circumstances or conditions. The nature of science is that *exactly* the same thing does *not* always occur at the same time in exactly the same way in the universe. Therefore, scientists, through experimentation, seek a statistical average on which to make their predictions.

There are two general classification of theories. One covers a large range of ideas or concepts, often referred to as *breadth* or *broad* theory. Historically, broad theories such as myths were used by ancient people in attempting to understand their world by using stories and folktales to explain observed phenomena. Scientists are still attempting to reduce all scientific theories, principles and laws into a unified field theory (UFT), grand unification theory (GUT), theory of everything (TOE), or come up with the "final answer." However, theories based on the laws of gravity, time, conservation, symmetry, and so forth are not yet explained in ways that can be incorporated into a general, universal theory of everything. Nature is extremely complex, and humans are just beginning to understand it. (See Figure 1 and Figure 2.)

An important characteristic of a theory is that it must be stated in such a way that it could lead to a negative assessment of the experimental results. It cannot be ambiguous. A vague theory cannot be proved wrong, and it is possible to come up with almost any answer desired for an ambiguously stated theory. Confirming a theory requires specified conditions, an experiment, and measured results that are analyzed statistically. If the related facts indicate a high probability for its validity and reliability, the theory can then be said to be justified and acceptable. Even so, there must exist the possibility that the theory may be wrong.

A theory is only as good as the limited number of assumptions and generalizations postulated. The fewer astute assumptions incorporated into a theory, the more likely it will stand the test of time. This is known as Ockham's razor (William of Ockham, c. 1284–1349) which is a maxim that states, "Entities ought not to be multiplied, except from necessity." In other words, the number of unnecessary assumptions should be avoided in formulating theories and hypotheses. In modern vernacular, KISS (keep it simple, stupid).

In summary, a theory must exhibit the following conditions: (1) It must explain the law from which it was derived and be deduced from that law, (2) it must in some way be related to the law it challenges, and (3) it must be able to predict new, verifiable adjustments to the law or postulate a new law.

Scientific Hypotheses

The word *hypothesis* comes from the Greek, *hypothesis*, meaning "placed under" or "foundation." Somewhat similar to theories, hypotheses are more tentative statements than are theories about nature, which lead to deductions that can be experimentally tested. They are reasonable statements, measurable assumptions, and generalizations drawn from a series of observed and selected facts. The origins of hypotheses (or concepts) are immaterial; they can be derived from intuitions, dreams, or ideas arrived at by scientists knowledgeable

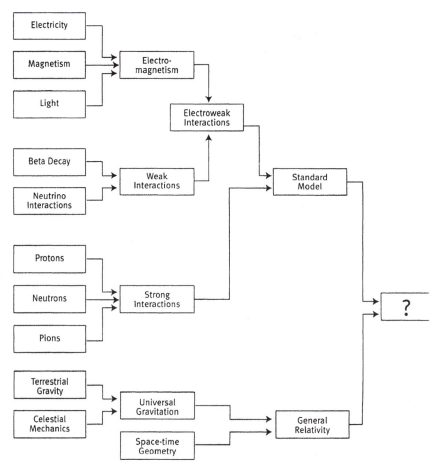

Figure 1: Unification of disparate phenomena within one theory has long been a central theme of physics. The Standard Model of particle physics successfully describes three (electromagnetism, weak and strong interactions) of the four known forces of nature but remains to be united definitively with general relativity, which governs the force of gravity and the nature of space and time. Figure by Johnny Johnson from "A Unified Physics by 2050?" by Steven Weinberg, *Scientific American*, December 1999, p. 70. © 1999 Scientific American, Inc. All rights reserved.

of the subject. What matters is that they must systematically be tested to determine their consequences. Like theories, hypotheses are products of the imagination of scientists, but they are not wild speculations, and they are accepted only when tested and confirmed by additional observations and experiments by other investigators.

A viable scientific hypothesis must be stated in such a way that it has some chance of being disproved, as well as proved, to conform to the observed facts

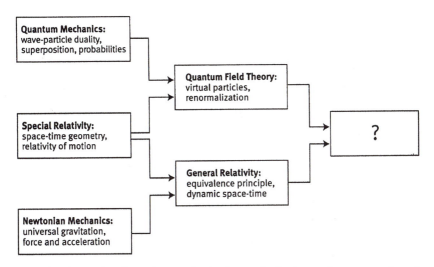

Figure 2: The profoundest advances in fundamental physics tend to occur when the principles of different types of theories are reconciled within a single new framework. We do not yet know what guiding principle underlies the unification of quantum field theory, as embodied in the Standard Model, with general relativity. Figure by Johnny Johnson from "A Unified Physics by 2050?" by Steven Weinberg, *Scientific American*, December 1999, p. 71. © 1999 Scientific American, Inc. All rights reserved.

(just as with a theory). At the same time, just because something cannot be disproved does not mean that it is true. It is the responsibility of the person advancing the hypothesis (or theory) to verify his or her case—not the responsibility of someone else to disprove it. Scientific hypotheses can answer questions for which answers can be achieved only by observing, testing, measuring, using statistics, and analyzing related data. The results become valid and reliable only when the process can be repeated by others who obtain similar results. This is why science can deal only with answerable questions. Most questions are not answerable as a scientific hypothesis because answers cannot be measured, or effectively analyzed (e.g., "What is the secret of success?" or, "How can I become more popular?") Being able to formulate questions that can be answered by controlled experiments or statistical analysis is required by all who wish to investigate nature. A scientific hypothesis (also theory, principle, or law) must be changed if new observed verifiable facts or experiments contradict the original statement(s). This self-correcting process is one of the basic reasons that science may some day triumph over ignorance.

Scientific Concepts

A concept is one step above a specific idea. Related to ideas are assumptions (notions) based on beliefs in or about something, most often accepted without proof. What counts in science is how ideas are developed into viable concepts, hypotheses, and theories that describe nature. Very few good ideas postulated during all of human existence have resulted in the development of basic physical laws. What matters in science is conceptualizing an idea into a viable hypothesis that will support existing, or the development of, new theories, principles or laws.

Causality

Causal laws are considered those that can predict and explain empirical and theoretical laws that describe the universe, usually referred to as "cause and effect." Some science philosophers classify causal principles as (1) empirical generalizations from facts based on other facts, (2) a rational interpretation of a required connection between two or more events, and (3) a useful or pragmatic explanation of science. This is what we usually consider as a specific cause (or causes) leading to a specific effect (or effects) but may not necessarily be a direct, observable connection between the two.

During Sir Isaac Newton's time, science was considered a system to describe a mechanistic, deterministic, and reductionist world. Determinism and predictability are not the same. *Determinism* deals with how nature behaves, particularly nature as we know it within the solar system; it depends on the laws of nature. *Predictability* is based on what scientists are able to observe, measure, and analyze as related to outcomes for specific events. To understand complex nature better, scientists attempt to reduce its complexity to more manageable and understandable laws, principles, and theories. We live our everyday lives in a Newtonian mechanistic solar system based on logic and mathematics, whereas the Einsteinian universe is based on theories of relative space and time that are not as easy to apply on earth as Newtonian physics because the setting for events on earth is very small as compared to the relational aspects for the frames of reference in space and time of the universe. Einstein's relativistic physics rendered some of Newton's laws of motion only approximations. Newton's laws do not hold up when considered for the immense universe consisting of great energies and velocities in different frames of reference over very great distances of space and time. Historically, all effects or events were assumed to have a cause, or possibly several causes, or to be co-events. We now know that many natural events are described and predicted by statistical probabilities, not mathematical certainties. This is true for very large events in the universe, as well as the very small events as related to subatomic particles and energy. These very small events led to quantum theory and indeterminacy (uncertainty principle), resulting in some problems with the cause-and-effect concept for accepted physical laws.

Scientists do not think in terms of *possible* or *could*, or *impossible* or *couldn't*, but rather in terms similar to *likely* or *credible* (*probable*), or *unlikely* or *incredible* (*improbable*). This of course, makes the use of statistical methods such as probability theory a powerful tool. Probability is statistically measurable; possible or impossible cannot be measured.

No one ever expects an effect to precede a cause. Scientists would say that such a situation is unlikely (very low degree of probability), but we never know for sure. In other words, the probability is practically zero. There also seem to be situations in nature where events, and sometimes co-events, occur without obvious causes, or there are causes and events we have not yet been able to detect and measure. This is one reason that the cause-and-effect relationship is not highly thought of by some scientists.

The nature of the universe is extremely complex. Humans have been trying to decipher rationally its mysteries for only a very short period of its and their existence. Scientific facts, for all practical purposes, are accepted as true but never final. What we accept today as a scientific fact related to a law, principle, or theory may very well be changed or discarded tomorrow. Science is a discipline that is "becoming," and the fields of relativity and quantum theory will drive future directions for our understanding of nature. We do not know when or if scientists will arrive at a final answer. They make reasonable guesses (predictions) about nature, based on conditional statements and measurable facts related to specific situations. If the facts do not hold up, then the guess or prediction should be revised or dropped. Predictions made in modern science are based on the analysis of what is observed and measured, not what is hoped for; and physics is the reduction of the complex nature of the universe by making, analyzing, and confirming approximate guesses.

A

Abbe's Theory for Correcting Lens Distortions: Physics: *Ernst Abbe* (1840–1905), Germany.

The equation for Abbe's theory is: *u'/sin U' = u/sin U, where u and u' are the angles for the entering and exiting of rays from the object to the image, as in a microscope.*

The *Abbe sine condition* is a mathematical concept used to make lenses that produce sharper images and less distortion. It is a means to eliminate spherical aberration of an optical system in order to produce an **aplanatic lens** (corrected lens). In 1886 Abbe used his mathematical approach to develop apochromatic lenses (corrected for chromatic and spherical aberrations) to eliminate primary and secondary color distortions. The *U* and *U'* are the corresponding angles of any other rays transmitted. Abbe's contributions to the field of optics led to the improvement of all of today's optical instruments, including sharper images and less color distortion for cameras, microscopes, refracting telescopes, spectroscopes, and so forth. *See also* Newton.

Abegg's Rule and Valence Theory: Chemistry: *Richard Abegg* (1869–1910), Poland.

Chemical reactions are the result of electron transfer from one atom to another.

Richard Abegg arrived at this rule years before the existence of **valence** was established for atoms. He further theorized that *the attraction of electrons for atoms of all elements has two distinct types of similarities or valences, i.e., a "normal affinity" (valence) and a "counter affinity" (countervalence).*

This theory that two related valences always add up to eight, published in 1869, is responsible for the octet rule as related to the **Periodic Table of Chemical Elements**. (See Figure S2 under Sidgwick.) Abegg's early theories of valence became valuable for later chemists and were used to explain chemical reactions and the organization of the Periodic Table. *See also* Mendeleev; Newlands; Sidgwick.

Adams' Concept of Hydrogenation: Chemistry: *Roger Adams* (1889–1971), United States.

(1) *Additional hydrogenation will take place when hydrogen is added to double bonds of unsaturated molecules of organic substances such as liquid fats and oils.* (2) *Hydrogenolysis hydrogenation will take place when hydrogen breaks the bonds of organic molecules, permitting a reaction of hydrogen with the molecular fragments.*

The first type of hydrogenation led to the formation of solid or semisolid fats from liquid oils, and the second process is used in hydrocracking petroleum or adding hydrogen to coal molecules to increase its heat output. In the early 1900s Roger Adams' ideas resulted in the successful hydrogenation of unsaturated organic compounds by catalyzing them with finely powdered platinum and palladium metals under heat. It is similar to the process of reduction in inorganic chemistry that led to the hydrogenation of many of our fuels and foods, where liquid oils are converted to semisolid oils or fats. Examples are hydrogenated peanut butter, hydrogenated vegetable oils (margarine), and forms of solid petroleum fuels and medications. The second type of hydrogenation led to an increase in the production of petroleum products from crude oil, such as gasoline.

Adhemar's Ice Age Theory: Astronomy: *Joseph Alphonse Adhemar* (1797–1862), France.

Since Earth tilts 23.5 percent from its ecliptic (the orbital plane of the earth around the sun), the Southern Hemisphere receives about 200 fewer hours of sunlight per year. Therefore, more ice accumulates on Antarctica than at the North Pole.

In 1842, Joseph Adhemar proposed his theory, the first to provide a reasonable answer as to the cause of the ice ages. His theory is based on evidence gained from astronomical events. He realized earth is just one focus in its elliptical orbit around the sun, which means that earth is farther away from the sun in July. And due to the tilt of earth, the South Pole receives fewer hours of sunlight. In addition, earth's axis does not always point in the same direction in space; rather, it slowly rotates in a small circular orbit approximately every 26,000 years (called PRECESSION). These astronomical factors, plus the inclination of earth, cause the winters to be slightly longer for the Southern Hemisphere. According to Adhemar, the Antarctic ice sheet built up, resulting in the ice age. However, many scientists today do not completely agree with his theory. *See also* Agassiz; Kepler.

Agassiz's Geological Theories: Geology: *Jean Louis Rodolphe Agassiz* (1807–1873), Switzerland.

Agassiz's Glacier Theory: *The movement of glaciers created scratch marks in rocks, smoothed over vast areas of the terrain, gouged out great valleys, carried large boulders over long distances, and left piles of dirt, soil, and debris called moraines.*

Jean Agassiz was a keen observer of geological phenomena and believed there was evidence of glaciers where none now exists. A native of Switzerland, his theory, published in 1840, was based on observations of the glaciers in the mountains near his home. He further speculated that ice sheets and glaciers formed at the same time in most of the continents.

Agassiz's Second Glacier Theory: *Glacier ice sheets had movements that included advancements as well as periods of retreat, which correspond to the ice ages.*

Jean Agassiz's work led to the concept that glaciers were the *result* of the ice age. As the ice sheets that covered Earth melted in the warmer zones, great deposits of compact ice were deposited in the colder regions of the Northern and Southern Hemispheres. These ice masses, now called *glaciers*, moved slowly toward the warmer areas, and they continue to move even today. Periodic ice ages, some much smaller than the original frozen Earth period, formed, advanced, and retreated over many millions of years. Ice ages on Earth are considered normal cyclic occurrences. Previous to Agassiz's glacier theories, scientists assumed glaciers were caused by icebergs. In fact, icebergs are caused by the calving of huge chunks off the edges of glaciers that extend into lakes and oceans. Agassiz, sometimes known as the Father of Glaciology, also made contributions to evolutionary development through his study and classification of the fossils of freshwater fish.

Agassiz's Theory of Fossils and Evolution: *The lowest forms of organisms were found in rock strata located at the lowest levels of rock formations.*

Before Agassiz devised his theory, William Smith (1769–1839) proposed that fossils found in older layers of sedimentary rocks were much older than the more modern-appearing fossils located in sediments laid down in more recent geological times. This concept answered some of the questions about the evolution of species because the ages of fossil plants and animals now could be determined by their placement in the rock strata and led to Agassiz' more formal theory on the subject. Agassiz at first did not believe in evolution, but later did accept it while still rejecting Darwin's theory of natural selection because it required long, gradual periods of change. Rather, Agassiz accepted and expanded on Georges Cuvier's catastrophism theory, which postulated periods of rapid environmental changes as the basis of evolution. *See also* Adhemar; Charpentier; Cuvier; Darwin; Eldredge; Gould; Lyell; Raup.

Agricola's Theories of Earthquakes and Volcanoes: Geology: *Georgius Agricola* (1494–1555), Germany.

Agricola's original name was Georg Bauer, but he preferred to be known by his chosen Latin name, Georgius Agricola. In the 1540s Agricola's studies led to several geological theories based on his observations of minerals and stratified layers of rock.

Agricola's Theory of Stratification: *Stratified forms of rocks are the arrangement and relationships of different layers of sedimentary rocks as formed during earthquakes, floods, and volcanoes.*

The planes between different strata (layers) assist in determining not only the source of the sedimentary rocks but also the area's local history. In addition, the fossil content found in different strata provides a record of the biological and geological history of Earth. Today, the study of stratification is called *stratigraphy*, which is the branch of geology that studies the different layers of rock. His stratification system, though primitive, proved useful as one means to identify the location and sources of petroleum.

Agricola's Theory for Earthquakes and Volcanoes: *Earthquakes and volcanoes are caused by subterranean (below ground level) gases and vapors originating deep in Earth, where they are heated and then escape to the surface.*

A keen observer and practicing physician specializing in miners' diseases, Agricola's main interest was minerals, known in those days as metals. In 1546 he was one of the first to classify minerals according to their physical properties, such as color, weight, and texture. He believed these minerals/metals originated deep in the underground and were brought to the surface by earthquakes and volcanos. Agricola was also known as a paleontologist for his work with objects found in the soil, including fossils, gemstones, and even gallstones. He is sometimes referred to as the father or founder of mineralogy and the science of geology, since he was among the first to describe fossils as once living organisms. Although later scientists disputed some of his ideas and subsequently proposed revised theories, his publications were used in the field of geology for over 200 years.

Airy's Concepts of Geologic Equilibrium: Geology: *Sir George Biddell Airy* (1801–1892), England.

Airy, along with other scientists, proposed two theories of equilibrium as expressed in geological structures.

Airy's First Theory of Geologic Equilibrium: *Mountains must have root structures of a lower density in proportion to their heights in order to maintain their Isostasy (equilibrium).*

Isostasy is Airy's theory that there is a proportional balance between the height of mountains as compared to the distribution of the root structure or mass underneath the mountain. He claimed this equilibrium resulted in a balance of **hydrostatic** pressure for the formation of mountains. It became an important concept in geology and aided in the exploration of minerals and gas and oil deposits, although it has been revised.

His second theory deals with internal water waves and is based on ideas expressed by Vagn Walfrid Ekman (1874–1954).

Airy's Second Theory of Equilibrium: *In areas where the sea is covered with a thin layer of fresh water, energy is generated by internal waves and is radiated away from ships, which subsequently produces a drag on ships.*

Because this slows the ship's progress, it is known as *dead water*. One area where freshwater and seawater are at different levels, which causes these internal waves to drag on ships, is the entrance to the Mediterranean Sea at Gibraltar. A story from World War II relates that since the water in the Mediterranean Sea has a higher salt content than the water in the Atlantic Ocean, the Mediterranean's water is more dense and thus flows out past the Strait of Gibraltar into the Atlantic near the bottom. At the same time, the less salty (less dense) Atlantic water flows into the Mediterranean near the surface. When submarines wished to avoid detection, they drifted quietly into the Mediterranean near the surface past the enemy. When submarines wished to leave without being detected, they drifted quietly near the bottom as the denser saltwater carried them past the protected Gibraltar and into the Atlantic. By using the difference in density of saltwater and freshwater, submarines could apply the "Airy waves" concept to their advantage. The success of this tactic remains unknown.

Al-Battani's Theories: Astronomy: *Abu Abdullah Al-Battani* (c.858–929), Iraq.

Abu Abdullah Al-Battani developed various theories after improving measurements completed by Ptolemy of Alexandria.

Al-Battani's Theory of Solar Perigee: *Solar perigee equations demonstrate slow variations over time.*

Al-Battani determined that the sun's perigee (the point at which the sun is closest to earth in earth's elliptical path around the sun) is greater than Ptolemy's measurement by a difference of over 16°47'. Although Al-Battani admired Ptolemy, he improved on several of Ptolemy's calculations, including the ecliptic of earth's orbit to its equatorial plane.

Al-Battani's Theory of the Earth's Ecliptic: *The inclination of the angle of earth's equatorial plane to its orbital plane is 23°35'.*

This figure is very close to the current measurement. The ecliptic for earth can also be thought of as the apparent yearly path of the sun as earth revolves around it. In other words, it is the angle of tilt of earth to its solar orbital plane that is the major cause of our four seasons—not how close earth is to the sun. The Northern Hemisphere is tilted more toward the sun during the summer, thus receiving more direct sunlight for more daylight hours than in the winter, when the Northern Hemisphere is tilted away from the sun. The situation is reversed for the Southern Hemisphere.

Al-Battani's Theory of the Length of the Year: *The solar year is 365 days, 5 hours, and 24 seconds.*

Al-Battani's calculations are very close to the actual figure accepted today 365.24220 days. This led to more exact recalculations for the dates of the spring and fall **equinoxes**. Both the dates for the equinoxes and the accuracy for the length of the year were important for various religions of the world that base many of their holy days on the seasons and these particular dates.

Al-Battani's Concept of the Motion of the Moon: *It is possible to determine the acceleration of the motion of the moon by measuring the lunar and solar eclipses.*

Al-Battani was able to time the lunar and solar eclipses and thus able to extrapolate this figure to calculate the speed of the moon in its orbit. He also devised a theory for determining the visibility of the new moon. Albategnius, an eighty-mile plane surface area on the moon surrounded by high mountains, is named for Al-Battani.

Al-Battani's Concepts of Two Trigonometric Ratios: (1) *Sines* (which he derived) *were demonstrated as more practical and superior to the use of Greek chords.* (A chord is a line segment that intersects a curve only at the end of the curve.) (2) *Cotangents are the reciprocal of tangents.*

Al-Battani devised tables for the use of sines, cosines, tangents, and cotangents, which are invaluable for modern algebra and trigonometry. Copernicus credited Al-Battani with advancing astronomy based on his work in trigonometry and algebra. For many years, Al-Battani's contributions to science and mathematics were considered preeminent and advanced the cause of knowledge over the next several centuries.

Ampère's Theories of Electrodynamics: Physics: *André-Marie Ampère* (1775–1836), France.

In the early 1820s Ampère based his theories of electrodynamics on how electric currents influence each other and how they interact with magnetism.

Ampère's Theory of Flowing Current and Magnetism: *When two parallel wires carry current in the same direction, they will attract each other. And if the current flows in opposite directions in the two parallel wires, they will repel each other.* (See Figure A1.)

He related this to the two poles of **magnets**, which led to his famous laws developed in 1825.

Ampère's Law, Part I: *The force of the electric current between two wires (or conductors) will exhibit the inverse square law, which states that the force*

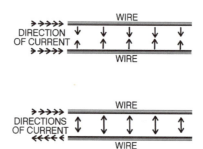

Figure A1: Ampère's Law. For both examples, the strength of the magnetic fields generated between the wires decreases with the square of the distance between the wires.

decreases with the square of the distance between the two **conductors**, *and that the force will be proportional to the product of the two currents.*

Ampère's Law, Part II: *When there is an electric charge in motion, there will be a magnetic field associated with that motion.*

Ampère's law is related to induction, which can be expressed as the equation: $d\beta = k\,A\,dl\,\sin\,\theta/r^2$, where A is the current and k is a proportional factor based on either the cm (centimeter) or m (meter) units of the SI system. $d\beta$ is the increase in the strength of the magnetic field due to an infinitesimal increment dl in the length of a wire carrying the current A; r is the distance from the element of wire to the part where the field is measured.

Ampère devised another law related to the magnetic effects of flowing electrical currents in curved wires.

Ampère's Circuital Law: *The magnetic intensity of a curved or enclosed loop of wire is the sum of the current and can be determined by considering the sum of the magnetic field for each segment of the loop.* (This is similar to Gauss's magnetic law for closed surfaces.)

The equation for the circuital law is: $B(r) \times 2\pi\,r = \mu A$. (B is the magnetic field at the center of the loop, r is the radius of the loop, π and μ are constants, and A (or I) is the current.) This law measures the strength of the magnetic field in a **solenoid** and determines the strength of the field in electromagnets used in electric generators and motors.

The ampere or amp named after Ampère is given the symbol of A or I.

Ampère Rule: *The unit of electric current flowing through parallel, straight, long wires in a vacuum produces a force between the wires of* 2×10^{-7} N (newtons) *for every meter of wire through which the current flows.*

In other words, it is the measure of the "amount" of electrical current flowing through a wire. As an analogy, consider amps similar to the "amount" of water flowing every second through a pipe. Another way to think of an amp (A) is to count the number of **electrons** that cross a particular point while flowing though a wire. The rule also states: *One amp equals* 6×10^{18} *electrons passing this point every second.* Electrical appliances are rated according to the number of amps (current) they use (e.g., a TV set uses 3 to 5 amps, a small motor about 2 to 5 amps, a 100 watt light bulb about 1 amp, an electric stove between 10 and 25 amps, or more).

Ampère's contributions advanced the development of many practical industrial devices that make our lives more enjoyable and easier. For example, he suggested his discovery could be used to send signals, which, over time, became a reality (e.g., telegraph, the radio, television). Others, including Faraday and Oersted, used Ampère's laws to construct the dynamo (electric generator) and the electric motor. A more recent application is the experimental nuclear fusion project to generate heat for the production of electricity. This process requires very strong magnets to produce the pinch effect, which will contain and concentrate the hot plasma gases required for the application of nuclear fusion. However, controlled nuclear fusion to produce electricity has yet to be developed to the point where it is practical. Strong electromagnets are also important for the operation of particle accelerators and magnetic resonance imaging (MRI) medical diagnostic instruments. *See also* Faraday; Galvani; Oersted; Ohm; Maxwell.

Anderson's Positron Theory: Physics: *Carl D. Anderson* (1905–1991), United States. Carl Anderson shared the 1936 Nobel Prize with Victor Franz Hess, who discovered cosmic rays.

Cosmic rays passing through a cloud chamber produce tracks of negative particles deflected in one direction (electrons), while at the same time producing tracks of particles with equal curvature in the opposite direction with equal mass, deflected by a magnetic field. These particles can be only positive-type "electrons." Thus, they are a new elementary particle named positrons.

In 1932 Carl Anderson's concept of a positive electron (positron) was verified by Patrick Blackett (1897–1974) and Giuesppe Occhialini (1907–1993) who determined, according to Paul Dirac's theory, that the **positron** was the equivalent but opposite charge to the electron. They considered it **antimatter** instead of normal matter. Anderson later discovered what is called the *mu-meson*, which was predicted by Hideki Yukawa. It is now called the **muon**, whose nature and role in nuclear physics are not yet completely understood. Anderson's use of the **cloud chamber** and his discovery of two new particles opened the path to the exploration and understanding of numerous subnuclear particles. *See also* Dirac; Yukawa.

Archimedes' Theories: Mathematics: *Archimedes of Syracuse* (c.287–212 B.C.), Greece.

Archimedes was an accomplished theoretical and applied mathematician who developed experiments to test his ideas and then expressed their results mathematically. Only a few of his many theories and accomplishments will be explored.

Archimedes Theory of "Perfect Exhaustion" (Calculation of Pi): Archimedes was not the first to recognize the consistency of the ratio of the diameter to the circumference for all circles or to attempt the calculation of **pi**. Ancient stone age people were intrigued by objects of differing shapes and how their dimensions were related. They realized straight lines do not exist in nature and recognized curved lines in the shape of rocks, plants, animals, and other objects. By about 2000 B.C. people recognized and roughly calculated the relationship of a circle's measurement in the sense that the larger the circle, the greater is its circumference. By the time of the ancient Greeks, mathematicians understood this ratio was consistent for all circles since they measured and compared the diameters and perimeters of various circles. Soon after, this constant irrational number was given the Greek symbol π (pi).

Using his knowledge of geometry of many-sided plane figures, such as squares and multiple **polygons**. Archimedes proposed his theory of *perfect exhaustion*, which he demonstrated by drawing a circle and inscribing several polygons on both the inside and outside of the circle. At first, he used polygons with just a few sides. Later he used multiple polygons with as many as ninety-six or more sides. This is often referred to as *perfect exhaustion* because Aristotle used polygons with larger numbers of sides. Theoretically, a polygon with an infinite number of sides could be used. Through the use of geometry and fractions, Archimedes measured the inside polygons and compared them with the measured outside polygons. He concluded that the polygons touching the circle on its outside circumference (perimeter) were slightly larger than pi and that the polygons touching the inside of the rim of the circle were slightly smaller than pi. Therefore, pi must be a value somewhere between these two measurements. His value for pi was 3.14163, which he calculated as the figure between the inner and outer polygons ($3_{10/71} < \pi < 3_{1/7}$). His figure for pi, used for many hundreds of years, was developed by using Euclidean plane geometry, which has physical limitations for this purpose. Later mathematicians used algebra, which enabled the calculation of a more accurate value for pi. With the invention of fast computers, pi has been run off to several hundred thousand decimal places in a few hours. Yet one could run off pi on a computer forever and still never reach a final number to make pi come out even, because it is an **irrational number**.

Archimedes Theory for the Volume of Spheres: *The volume of a sphere is two-thirds the volume of a cylinder that circumscribes (surrounds) the sphere.*

It is said that Archimedes wanted this theorem inscribed on his tombstone. Historically, measuring the volume of a sphere was difficult, while measuring the volume of a cylinder was easy. Therefore, if one knew the volume of a cylinder that surrounded a sphere, its volume could be determined.

Archimedes' Theory of Levers: *The mechanical advantage of a lever is due to the ratio of the weight (load) to the action (effort) required to move the load, which is determined by measuring the distance the effort moves from the central point (fulcrum) divided by the distance the load moves from the central point.*

The simple lever has been used by humans since prehistoric times. How people learned to take advantage of this simple lever is unknown, but evidence exists that ancient people were aware of the advantage of using sticks for digging and moving heavy objects. Archimedes was the first to calculate the ratio of the distance between a force and a weight, separated by a fulcrum. The placement of the fulcrum in relation to the force and weight determined the ratio for the mechanical advantage. Archimedes used his knowledge of geometry and mathematics to calculate the mechanical ratio for several simple machines. For the simple lever, he believed the advantage was the ability to move very heavy loads with little effort. Most of his demonstrations of mechanics dealt with the simple lever. His major demonstration was the raising of a large ship by pushing down on one end of a lever that he had designed.

Archimedes' Concept of the Inclined Plane: *It is easier to move a load along a long, sloping ascent of a given height than it is to move a load of the same weight along a shorter but steeper ascent to the same height.*

Archimedes knew the mechanical advantage of rolling objects up a long, inclined plane of a given height rather than lifting them vertically for the same height. He applied the concept of an inclined plane as a means of raising water from a well to the surface. He wrapped an inclined plane device around a central shaft to form a "water screw," which was placed with one end in the well and the other on the surface where the water was to be used. When turned by a crank handle, this "helical pump" enabled one man to lift water more efficiently than with any other pump then known. Remarkably, it is still used, 2,300 years later, in Egypt and other parts of the world. Archimedes also developed catapults, cranes, pulleys, and optical devices that consisted of a series of shiny metallic mirrors that reflected and concentrated rays of the sun. All of the devices are said to have been used to defend his city of Syracuse from Roman invaders. Although not the first to use his knowledge of physics and mechanics in the name of war, Archimedes was one of the most successful.

Archimedes' Concepts of Relative Density and Specific Gravity: *The compactness (amount of matter) of an object is related to the ratio of its weight divided by its volume.*

Archimedes used his concept of buoyancy to measure the relationship between the weight and volume of an object. No discussion can omit the famous story of how Archimedes gained insight into the concept of density and specific gravity. As the story goes, King Hieron asked Archimedes to ascertain whether a gold crown he commissioned from a goldsmith was pure gold or whether silver was substituted for some of the gold. Archimedes pondered the question while taking a public bath. He lowered himself into a bath filled to the brim. As he sank deeper into the bath, more and more water spilled over the sides. He immediately grasped the significance of this phenomenon, jumped out of the bath, and ran naked down the street shouting *"Eureka! Eureka!"* (I have found it! I have found it!) He proceeded to fill a bucket to the brim with water into which he lowered the crown, catching and measuring the volume of the water that overflowed. He did the same with equal weights of gold and silver. Since gold has a greater density than silver, the ball of gold was smaller; thus less water spilled over the edge of the bucket. Once he could measure the volumes of the water representing the volumes of the gold, silver, and the crown, all he needed to do was to divide the figures obtained for the volume of each item into their weight and calculate a ratio representing their comparative densities. He could then determine how much of the crown was gold and how much was silver. (Supposedly, the crown was not pure gold, and the goldsmith was executed.)

His principle led to the expression of density as the weight (mass) of an object divided by its volume ($d = m/v$). Specific gravity is the ratio of an object's density to that of some standard. For liquids, water is the standard for specific

gravity of 1.0 at 15°C. For gases, dry air is used as the standard pressure and temperature. Specific gravity is easier to use than density for making calculations because it is the same value in all systems of measurement.

Aristotle's Theories: Physics: *Aristotle of Macedonia* (384–322 B.C.), Greece.

In the estimation of many historians, Aristotle was one of the most influential humans who ever lived. He was a philosopher concerned with classes and hierarchies rather than a scientist concerned with observations and evidence. His philosophy, methods of reasoning, logic, and scientific contributions are still with us and continue to be influential. Much of Aristotle's philosophy is related to his four causes: (1) *The matter cause, which makes up all material, including living organisms*; (2) *The form cause of species, types, and kinds of things*; (3) *The efficient cause of motion and change*; (4) *The final cause of development or the final goal of an intended activity (maturity).* He related these "causes" to both inanimate and animate phenomena. Only a few of Aristotle's theories related to limited areas of science will be presented.

Aristotle's Topological-Species Theories: *An ideal form is a living group in which each member resembles each other but the group is distinct in structure from members of other groups.* His concept can further be stated as: *Each living thing has a natural built-in pattern that, through reproduction, growth, and development, leads to an individual type **(species)** similar to their parents.* Aristotle also believed that all living species reproduce as to type (e.g., humans beget humans, cattle beget cattle), but he considered a possible exception to this theory when applied to the lowest of species. He organized species from lowly flies and worms at the bottom, then lower animals, up to mammals, and then humans at the top. Aristotle classified everything and endeavored to write all as a unified theory of knowledge, which preceded Einstein's long-sought grand unification theory by many centuries.

Aristotle's Theory of Spontaneous Generation: *Flies and low worms are generated from rotting fruit and manure.* He based this theory on observations, not experimentation. Later scientists, through the use of simple experimentation, demonstrated "spontaneous generation" of life does *not* exist, at least as expressed by Aristotle and some former scientists. *See also* Leeuwenhoek; Pasteur; Redi.

Aristotle's Theory of Taxonomy (classification) of Living Things: *Nature proceeds from tiny lifeless forms to larger animal life, so it is impossible to determine the exact line of demarcation. Reproduction identifies those giving live birth (viviparous) as being mammals and humans, while those laying eggs (oviparous) are subdivided into birds and reptiles, and fish and insects.* Aristotle developed an elaborate classification system of nature later called *Aristotle's ladder of nature.* It listed inanimate matter at the bottom, progressing upward from lower plants, higher plants, minor water organisms, shellfish, insects, fish, reptiles, whales, mammals, and finally on the top rung of the ladder, humans.

Aristotle's Three Classes of Living Things: (1) *Vegetable, which possessed a nutritive soul*; (2) *animals, who were able to move and thus had a sensitive soul*; and (3) *humans who had intelligence and thus a rational soul, and who also possessed souls of all the types of creatures.*

One of Aristotle's classifications was that male humans had more teeth than did females. As a philosopher concerned with the meaning of classes and hierarchies rather than a scientist concerned with observations and evidence, neither he nor anyone else at that time bothered to count the teeth in men and women. Regardless, "man" was at the top class with a rational soul.

Aristotle's Concept of Reproduction: *An invisible "seed" of the most rudimentary structure was imparted by the male to join a female egg to produce an offspring of the same species.* Some philosophers and scientists of Aristotle's day (and later) believed the "seed" (sperm) was a tiny, invisible person or animal that grew larger once it joined with a female egg. Aristotle concluded this by observations made while dissecting and studying fertile chicken eggs at different stages of embryonic development. He rejected most theories about reproduction proposed by previous philosophers, which included that the sex of an embryo is determined by how it was placed in the womb, a seed originates as a whole body, and the embryo contains all of the adult body parts (preformationism).

Aristotle's Laws of Motion: (1) *Heavy objects fall faster than lighter ones, and the speed of descent is proportional to the weight of the object.* (2) *The speed of the falling object is inversely proportional to the density of the medium through which it is falling.* (3) *An object will fall twice as fast as it proceeds through a medium of half its density. Thus a vacuum cannot exist because the object would proceed at an infinite speed.* This law is one of the few examples indicating Aristotle's concern with the quantitative nature of things. Unfortunately, he did not verify his insights by experimenting and making measurements. Although Aristotle's laws of motion were incorrect, they were accepted for many years and provided the background for Galileo and Newton in which to revise his original concepts.

Another of Aristotle's laws of motion stated: *Violent (forced) motion will always be displaced by natural motion that ends in a state of rest. The speed of a moving object is directly proportional to the force applied to it.* In simple language this means if you cease pushing an object, it will stop moving. Philosophers and scientists, before as well as after Aristotle's time, could not accept the concept of action at a distance, such as gravity. There had to be something in contact with the object that would force the object to move, and it could not just be "spirits," as some believed, but rather something physical. To Aristotle, all motion was self-explanatory because all bodies sooner or later came to their natural place of rest in the universe. He explained his theory somewhat in this way: Once "impulse" was given to a stone by throwing it up in the air, this impulse was transferred to the air in tiny increments, which kept pushing the stone up. These "air impulses" pushing the stone upward became weaker as the stone rose, and now the natural motion of the stone returned it to the ground in a straight line, and finally to its natural state of rest. When the "impulse" completely stopped, so did the object's motion.

Aristotle applied his concepts of motion to his observations of heavenly bodies. His theory states: *Heavenly bodies move in perfect circles rather than in straight lines as bodies do on earth. Thus, heavenly bodies are not composed of the four earth elements but rather a fifth element called aether.* This concept that heavenly bodies and bodies on earth obey separate laws was followed by scientists until Newton's time. Celestial bodies were "pure," while those on earth were subject to death and decay. Aristotle's theory of the prime mover, impulse, and motion came very close to the modern physical law of conservation of momentum. *See also* Galileo; Newton.

Aristotle's Concept of Infinity: *Since the universe is spherical and has a center, it cannot be infinite. An infinite thing cannot have a center, and the universe does have a center (the earth). Therefore, infinity does not exist.*

Most scientists of Aristotle's day believed the universe was composed of crystalline concentric spheres with the earth at the center; therefore, the universe was finite. Somewhat the same argument was used to negate the existence of a void, or vacuum. Accepting concepts such as infinity and vacuum was beyond the philosophical reasoning of people in Aristotle's time. It was not until the sixteenth century, when Copernicus provided credible evidence that the earth was not the center of the **universe**, that this geocentric concept was overcome.

Aristotle's Theory of the Matter and the Aether: *Since all celestial bodies move in perfect circles, there must be a perfect medium for this to occur.*

This perfect medium that enables circular motion is known ·as the **aether**, which also has circular motion. Aristotle accepted the classification of elements as devised by Empedocles (c.490–c.430 B.C.) and others that placed all things into four elementary groups: earth, water, fire, and air. He saw the need for a fifth class of matter when addressing the heavens. The concept of aether (or *ether*, the Greek word for "blazing") was used by scientists until Newton's time. In its more sophisticated form, it was referred to as the "fabric of space." The concept of an ether existed into the days of early radio. It was popular to believe that radio signals (and other electromagnetic waves) were transported by something in space similar to the way sound is carried by air. In Aristotle's time, people did not believe the sun's heat could reach earth without some form of "matter" transporting it. *See also* Maxwell.

Arrhenius' Theories, Principles, and Concepts: Chemistry: *Svante August Arrhenius* (1859–1927), Sweden. Arrhenius was awarded the Nobel Prize for Chemistry in 1903.

In 1883 Svante Arrhenius proposed two related theories of dissociation. One deals with what occurs when substances are dissolved in solutions; the other explains what happens when a current of electricity is passed through a solution.

Arrhenius's Theory of Solutions: *When a substance is dissolved, it is partly*

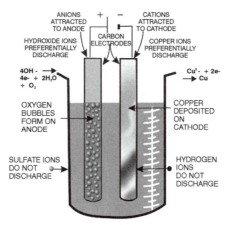

Figure A2: The positive ions are attracted to the negative cathode, and the negative ions are attracted to the positive anode. The ions lose their electrons to form neutral atoms through a discharge of the current between the electrodes and through the electrolyte, which is an ionized compound in solution that can carry electricity.

converted into an "active" dissolved form that will conduct a current. Arrhenius's theory is based on the concept that **electrolytes** in solution dissociate into atoms. (See Figure A2.) *See* Faraday.

Arrhenius's Theory of Ionic Dissociation: (1) *When an electric current is passed through molten salt (sodium chloride, NaCl), it is dissociated into charged ions of Na^+ and Cl^-. (2) Positive charged Na^- ions are attracted to the negative pole (**cathode**) and deposited as neutral atoms of sodium metal, and the negatively charge ions of Cl^- are attracted to the positive pole (**anode**) and changed back to neutral atoms of chlorine, as the gas molecule Cl_2.*

Once the atoms are "dissociated" into **ions** in a liquid and become an electrolyte solution, an electric current can pass through the solution, producing an attraction of negative ions to the negative pole within the completed electric circuit.

These theories are related to electrolysis and are important to many industrial processes today, including electroplating. For instance, ionic dissociation is one way to produce chlorine and sodium from common salt (NaCl). The dissociation of substances, such as $NaNO_3$, separates the compound molecule into Na (sodium metal) and NO^{---} (the negative nitrate ion). A similar process occurs when electroplating gold and chromium, and other metals. *See also* Faraday.

Arrhenius' Principle of Acid-Base Pairs: *When an acid splits, it will yield hydrogen ions (H^+). When a base breaks apart, it will yield hydroxyl ions (OH^-).* Arrhenius extended his concepts about ionic dissociation to include his theory related to acids and bases. In general terms, this principle has been broadened to make it more useful for chemists, who still speak of *acid-base pairs.* This broadened principle refers to the transfer of a proton from one molecule to another: *The molecule that gave up the proton is the **acid**, and the one receiving the proton is a **base**.* The process is used in industry to produce both acids and alkaline (basic) chemicals.

Arrhenius' Rate Law: *The rate of a chemical reaction increases exponentially with the absolute temperature.*

Arrhenius and others based this phenomenon on their observations that when the temperature rises, the rate (speed) of chemical reactions increases; conversely, when the temperature cools, reactions slow down. They noted this for such things as spoilage and decomposition of fruits and vegetables in hot climates, while their usefulness can be extended by refrigeration. Also, bread rises faster in a warm environment, substances dissolve faster in warm water, and so forth. Arrhenius' rate law can be expressed mathematically: Rate = A $\exp(-B/T)$, where A and B are constants that differ from one reaction to another. The value of this equation is its use in generalizing the concept based on the fact that most chemical reactions occur at room temperatures of about 20°C, and a rise of 10°C will double the rate of the reaction.

Arrhenius Theory of Panspermia: *Life came to earth as a bacterial spore or other simple form from outside the solar system.*

After the theories for spontaneous generation were discredited by Redi and Pasteur, other theories for the origin of life were postulated, including theories of life arriving on earth from outer space. One was proposed by Arrhenius, who called his theory for the beginning of life on earth *panspermia*. This theory is gaining some new proponents since it was reported that simple organic molecules were found in some meteorites, possibly from Mars, that landed on the ice sheets of Antarctica. A modern version of the theory is that prebacteria-type organisms capable of reproducing are universal and develop within a suitable environment anywhere in the universe. *See also* Hoyle; Pasteur; Redi; Spallanzani; Struve.

Arrhenius' Theory for the Greenhouse Effect: *The percentage of carbon dioxide in the upper atmosphere regulates the temperature, which may be the cause of the ice ages.*

Over eighty years ago, Arrhenius was one of the first to relate carbon dioxide to global climate changes. Although he was unable to establish an exact relationship between carbon dioxide and atmospheric temperatures, he considered both the cooling and warming effects of CO_2 as evidence for the cause of the past ice ages. His theory is based on the belief that CO_2 in the atmosphere does not absorb the energy from the sun that arrives on the surface of earth in the form of light and infrared (heat) radiation. In addition, the energy radiated from earth is in the form of infrared radiation that is absorbed by CO_2, acting as a blanket thus creating a *greenhouse effect*. This theory is still controversial; however, evidence indicates that there is a slight increase in the levels of atmospheric CO_2 which may, or may not, have a slightly more warming than cooling effect on earth (about 1.5°C in the twentieth century). The increase in water vapor plus methane from industry, decaying of organic matter, and animal flatulences also contribute to a greenhouse effect. Another theory states the massive ice age 600 million years ago, when earth was a complete ice planet, ended as trillions of simple organisms produced enough carbon dioxide to create a greenhouse effect. This may have resulted in the melting of the ice that covered earth, thus permitting the evolution of higher forms of life. Carbon dioxide also increases plant growth on earth, which provides food for a more diverse animal kingdom. *See also* Rowland.

Aston's Whole Number Rule: Chemistry: *Francis William Aston* (1877–1945), England. Francis Aston received the 1922 Nobel Prize in chemistry.

The mass of the oxygen isotope is defined as a whole number, and all the other isotopes of elements have masses that are very nearly whole numbers.

Aston invented the mass spectrograph, an instrument that uses electromagnetic focusing to separate **isotopes** of the same element by slight differences in their **atomic weights**. The **nuclei** of atoms of a specific element are composed of both positive **protons** and neutral **neutrons**. The number of protons determines the chemical identity of the element, which does not change, while the number of neutrons can be more or less than the number of protons. This explains the formation of isotopes of the same element with different atomic weights. This difference in weight is very slight and was not successfully detected until the development of the spectrograph. Aston used his instrument to measure this minute difference in atomic weights and to separate and identify 212 isotopes of nonradioactive elements. From his research he devised the *whole number rule*, which advanced the fields of inorganic and nuclear science.

Atomism Theories: Physics. Theories and Scientists listed Chronologically.

Theories of atomism go back to the fifth century B.C. when philosophers conceived the idea that all matter was composed of tiny, indivisible particles. In ancient times these classical theories were philosophical and were deduced by reason and logic, not by empirical or experimental evidence. The word *atomos* is derived from two Greek words: *a*, which means "not," and *tomos*, which means "cut." In other words you cannot cut it, or it is indivisible. Several examples of atomic theories follow:

Leucippus' Atomic Theory: *Leucippus of Miletus* (c.490–430 B.C.), Greece. *All matter is composed of very minute particles called atomos. They are so small that there cannot be anything smaller, and they cannot be further divided.*

Not much is known about Leucippus, but he was the first to be credited with originating the atomic theory, giving the concept the name *atom* and describing the indestructible nature of atoms. One of Leucippus's students was a philosopher named Democritus, who is also credited with the "atomic concept."

Zeno's Paradox: *Zeno of Elea* (c.495–430 B.C.). One of Zeno's theories stated that *conclusions could be reached by reason even when there was contradictory sensory evidence.* He used paradoxes to present his hypotheses that motion and distance could be divided into smaller units forever. His most famous paradox was used by other philosophers and scientists to explain the concept of the division of matter into smaller and smaller particles but never reach a final indivisible particle. *See also* Zeno.

Democritus' Theory of Atoms: *Democritus of Abdera* (c.460–370 B.C.), Greece. It is assumed that Democritus and others who followed questioned Zeno's paradox as a rational way to look at nature in the sense that the division of space and motion could be divided indefinitely, and perhaps there was a final limit to the point of indivisibility. Democritus further developed the atomic theory of his teacher, Leucippus. He and other philosophers considered what would happen if a person took a handful of dirt and divided it by half, then divided that half into half, and continued dividing it by halves. Soon a point would be reached at which a single tiny speck of dirt that could no longer be divided was all that remained. The result was considered, on a philosophical basis, the indivisible atom of dirt. Democritus also theorized, *these tiny "atoms" of matter unite to form larger masses, and the large mass could fly apart and the smallest particles would still be the tiny atoms*. This led to his theory that *nothing can be created out of nothing*, which was the precursor to the basic physical law of conservation of matter and energy.

Aristotle's Theory of the Atom: *Aristotle* (384–322 B.C.), Greece. Aristotle recorded much of the philosophy of Democritus. He also credited Democritus with the concept of the indivisible atom and accepted it as a rational, logical, philosophical explanation. *See also* Aristotle.

Epicurus' Theory of the Atom: *Epicurus of Samos* (c.341–270 B.C., Greece. Epicurus kept the atomism theory current by demonstrating how it could be the basis for perceiving reality and eliminating superstitions—the Epicurean concept of "just being happy" and living a good life without fear. Later, the Romans adopted this philosophy of the "good life." The modern word *epicurean* is derived from Epicurus, whose theory stated that *atoms were forever in constant motion, perceivable, and thus "deterministic."* Although he disputed Democritus' concept of atoms as having "free will," Epicurus was the first to suggest atomic or molecular motion, which later developed into the concepts of kinetic energy, heat, and **thermodynamics**.

Lucretius' Theory of Atoms: *Titus Lucretius Carus* (c.95–55 B.C.), Italy. Lucretius was a follower of Epicurean philosophy that life's goal should be to avoid misery. His theory was the last of the ancient classical period: *There is a natural origin of all things in the universe, including the heavens, physical objects, and living things, and all things, including living organisms, are composed of atoms of different substances*. Although atoms and molecules could not be seen at this time, his ideas preceded both the cell theory and the theory for the chemical basis of **metabolism**. He also preceded Charles Darwin by many centuries with his philosophy that all living things struggle for existence, which is one of the principles of evolution, more accurately stated as "natural selection." Up to this time, the indivisible atom was a concept that usually included inorganic matter. Lucretius is credited as being one of the first to write about the atomic structure of living things, including humans, based more on divine knowledge and his philosophy than on **empirical** evidence.

These ancient classical theories, concepts, and philosophies of atomism were mostly ignored and unexplored for over fifteen hundred years. The more modern atomic theories that developed during later periods are presented alphabetically under the names of the scientists. *See also* Bohr; Boyle; Gassendi; Heisenberg; Rutherford; Thomson.

Avogadro's Law, Hypotheses, and Number: Chemistry: *Amedeo Avogadro* (1776–1856), Italy.

Avogadro's Hypothesis: *If the density of one gas is twice that of another, the atomic mass of particles of the first gas must be twice that of the particles of the second gas.* This relationship between the density and the number of particles in a given volume of gas opened the field of quantitative chemistry, which became a more exact science because it involved the analysis of measurements made of observed phenomena. It enabled molecular weights of different substances to be compared by weighing and measuring the combining substances. Using his hypothesis to compare molecular weights of the oxygen and hydrogen, Avogadro established that it required two hydrogen atoms to combine with one oxygen atom to form a molecule of water. Avogadro also hypothesized that *gases such as oxygen and nitrogen must be composed of two atoms when in their gaseous phase.* He named the particle composed of more than one atom **molecule**, which is Latin for "small mass." This concept led to the structure of the diatomic molecule for gases (e.g., H_2, O_2, Cl_2).

Avogadro's Law: *Equal volumes of gases at the same temperature and pressure contain the same number of molecules, regardless of the physical and chemical properties of the gases.* This is true only for a "perfect gas." Avogadro knew that all gases expand by equal amounts as the temperature becomes greater—assuming that the pressure on a gas remains unchanged. Through some insight on his part, he realized that if the volume, pressure, and temperature were the same for any type of gas, the number of particles of each of the gases, existing under the same circumstances, would be the same. His reasoning that all gases under the same physical conditions have the same number of molecules was based on the fact that all gas molecules have the same average kinetic energy at the same temperature. Other physicists of his day called this unique law *Avogadro's hypothesis.*

The scientists of this period of history did not completely understand this concept and its relationship to the atomic weights of elements. This delayed the use of Avogadro's theories and principles for about five decades until they were rediscovered and applied to modern chemistry. In 1858 the Italian chemist Stanislao Cannizzaro, used Avogadro's hypothesis to show that molecular weights of gases could be definitely determined by weighing 22.4 liters of each gas; thus the results could explain molecular structure. *See also* Cannizzaro.

Avogadro's Number: *6.023 × 10²³ is the number (N) of atoms found in 1 mole of an element.*

In other words, 1 mole of any substance, under standard conditions, contains 6.023×10^{23} atoms. Avogadro's number provides scientists a very easy and practical means to calculate the mass of atoms and molecules of substances. As an example, the number of atoms in 12 grams of the common form of carbon 12 equals the *atomic weight* of carbon 12 (6 protons + 6 neutrons in the nucleus); thus 12 grams of carbon is equal to 1 mole of carbon. This is expressed as the constant N and applies to all elements and also to molecules of compounds.

Scientists assigned the simplest atom, hydrogen, an atomic weight of 1, which then results in the weight of 2 for a diatomic molecule of hydrogen gas (H_2). It was determined that at 0°C, under normal atmospheric pressure, exactly 5.9 gallons (or 22.4 liters) of hydrogen gas weigh exactly 2 grams (22.4 liters of any gas, under the same conditions, equals its atomic or molecular weight in grams). This established that the atomic weight of any element, expressed in grams, is 1 mole. Avogadro's number is one of the basic physical constants of chemistry. Thus, 22.4 liters of any gas weighs the same as the molecular weight of that gas—1 mole. Two other examples: one molecule of H_2O has a weight of 18 (2 + 16), so 18 grams of water is equal to 1 mole of water; sulfuric acid, H_2SO_4, has the molecular weight of 98 (2 + 32 + 64), so 1 mole of sulfuric acid equals 98 grams of H_2SO_4.

Using this constant makes chemical calculation much easier. All that is needed to arrive at a mole of a chemical is to weigh out, in grams, the amount equal to its atomic or molecular weight. These examples can be changed to kilograms by multiplying grams by 1000, but they are still equivalent as a molar amount.

Using the *kinetic theory of gases* and the *gas laws*, it is now possible to calculate the total number of molecules in 22.4 liters (1 mole) of a gas. The figure turned out to be six hundred billion trillion, or 600 followed by 21 zeros—more exactly 6.023×10^{23}.

B

Baade's Theory of Star Types: Astronomy: *Wilhelm Heinrich Walter Baade* (1893–1960), United States.

Population I stars are like our sun and are found in the disk portions of galaxies. Population II stars are found in the "halo" region of galaxies.

As a result of his observations at the Mt. Wilson Observatory located in Pasadena, California, Baade developed the concept of two different types of stars and his theory of galactic evolution, which was based on the following characteristics of the two star populations:

Population I Stars	*Population II Stars*
a. Population I stars are younger halo stars (formed more recently than Population II).	a. Population II stars are older disk stars (formed early in galactic history).
b. Thus, Population I stars have more "heavy metals."*	b. Population II stars have fewer "heavy metals."*
c. Population I stars have lower velocities as compared to our sun (disk stars).	c. Population II stars have random orbits and higher velocities than Population I stars.
d. Orion Nebula is an example of a younger Population I disk with more metals than sun.	d. Stars in the "galactic bulge" are old, but received heavy elements from **supernovas**.

*"Heavy metals" refer to all elements heavier than hydrogen and helium.

Based on the work of Henrietta Leavitt (1868–1921) and Edwin Hubble that determined the relationship of periodicity and luminosity of Cepheid-type stars that vary in brightness, Baade further theorized: *The period/luminosity relationship was valid only for Population II-type Cepheid stars.* His work, combined with the results of other astronomers, led to methods for determining the distance (in light-years), size, and age (in billions of years) of Andromeda and other galaxies. Baade concluded that the Milky Way **galaxy** was larger than average, but, by far, it is not the largest (or oldest) galaxy in the universe.

Baade is also credited with discovering two minor planets (more likely **asteroids** than planets) that follow very elliptical orbits. One extends from the asteroid belt (between Mars and Jupiter) to beyond Saturn. He named it Hidalgo, Spanish for a person of noble birth. The other he named Icarus after a mythical Greek god. The "planetoid" Icarus has an elliptical orbit inside Mercury's orbit and sweeps past earth. *See also* Zuckerandl.

Bacon's Concept of Inductive Reasoning: Philosophy: *Francis Bacon*, 1st Baron Verulam, Viscount St. Albans (1561–1626), England.

Francis Bacon was a philosopher and writer whose book *Novum Organum* has influenced every scientist since his day by introducing the logic of induction and devising his "scientific method," which in essence states:

1. *Approach the problem without prejudices; proceed with inquiry.*

2. *Observe situations accurately and critically.*

3. *Collect relevant facts and data from observations; make measurements.*

4. *Infer by use of analogies based on characteristics of observed facts.*

5. *Draw general conclusions from the specific to the general.*

6. *Correct initial conclusions with new insights. Truth comes from error—not confusion.*

His *inductive method* was a great improvement over the Aristotelian *deductive method* and the ancient philosophical thought processes of arriving at conclusions. Bacon's inductive reasoning improved the way scientists observed and experimented. It also improved the process of establishing scientific hypotheses that could lead to new theories, principles, and laws of nature. Of great importance was his idea of generating tentative conclusions that could be addressed and corrected by further scientific investigations. Bacon was the first to observe that the coastlines on both sides of the Atlantic Ocean (Europe/Africa and North/South America) seemed to fit each other. Years later, this concept was developed into the theory of continental drift by Suess, and Wegener. *See also* Ewing; Hess; Suess; Wegener.

Baeyer's Theory of Compound Stability: Chemistry: *Johann Friedrich Adolph von Baeyer* (1835–1917), Germany. He was awarded the 1905 Nobel Prize in Chemistry.

Chemical compounds (molecules) are less stable the more they depart from a regular tetrahedral structure.

A regular tetrahedron is a four-sided (faces) polyhedron. Each face is a triangle. It has four verticals and six line segments that join each pair of verticals. It may also be described as an analog of a three-dimensional triangle. This is the typical structure for some crystals and carbon **compounds**, where the carbon atom provides a **covalent bond** to each of the four corners of the tetrahedral atom to other elements. (See Figure V3 under Van't Hoff.) Think of the tetrahedron structure of the carbon atom as having one electron at each of the four corners. Each of these electrons can be shared with the outer electrons of other atoms to form a wide variety of structures, such as long chains of carbon atoms with branches or rings whose "skeleton" is formed by connecting carbon bonds. This unique tetrahedron structure for carbon makes it important for the formation of the many different types of organic molecules that make up plants and animals. *See also* Kekule; Van der Waals; Van't Hoff.

Bahcall's Theory for the Solar Neutrino Model: Astronomy: *John Noris Bahcall* (1934–), United States.

The sun produces 10^{36} neutrino events every second (solar neutrino units or SNU) at a density (flux) of 8 SNU.

A **neutrino** is an elementary particle classed as a **lepton** (somewhat like an electron) that has zero mass at rest as well as a zero electrical charge. It has other characteristics that make it useful for studying other minute particles in accelerators. The solar neutrino model presents several problems related to the number and types of emissions of neutrinos from the sun. John Bahcall and several other theorists predicted that neutrinos, which are considered weightless, will strike the earth and not be absorbed as are some of the other heavier particles created by the fusion reaction taking place in the sun. His theory was tested by others but did not seem to hold up very well. This prompted Bahcall to consider several options to his theory. One was that the sun was going through a passive phase and that only over a long period of time (cycles) would his predictions be accurate. Another consideration was that the neutrinos were decaying before they reached earth, thus causing a lower count than his predictions. Still another consideration was that perhaps the entire solar neutrino model was wrong, which caused a false count in his predicted neutrino rate and density. The problem is still not settled and is now left up to the development of better instrumentation or revised and improved theories to account for the extent of neutrino production by the sun. More recent speculation involves the vast amounts of dark matter (about 90 percent of all matter in space) that may be composed of neutrinos left over from the big bang. It is now estimated that one type of neutrino, called the *electron neutrino*, is not exactly massless, but has a tiny mass of about 0.5 eV to 5.0 eV, which is less than 1 millionth the mass of a regular electron. *See also* Bethe; Birkeland; Fermi; Pauli.

Bakker's Dinosaur Theory: Biology: *Robert Bakker* (1945–), United States.

Dinosaurs were warm-blooded, similar to mammals, and not cold-blooded as are reptiles.

Robert Bakker based his theory of warm-blooded dinosaurs on the following evidence: (1) Bones of warm-blooded animals, such as mammals and birds (including dinosaurs), have blood vessels, while cold-blooded reptiles' bones exhibit growth rings; (2) cold-blooded animals cannot withstand large variations in climate, such as the cold northern parts of the United States and Canada, yet dinosaur fossils have been found in cold northern climates; and (3) the *prey ratio* is much higher for warm-blooded mammals than reptiles—that is, the food consumption for warm-blooded animals is many times higher (per unit of body weight) than it is for cold-blooded reptiles. Fossil evidence suggests dinosaurs consumed vast amounts of both plant and animal foods.

From these data, Bakker concluded that dinosaurs were more closely related to birds than reptiles, both having a common ancestor known as *thecodonts*, which means "animals with teeth embedded in the jaws." Bakker's theory created much discussion in the field of paleontology and raised concerns about some of the concepts of evolution. Not all scientists agree that dinosaurs were warm-blooded. His theory is still being debated.

Baltimore's Hypothesis for the Reverse Transfer of RNA to DNA: Biology: *David Baltimore* (1938–   ), United States. David Baltimore, Howard Martin Temin, and Renalto Dulbecco jointly received the 1975 Nobel Prize for physiology or medicine.

A special enzyme, called reverse transcriptase, will reverse the transfer of genetic information from RNA back to DNA, causing the DNA possibly to provide information to protect cells.

Previous work with **DNA** and **RNA** indicated that genetic information could be passed from DNA to RNA but not the other way around. David Baltimore and Howard Temin independently announced that the enzyme reverse transcriptase enabled RNA to pass some genetic information to DNA, which possibly could aid cells to fight off cancer and other diseases, such as HIV/AIDS.

Baltimore also worked on the replication of the polio virus and continues research on the HIV **retrovirus**, which was identified by other biologists. Baltimore and other virologists hope their research will lead to a better understanding of the relationship between the HIV retrovirus and AIDS. Biomedical researchers are attempting to find an effective vaccine that will prevent the damage the HIV virus does to the immune system or prevent AIDS by immunization. *See also* Dulbecco; Gallo; Montagnier; Temin.

Barringer's Impact Theory of Craters: Geology: *Daniel Moreau Barringer* (1860–1929), United States.

Craters were formed on the planets (including earth) and the moon by the impact of large extraterrestrial objects such as meteors, asteroids, and comets.

Following is the evidence Barringer developed for his theory:

1. The large amount of silica powder found at crater sites could be formed only by very great pressure.

2. In the past, large deposits of meteoritic iron "globs" were found at the rims of craters. Most of it was removed many years ago by humans.

3. Rocks from deep in the craters are mixed with meteoritic material.

4. There is no evidence of volcanoes at crater sites. Therefore, they could be ruled out as a possible cause of impact craters.

Barringer's impact theory for craters is based on his study of the famous meteor crater (also referred to as the Barringer meteorite crater) located near Flagstaff, Arizona. Estimated to be 20,000 to 25,000 years old, it is almost 1 mile across and 600 feet deep. As meteorite impact craters go, it is considered small.

Barringer was not the first to study this crater or come up with theories of crater formation on earth. He did, however, establish the impact theory for craters, which is generally accepted within the scientific community. He did this despite having at one time agreed mistakenly with the theory that the meteor crater was the result of the impact of a meteor of the same size as the crater itself. (The current estimation of the size of the meteorite that impacted to create the Barringer crater is about 35 feet in diameter. It was a very dense iron meteorite weighing about 10,000 tons.) After Barringer found small pieces of nickel-iron rocks in the area, he spent a great deal of money establishing a mining company to extract the meteorite iron thought to be at the bottom of the crater. However, he was unsuccessful in finding significant deposits. Today, his theory is still the best explanation for most craters, including the Barringer meteorite crater found in Arizona.

Beaumont's Theory for the Origin of Mountains: Geology: *Jean Baptiste Armand Leonce Beaumont, Elie de* (1798–1874), France.

Mountains were rapidly formed by the distortion of molten matter as it cooled in the earth's crust.

Jean Beaumont's theory is an explanation for the formation of mountains consisting mainly of basalt rocks, but not sedimentary shales or layered limestone. His theory is still considered viable by some biologists and geologists, particularly by those who believe in the concept of *catastrophism*—theories that deal with the different types of catastrophic events that occurred in the past. These include earthquakes and volcanoes, which possibly are responsible for the formation of mountains, as well as catastrophic meteor impacts and major climate changes. The major evidence in support of Beaumont's theory is that "roots" of mountains are less dense than the rocks found at their higher elevations.

Modern theory for the origin of mountains is based on the concept of the earth's crust being raised above the surrounding area by the warping and folding of surface rock into layers. Another modern concept is *plate tectonics*: large plates on the ocean floor and under the continents move and crash into each other over eons. This plate movement, at a depth of 25 to 90 miles, has been ongoing for the past 2.5 to 3 billion years and still continues. The crashing together of the edges of these plates causes the development of earthquake fault lines similar to those located in California, eastern Europe and Asia. This process also is responsible for building the global distribution of mountains, as well as causing earthquakes and volcanoes. Mountains are formed in either a ring configuration, as in the Olympic Mountains in Washington State, or, more often, in ridges linked together, as in the Sierra Nevada range. A third type are groups of ranges similar to the Rocky Mountains in the western United States, the Andes in South America, the Alps in Europe, and the Himalayas in Asia. Beaumont's theory, while not completely wrong, is too limited as a geological concept for the origin of mountains. *See also* Buffon; Cuvier; Eldredge; Gould.

Becquerel's Hypothesis of X-Ray Fluorescence: Physics: *Antoine Henri Becquerel* (1852–1908), France. Antoine Becquerel shared the 1903 Nobel Prize for physics with Marie and Pierre Curie.

The exposure of fluorescent crystals to ultraviolet light will produce x rays.

Antoine Becquerel's concept is an excellent example of how his hypothesis, which proved false, later resulted in a discovery of great importance.

Wilhelm Roentgen discovered x rays in 1895. Becquerel believed he could produce x rays by exposing his **fluorescent** crystals (salts) to sunlight (**ultraviolet** radiation). He placed his crystals on a photographic plate covered in black paper and then exposed both to sunlight. His original hypothesis assumed that the photographic plate had been darkened by what he incorrectly thought was exposure to x rays passing through the paper from the crystals. He inadvertently left an unexposed, wrapped, photographic plate in a desk drawer with some of his fluorescent crystals on top of the plate. To his amazement, when the plate was developed, it was darkened as if it had been exposed to something coming from the crystals—obviously not ultraviolet light, since it was stored in a dark drawer. Since neither the plate nor the crystals were exposed to the sunlight, he concluded that his original hypothesis was incorrect. He now hypothesized that the crystals gave off some form of penetrating radiation (later identified as radiation of short wavelengths with an electrical charge such as beta and gamma rays). He continued to experiment and found that the radiation could be deflected by a magnet and thus must consist of minute charged radiation particles. Becquerel is credited with discovering *radioactivity*. *See also* Curies; Roentgen; Rutherford.

Bernoulli's Principle: Physics: *Daniel Bernoulli* (1700–1782), Switzerland.

The sum of the mechanical energy of a flowing fluid (the combined energy of fluid pressure, gravitational potential energy, and kinetic energy of the moving fluid) remains constant.

Daniel Bernoulli's principle is related to the concept of energy conservation of ideal fluids (gases and liquids) that are in a steady flow. This principle, used by mathematicians and engineers to explain and design many machines, further states that if fluid is moving horizontally with no change in gravitational potential energy and if there is then a decrease in the fluid's pressure, there will be a corresponding increase in the fluid's velocity (or vice versa). One example of this aspect of the principle is the design of airplane wings. The air flowing over the upper curved surface of the wing moves faster than the air passing the underside of the wing, creating a pressure differential. The air must travel faster over the curved top of the wing and thus the pressure is less (i.e., the air molecules are spread further apart). While on the underside of the wing, the air flows slower (molecules closer together) and thus exerts greater pressure. This causes the wing to be "pushed" up, keeping the aircraft in flight. This upward pressure on the wing is called *lift*, but it might be more appropriate to call it *push*.

A similar part of the Bernoulli principle states for example, that if there is a partial constriction in a pipe or air duct, the velocity of the fluid (gas or liquid) will increase as the pressure increases. This is known as the *Venturi effect* and can be demonstrated by the narrow nozzle of a garden hose that constricts, and thus speeds up, the flow of water as the water pressure forces the same amount of water through a smaller opening. A spray bottle or atomizer works on the same principle. It is named after G. B. Venturi (1746–1822), who first described the effect by constrictions on water flowing in channels.

Berzelius' Chemical Theories: Chemistry: *Baron Jons Jakob Berzelius* (1779–1848), Sweden.

Berzelius' Electrochemical Theory: *Molecules that make up compounds carry either a negative or positive electric charge.*

Baron Berzelius's work with the electrochemical nature of chemical reactions led to his concept of **catalysts** being related to the speeds at which chemical reactions take place. He not only gave this concept the name catalyst but also coined the names *protein* and *isometric*. However, his positive-negative, or, as it became known, his dualistic view of compounds, did not hold up very well for future theories related to organic chemistry.

Berzelius' Theory for Chemical Proportions: *The proportion of chemicals in a reaction is related to the masses (atomic weights) of the molecules involved in the reaction.*

This theory allowed Berzelius to develop an accurate table of atomic weights for elements and molecules. Somewhat oddly, despite understanding the proportional relationship of atomic weights in chemical reactions, he did not accept Avogadro's hypothesis or number. Even so, his accurate measurements of atomic weights of chemicals were one of his most important contributions to chemistry.

Berzelius' Theory of Radicals: *Groups of atoms can act as a single unit during a chemical reaction and have at least one unpaired electron.*

Baron Berzelius called these *radicals* because of the nature of these groups of atoms to act as a singular electrically charged unit (ions), for example, OH^-, $SO4^{--}$, $NH4^+$, and NO_2^-, all of which have a charge. They are short-lived, highly reactive charged particles that can initiate a chemical reaction by splitting molecular bonds. It was later discovered that ionizing radioactivity can cause illness, including radiation poisoning, and death. Other causes for the formation of free radicals in the tissues of living organisms are not completely understood, but their formation is related to normal metabolism within organisms. The role of free radicals as they affect cells and accelerate the aging process in humans is currently being studied.

Berzelius also developed the system for naming elements by the first one or two letters of their names (e.g., H for hydrogen, O for oxygen, S for sulfur, Na for Natrium (sodium), Co for cobalt, and Ra for radium). *See also* Avogadro; Dalton; Dumas.

Bethe's Theory of Thermonuclear Energy: Physics: *Hans Albrecht Bethe* (1906–   ), United States. Hans Bethe was awarded the 1967 Nobel Prize for physics.

Carbon-12 atoms found in all stars undergo a series of catalytic reactions that convert hydrogen nuclei (^+H protons) into helium nuclei through the process of a thermonuclear reaction, releasing 17.5 million electron volts of energy (17.5 MeV).

Bethe's theory became known as the *carbon cycle*. Others had previously determined the sun is composed of about 85 percent hydrogen and 10 percent helium. Bethe postulated his thermonuclear theory as the explanation for the tremendous, long-lived source for the energy produced by the stars, including our sun. Although the thermonuclear reaction involving just one carbon-12 atom and a few hydrogen protons will not produce much energy, the stars have an enormous quantity of hydrogen. This reaction has continued over billions of years and produces prodigious amounts of energy. One unsolved problem was why the reaction did not take place faster and blow up the stars similar to a hydrogen (fusion) bomb. Herman Ludwig Ferdinand von Helmholtz suggested that gravitational forces slowed the contraction of hydrogen to keep the system running. This theory did not hold up. Sir Arthur Stanley Eddington suggested that the hydrogen-to-helium reaction could be sustained in the stars if their centers contained very high-temperature gases that would force the nuclei together. Experiments on earth with high-pressure and temperature-heavy hydrogen plasmas to replicate the fusion process of the sun indicated that Bethe's and Stanley's theory is correct. (See Figure O1 for heavy hydrogen under Oliphant.)

The essence of this thermonuclear reaction is that four protons (4 $^+$H) are converted into a helium nucleus ($^{++++}$He), with the carbon-12 atom acting as a catalyst that is not consumed. The hydrogen involved are isotopes—deuterium (D) and tritium (T)—which are forms of heavy hydrogen. The reaction can be written as: D + T + $e \rightarrow {}^4$He + n + 17.5 MeV of energy, where e is an input of energy required to start the reaction and n is radiation. This process led others to develop the "fusion" H-bomb, which is many times more destructive than the nuclear "fission" atomic bomb but produces less harmful radiation. For the past half-century, research had continued to attempt to achieve a similar controlled thermonuclear reaction to produce prodigious amounts of controlled energy. *See also* Hoyle; Gamow; Teller.

Birkeland's Theory of the Aurora Borealis: Physics: *Kristian Olaf Bernhard Birkeland* (1867–1917), Norway.

The aurora borealis is caused by rays (charged particles) from the sun that are trapped in earth's magnetic field and concentrated at the polar regions.

The aurora is a curtain-like, luminous, greenish-white light produced by upper atmospheric atoms and molecules that become ionized after being struck by electrons, thus emitting radiation. It is a large-scale electrical discharge affected by the solar wind and earth acting as a **magnetosphere** "generator" that concentrates the aurora at the polar regions.

Kristian Birkeland studied this phenomenon for some time and arrived at his theory from his knowledge of cathode rays recently produced and named by Eugen Goldstein (1850–1930). He recognized the relationship of the glowing charged particles in cathode rays whose directions could be altered by magnetism.

Birkeland made another important contribution involving the great worldwide demand for nitrogen fertilizer. The supply was limited to guano (bat dung found in caves) and some natural nitrogen compounds. However, the atmosphere is about 78 percent nitrogen and could be an unlimited source. Birkeland and Samuel Eyde (1866–1940) developed a process by which air was passed through an electric carbon arc and produced nitrogen oxides, which were then dissolved in water to form nitric acid. The nitric acid reacted with lime to add calcium to produce calcium nitrate, an excellent fertilizer. It required great amounts of electricity to operate the electric arc, and shortly afterward, the Haber process superseded the electric arc as a means of fixing atmospheric nitrogen. *See also* Haber.

Black's Theories of Heat: Chemistry: *Joseph Black* (1728–1799), Scotland.

There are three aspects to Black's theory. One deals with solids changing to liquids (fusion), one involves the change of liquids into gases (vaporization), and the third relates the capacity of heat required to a specific temperature change of a given mass.

Black's Theory of Latent Heat of Fusion: *The heat of fusion is the heat capacity required to change 1 kilogram of a substance from a solid to a liquid without a temperature change.*

Black's Theory of Latent Heat of Vaporization: *The heat of vaporization is the heat capacity required to change 1 kilogram of a substance from a liquid to a gas without any temperature change.*

Black's Theory of Specific Heat: *Specific heat is the amount of heat required to raise 1 kilogram of a substance by 1 degree kelvin.*

Joseph Black proposed these theories after experimenting and making many

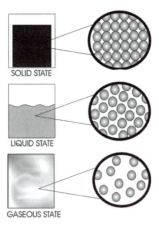

SOLID STATE

LIQUID STATE

GASEOUS STATE

Figure B1: States of matter: A solid state exists when the substance has a definite shape and volume and tends to maintains its shape and volume. A substance in the liquid state has a definite volume but no definite shape; it flows and takes on the shape of its container. The liquid state is between the solid and gaseous states. A substance in the gaseous state has a lower density than do solids and liquids, and it will expand to fill the extent of its container. The states of matter are related to the densities of matter and the kinetic energy of their constituent particles.

measurements involving changes in the states of matter (e.g., water to ice and boiling water to steam; see Figure B1). He was the first to distinguish between temperature and heat, a distinction that many still confuse today. *Temperature* is based on the law of thermodynamics and is the degree of hotness or coldness transferred from one body to another as measured in degrees Celsius, kelvin, or Fahrenheit. It is a measure of the average (mean) energy of the motion of molecules and atoms in a substance in internal equilibrium. *Heat*, on the other hand, is a form in which energy is transferred from one body to another. Heat always flows from a substance that contains more **energy** to one with less. Thus, the temperature of the first substance is reduced, and the second increased until equilibrium between the two is established.

During the late eighteenth and early nineteenth centuries, a number of physicists developed the science of heat, later named *thermodynamics*, the second law of which states that heat flows naturally from hot to cold, but never the other direction (entropy).

Black's work in several ways was the beginning of modern chemistry. More important, he was one of the first to measure chemical reactions quantitatively. The reaction of limestone with acid that produced an effervescence that he called "fixed air," which later proved to be carbon dioxide, was one of his most famous experiments. *See also* Carnot; Clausius; Joule; Kelvin; Maxwell; Mayer; Thompson; von Helmholtz.

Bode's Law for Planetary Orbits: Astronomy: *Johann Elert Bode* (1747–1826), Germany.

Bode's law is a numerical system for determining the average radii (distance) of a planet from the sun, calculated in astronomical units (AU). An AU is the average distance of earth from the center of the sun—approximately 93 million miles.

Start with a series of numbers where each one is twice the preceding number, viz., 3, 6, 12, 24, 48, 96 . . . , then add 4 to each number, viz., 7, 10, 16, 28, 52, 100 . . . , then divide the sum of each by 10. The answer is the mean radii of the planetary orbits in astronomical units (AU), which is the planet's mean distance from the sun—for example:

$$3 + 4 = 7 \div 10 = 0.7 \text{ AU}$$
$$6 + 4 = 10 \div 10 = 1.0 \text{ AU}$$
$$12 + 4 = 16 \div 10 = 1.6 \text{ AU}$$
$$24 + 4 = 28 \div 10 = 2.8 \text{ AU}$$
$$48 + 4 = 52 \div 10 = 5.2 \text{ AU}$$

Bode's law is really the mathematical expression of a concept proposed by Johann Titius (1729–1796) in 1766 or 1772. It is based on Titius's idea that a simple numerical rule governs the distance of planets from the sun. A few years later Bode proposed a useful combination of simple numbers that he claimed could predict the location of unknown planets. It is unknown if this is some true relationship of the nature of the solar system or just coincidence. Most astronomers of his day were unimpressed with his number sequence because the rule did not apply for the planets Neptune and Pluto. Bode's law predicted a planet between Mars and Jupiter, but none was found until Giuseppe Piazzi (1746–1826) discovered a very small (about 650 miles in diameter) asteroid-like planet, Ceres, in 1801. Ceres was located at 2.55 AUs from the sun in an area with many, many asteroids. Bode's law was finally accepted when he accurately predicted the location of a yet-to-be discovered planet. Using a telescope, William Herschel (1738–1822) located Uranus in 1781, exactly where Bode's numbers indicated it should be, at 19.2 AUs. Bode was given the privilege of naming this new planet, calling it Uranus after the Greek god of the sky.

Bohm's Interpretation of the Uncertainty Theory for Electrons: Physics: *David Joseph Bohm* (1917–1992), United States.

The electron has a definite momentum and position and is thus a real particle, with both wave and particle characteristics, but this duality is the result of new "pilot waves" that connect the electron with its environment.

David Bohm did not completely agree with the Heisenberg uncertainty principle or with then current interpretations of **quantum** theory. He considered Heisenberg's theory as presenting only a *description* of the behavior of an electron and not a *view* of the electron because it stated neither the position nor momentum (mass times velocity) of the electron could be determined at the same instant. Bohm claimed that this uncertainty does not represent the **deterministic** nature of reality—that is, an event cannot proceed its cause.

Bohm's pilot wave is not the classical or traditional explanation of quantum or indeterminacy theories. It can be measured only by complex mathematics, not by experimentation. Although Bohm's pilot wave interpretation maintains the concept of "real nature" as being deterministic, it is not as well accepted as is Bohr's interpretation. Today, most physicists accept the latter's theory. *See also* Bohr; Dehmelt; Einstein; Heisenberg; Planck.

Bohr's Quantum Theory of Atomic Structure: Physics: *Niels Hendrik David Bohr* (1885–1962), Denmark. Bohr was awarded the 1922 Nobel Prize for Physics.

Niels Bohr based his theory for the structure of the atom on Ernest Rutherford's famous experiment demonstrating that atoms were composed of very small, heavy, dense, positively charged central nuclei surrounded at some distance by very light, negatively charged particles, referred to as *electrons*. This concept of the negative electrons orbiting the positive nucleus was somewhat similar to planets orbiting around the sun. This classical mechanical-electrodynamic concept presented a problem in the sense that electrons carry a negative electrical charge, and according to the laws of physics, they should radiate energy as they orbit the nucleus, which would result in instability and cause them to spiral into the positively charged nucleus. Thus the conservation of momentum would be violated. Bohr solved this problem of the atom's potential instability by postulating that the circumference of the orbit must be equal to an integral number of wavelengths. The extension of this idea led to the development of quantum mechanics.

Bohr's Quantum Mechanics Theory for Atoms: (1) *Electrons reside in discrete energy levels (similar to the shells or orbits of Rutherford's model) in which they move. As long as they remain in their orbit, they do not emit radiation. Therefore, these energy levels (orbits) are stable.* (2) *Electrons move in stable orbits because they can only emit or absorb discrete radiation "packets" of energy that are equal to the difference between the original and the final energy levels of the electrons. The quanta "packets" of energy are absorbed or radiated when electrons change from one orbit to another.*

To account for the conservation of momentum, Bohr assigned specific values to orbits and later to suborbits. This led to his concept that when an electron emitted a quantum of energy (photon), it would move to a lower orbit (lower energy level). Conversely, when an electron absorbed a quantum of energy, it would move to an outer or higher orbit (energy level). This became known as a *quantum leap*. By using Planck's constant ($\hbar$) he measured the difference in radiation for these energy level changes by $\hbar v$, where v is the frequency of the radiation. These developments led Bohr to another principle.

Bohr's Correspondence Principle: *The quantum theory description of the Bohr atom relates to events on a very small scale, but corresponds to the older classical physics, which describes events on a much larger scale.*

This principle is based on electrons' obeying the principle of quantum mechanics but with limits similar to Newtonian classical mechanic. Thus, his model of the atom could exist only if electrons exhibited both wave and particle properties, which explains how electrons, as standing waves, could move in orbits without emitting radiation but still have particle characteristics. Bohr's next principle is related to the quantum nature of photons and electrons.

Bohr's Complementary Principle: *The electron can be behave in two mutually exclusive ways. It can be either a particle or a wave.*

The wave-particle duality was demonstrated by others and is accepted today as the duality nature of quantum particles. Bohr also was the first to theorize that an electron could enter a nucleus and cause it to be excited and unstable. This led to his next contribution.

Bohr's Theory of a Compounded Nucleus: *The nuclei of atoms are compounded or composed of distinct parts. The heavier the nuclei the more "parts" they contain and more likely to be unstable and break up.*

This led to his next theory.

Bohr's "Droplet Model" Theory: *The impact of a neutron (corresponding to a "droplet") on a very heavy nucleus can cause the heavy nucleus to be compounded and become unstable and fission or split into two parts, whose total mass almost equals the mass and charge of the original heavy nucleus.*

Later, Otto Hahn (1879–1968), who discovered *protactinium*, and Fritz Strassman (1902–1980) chemically identified fragmentary decay particles of uranium predicted in Bohr's model, but did not identify it as **fission**. This decay reaction, which occurs when "compounded" heavy nuclei break into two or more lighter radioactive nuclei, was called *nuclear fission* by Lise Meitner and Otto Frisch. These experiments were the first evidence for fission of the rare uranium radioactive isotope U-235, which ended as a small radioactive isotope of barium-56. This led to the use of another fissionable element, plutonium, used in atomic bombs, which Bohr assisted in developing. Among Bohr's other contributions was his early (1920) theoretical description of the Periodic Table of Chemical Elements which he based on his theories of atomic structure. *See also* Bohm; Dehmelt; Frisch; Hahn; Meitner; Planck; Rutherford.

Bok's Globules Theory of Star Formation: Astronomy: *Bart Jan Bok* (1906–1983), United States.

The small, circular "clouds" of matter that are visible against a background of luminous gas or by the light from stars are actually massive "globular-like clouds" of dust and gas that are in the process of condensing to form new stars.

Bart Bok identified these interstellar dark globules near the **nebula** Centaurus, referred to as IC 2944. Only recently has his theory been reexamined as a possible creation, regeneration, or rebirthing of stars to explain one of the theories for the origin of our **universe**; the ever-continuing universe. Bok and his wife, Priscilla, studied the Milky Way galaxy and were among the first to propose that it was a spiral structure, not just one big mass of stars. This was later confirmed by Water Baade when he observed bright stars in the "arms" of the Andromeda galaxy. *See also* Baade.

Boltzmann's Laws, Hypotheses, and Constant: Physics: *Ludwig Edward Boltzmann* (1844–1906), Austria.

Boltzmann's Law of Equipartition: *The total amount of energy of molecules (or atoms) is equally distributed over their kinetic motions.*

In other words, on the average, the energy of molecular motion is distributed with discrete degrees of freedom within an ideal gas. This led to Boltzmann's description of how *the total energy of a gas is distributed equally among the molecules in the gas, viz., heat.* This became known as the Maxwell-Boltzmann distribution equation, which is based on the Boltzmann constant: $k = R/N = 1.38 \times 10^{-23} \ J/K$, where k is the Boltzmann constant, R is the universal gas constant, N is the number of molecules in 1 mole of gas as per Avogadro's number, and J/K is joules per degree of kelvin.

Boltzmann Distribution Equation: *The probability exists that a molecule of a gas will be in energy equilibrium with both the position and movement of the molecule and will be within an unlimited ranges of values.*

This is another way of stating the energy distribution of gas molecules. It states that atoms and molecules should obey the laws of thermodynamics.

Boltzmann's Entropy Hypothesis: *The entropy (the measure of disorder in a system) in a given state is directly proportional to the logarithm of the number of distinct states available to the system.*

Entropy was the term given to the concept of the second law of thermodynamics. It is based on the fact that unless energy is added to a system, the system will always proceed to a state of disorganization and finally to a state of equilibrium. Boltzmann supplemented the mathematics related to thermodynamics using a statistical treatment to interpret the second law of thermodynamics, which in essence states that heat can only move toward cold, never the other way around. This hypothesis can be stated as the Boltzmann constant equation: $S/k = \log p + b$, where S is entropy and k is the Boltzmann constant, which has the value of 1.380×10^{-23} joule per degree kelvin. Boltzmann thought so much of this equation that he had it inscribed on his tombstone. Most of Boltzmann's theoretical work contributed to the science of statistical mechanics. He is also known as the Father of Statistical Mechanics. *See also* Carnot; Clausius; Kelvin; Maxwell; Mayer; Rumford.

Bonnet's Theory of Parthenogenesis: Biology: *Charles Bonnet* (1720–1793), Switzerland.

All organisms are preformed in miniature (homunculi) as little beings inside the female of the species, and the "germ" of a species is constant over time.

Charles Bonnet developed this theory of parthenogenesis after discovering the female of a species of a tree aphid reproduced without the aid of male sperm (parthenogenesis, or "virgin birth"). To overcome the objections to his theory, which implied all living organisms were unchanged from the beginning of time, he proposed the concept of catastrophism. Although he was the first to explain evolution in a biological context, he did not accept the extinction and changes of species as a gradual process. He was one of the first to propose catastrophism as the cause of biological evolution, where periodic catastrophic events on earth result in great extinction and changes in all the species. *See also* Buffon; Cuvier; Eldredge; Gould.

Born-Haber Theory of Cycle Reactions: Physics: *Max Born* (1882–1970), Germany. Born shared the 1954 Nobel Prize for physics with Walter Bothe.

The sequence of energy involved in the chemical and physical reactions that form lattice ionic crystals is related to the crystal's initial state (zero pressure at zero kelvin), and to the crystal's final state, which is also at zero pressure and zero K (e.g., for a gas of infinite dilution).

The Born-Haber cycle is better known by the early work of Max Born, which resulted in the mathematical theory referred to as the cycle explaining how chemical bonds are the result of sharing or transferring electrons between atoms. As a result, several scientists applied quantum mechanics to the concept of chemical **bonding**. Born and others used the hydrogen atom as a model, and it was soon obvious that quantum mechanics could explain almost all aspects of chemistry.

Max Born and Ernst Pascual Jordan (1902–1980) developed the mathematics for introducing *matrix mechanics*. Born was among the first to provide the mathematics to explain the possibility that particles can also behave like waves. This was about the time that the duality concept of light (photons) and other types of radiation was being discussed and debated. *See also* Bohm; Bohr; Dehmelt; Frisch; Hahn; Heisenberg; Meitner; Schrödinger.

Boyle's Law: Chemistry: *Robert Boyle* (1627–1691), Ireland and England.

There is an inverse relationship between the pressure and volume of a gas when the temperature remains constant. The equation for this law is: $P \times V = c$ (pressure times volume equals a constant inverse relationship).

Robert Boyle, an Irish chemist who later worked in England, used air pumps developed by Robert Hooke to experiment with the physical conditions of gases under differing pressures while maintaining constant temperatures. He published the results in his book, *The Sceptical Chymist*. In 1662 Boyle discovered air could be compressed, and as the pressure increased, the volume decreased. He demonstrated that if the pressure on a gas was doubled, its volume would be just one-half its original volume; if the pressure was increased by one-third, the volume would decrease one-third. Boyle also noted the opposite inverse relationship existed when he used a vacuum pump to decrease the pressure that increased the volume of air. This proved to be a classical inverse relationship, which seems to be a universal constant. It was an important conclusion because it helped explain the atomic nature of gases—that is, atoms (or molecules) of gases would spread farther apart when the pressure was decreased. Conversely, the atoms would be forced closer together if the pressure increased.

Boyle was an atomist who supported the original concept of matter first proposed by the ancient Greek Democritus. It took more than a century after Boyle's work before the modern atomic theory of matter was fully developed. It might seem ironic that a scientist who is considered one of the founders of modern chemistry was also an **alchemist** who spent much of his time attempting to transmute base metals into gold. *See also* Avogadro; Charles; Gay-Lussac; Ideal Gas Law.

Bradley's Theory of a Moving Earth: Astronomy: *James Bradley* (1693–1762), England.

Parallax exhibited by the stars indicates a movement of the Earth.

To prove his theory, Bradley set up a telescope in a stationary vertical position in order to observe the same spot in the sky each night. Over time he observed a slight displacement (**parallax**) of the image of the star Gamma Draconis. At first he thought this parallax of a star viewed over a period of time from the same place on earth could mean only the star moved in relation to earth. Bradley later realized it was not really the star's motion causing the parallax, because the pattern of the star's displacement was repeated every six months. Thus, it meant the observer and his fixed telescope on a moving Earth caused the change in the star's apparent position. This concept is based on Earth's orbiting around the sun every twelve months, which means earth is in a very different viewing position in relation to the star every six months. This creates an apparent displacement called *parallax*. The diameter of earth's orbit around the sun is approximately 186 million miles (twice the radius of 93 million miles). Every six months, Bradley's telescope was 186 million miles from where he viewed the star the previous six months. Parallax seemed to place the star in slightly different positions. Astronomers using the parallax concept can mathematically determine the distance of the closer stars to our solar system. Parallax does not work very well for distant objects located in deep space. Bradley also was one of the first to conceptualize that light has a finite, not infinite, speed. This, plus the movement of earth, were important concepts for compiling accurate observations and calculations of stars. *See also* Brahe; Copernicus; Galileo; Kepler.

Brahe's Theory of the Changing Heavens: Astronomy: *Tycho Brahe* (1546–1601), Denmark.

Since a new star did not exhibit any parallax and comets come and go, there must be changes in the heavens, proving earth with its orbiting moon is the center of the universe.

Tycho Brahe was an ardent proponent of Ptolemy's concept that earth was the center of the universe, but he did not accept the idea that the universe was static. Based on his observation, Brahe devised this theory of a changing universe but incorrectly believed earth was at its center. Brahe's concept of a changing universe was unique; until that time Aristotle's concept of a permanent, unchanging universe was accepted as fact. This belief changed when Brahe discovered a new supernova (exploding star), named *Tycho's star*. Brahe also discovered a large comet in 1577 that further supported his theory.

Brahe constructed several large sextants and quadrants for his direct sight viewing since telescopes had not yet been invented. He kept a journal of all his activities as well as astronomical tables, which later proved useful to his assistant, Johannes Kepler. Brahe's most important contribution was the accurate recording of his astronomical observations, which enabled Kepler to develop his three laws of planetary motion. *See also* Copernicus; Galileo; Kepler; Ptolemy.

Buffon's Theories of Nature: Biology: *Comte George Louis Leclerc de Buffon* (1707–1788), France.

Buffon's Theory of Ecology: *The animals of an area (ecology) are the product of the environmental conditions of the land where they developed.*

Comte Buffon based this theory on his concept that earth makes and grows the plants on which the animals depend; thus the region's plants determine the region's animals.

Buffon's Theory of Natural Classes: *Animals were classed not according to genera and species, but rather in a hierarchy of man, domesticated animals, savage animals, and lower animals.*

Taxonomy and species classifications based on structure and function were not yet fully developed. Therefore, Buffon classed animals according to major categories as he interpreted their status in life.

Buffon's Theory of Species: *Animals within a hierarchical group (species), and only those within that group, can reproduce themselves.*

Buffon based his theory on empirical evidence. He observed that animals of one group from his hierarchical classification of animals would breed only with others of their kind. He was unaware of hybrids or mutations.

Buffon's Theory of the Age of Earth: *Based on a series of stages as evidenced by geological history, earth is 78,000 years old.*

Buffon rejected biblical records that contended earth's age was 6,000 years. His estimate of the age of earth was based on fossils and geology. Buffon believed earth was originally a hot body that cooled off enough for people to exist and would continue to cool, at which time all life on earth would end. His extension of the age of earth led other scientists to examine fossils and geological evidence more closely, which provided a more accurate estimate of earth's age. Today, the universe is considered to be about 14 to 16 billion years old, with earth being formed about 4.5 billion years ago.

Buffon's Theory of the Origin of the Planets: *The formation of planets was the result of a collision between a large comet and the sun.*

Buffon is credited with providing an important naturalistic history of earth.

He based his concept of the origin of the solar system and planets on the more current explanation of natural forces where cosmic "dust" circulated to form the solar system and coalesce into the planets around the sun. He used the mechanics of motion as described by Sir Isaac Newton as well as his own empirical observations. *See also* Cuvier; Darwin; Wallace.

Bunsen's Theory of the Spectrochemistry of Elements: Chemistry: *Robert Wilhelm Bunsen* (1811–1899), Germany.

Each element, when heated, emits a unique electromagnetic spectrum that can be identified by careful spectrum analysis of the emitted light.

While assisting in an experiment of spectrum analysis, Robert Bunsen refined Michael Faraday's gas burner by adding a collar that could be adjusted to control the flow of air into the gas flame. This device greatly improved the burner by providing a hotter and steady flame. Ever since, it has been known as the Bunsen burner. In addition to the field of **spectroscopy**, Bunsen contributed to the fields of electrochemistry, electrodeposition of metals, and photochemistry. He is also credited with the discovery of the new elements rubidium and cesium. *See also* Berzelius; Faraday; Kirchhoff.

C

Cagniard De La Tour's Concept of "Critical State": Physics: *Charles Cagniard de la Tour* (1777–1859), France.

The critical state exists at the point where the temperature and pressure create equal densities between a liquid and its vapor. The vapor and liquid can be in equilibrium at any temperature that is below the critical point.

The term *critical state* is also known as *critical point.* (At the critical temperature there is no clear-cut distinction between the vapor [gas] and liquid states.) (see Figure B1 under Black.) Their densities are equal, and their two phases are also equal and considered to be one phase. The concept of critical state is also associated with **critical temperature**, which is based on the work of Louis Paul Cailletet, James P. Joule, and Thomas Andrews (1813–1885). Some examples of practical applications are boilers used in home heating and industry, steam engines, and frequently in generating electricity by steam turbines. By increasing the pressure, as in a pressure cooker, the liquid and vapor can be compressed with an increase in temperature, changing the critical state until pressure is released. Food then cooks faster under the increased pressure, using less applied heat than would unpressurized cooking. *See also* Cailletet; Joule; Kelvin.

Cailletet's Concept for Liquefying Gases: Physics: *Louis Paul Cailletet* (1832–1913), France.

As pressure on most gases increases, so does the boiling point. Therefore, reducing the temperature while increasing the pressure will liquify the gas at a lower pressure.

Cailletet's concept for liquefying gases was based on the work of others who developed the concept of critical temperature. Cailletet was the first to produce liquid oxygen by compressing and cooling oxygen gas, then rapidly releasing the pressure, which further reduced the temperature. Thus, the gas reached the critical temperature point at which the gaseous oxygen turned into liquid oxygen. This process is used today to produce various liquified gases (e.g., nitrogen and helium) and is used in experimental work in nuclear physics, chemistry, and cryogenics. Compressed and liquefied oxygen are essential in the health care of oxygen-dependent patients and in many industrial processes (e.g., steel production, smelting, rocket fuels, and welding). *See also* Cagniard; Joule; Kelvin.

Calvin's Carbon Cycle: Chemistry: *Melvin Calvin* (1911–), United States. Calvin was awarded the 1961 Nobel Prize in chemistry.

The path of carbon dioxide in the chemical and physical reactions of photosynthesis occurs only in the presence of chromoplasts of living plants.

Calvin exposed plants for a few seconds to radioactive carbon dioxide containing the isotope carbon-14. Carbon-14 has the same atomic number (6 protons) and is chemically similar to carbon-12, but the carbon-14 nucleus has two more neutrons than does nucleus of C-12. Since there is no chemical distinction between C-14 and C-12, the radioactive tracer carbon was assimilated in plant chloroplasts during the process of **photosynthesis** and could be traced with radiation detection devices. Since the process is very rapid, he worked quickly to mash the cells in order to separate the carbon-14 in boiling alcohol. He then separated and identified the components and products of photosynthesis by using paper chromatography. After many experiments, Calvin identified the cycle of absorption and the use of carbon dioxide by plants, leading to a better understanding of the roles of chlorophyll and carbon dioxide in the science of photosynthesis. We now know that most plants increase their rate of growth in an atmosphere rich in carbon dioxide. CO_2 is sometimes added to the inside air in greenhouses to accelerate plant growth.

Candolle's Concept of Plant Classification: Biology: *Augustin Pyrame de Candolle* (1778–1841), Switzerland.

There is a homologous or fundamental relationship of similarities for the parts of different types of organisms.

Several classification systems for plants and animals existed at the time of Candolle's work, but Candolle was the first to use the terms *taxonomy* and *classification* synonymously. His taxonomy (naming system) was based on the recognized similarity of various body parts of near relatives of different species, thus assuming they derived from common ancestors. This concept of relating taxonomy to species influenced Alfred Wallace and Charles Darwin in the development of their theories of organic evolution. Only six volumes of Candolle's twenty-one–volume taxonomy project were published before his death. His work, superior to that of Linnaeus, is still used today. *See also* Cuvier; Lamarck; Linnaeus.

Cannizzaro's Theory of Atomic and Molecular Weights: Chemistry; *Stanislao Cannizzaro* (1826–1910), Italy. The Royal Society of London awarded Cannizzaro the Copley Medal for his contributions to science.

The atomic weights of elements in molecules of a compound can be determined by applying Avogadro's law for gases, which states that gram-molecular weights of gases occupy equal volumes at standard temperature and pressure (STP).

Fifty years after Avogadro proposed his law and number (which was mostly ignored by other scientists), Cannizzaro recognized its utility for measuring the atomic weights of atoms of elements and the weights of gas molecules. He also determined that the theory could be applied to solids if their vapor density is unknown by measuring their specific heat. Although Cannizzaro credited Avogadro, it was Cannizzaro who finally convinced the scientific community that molecular weights of gases could be determined by measuring their vapor densities. The field of quantitative chemistry rapidly advanced once atomic and molecular weights could be accurately determined. Cannizzaro's theory provided a means for defining the molecular weights of many organic compounds, as well as clarifying the structure of complex organic molecules. *See also* Avogadro.

Cantor's Mathematical Theories: Mathematics: *Georg Ferdinand Ludwig Philipp Cantor* (1845–1918), Germany.

Cantor's Theory of Infinity: *One can without qualification say that the transfinite numbers stand or fall with the infinite irrational; their inmost essence is the same, for these are definitely laid-out instances or modifications of the actual infinite.*

This theory is related to Cantor's axiom, which states that if you start with a single point on a two-dimensional surface and continue to add points on each side of the original point, they will continue to extend out in both directions. By adding point-to-point, there will be no one-to-one relationships between the two directions of points, and the lines they form can be extended forever. Another way to look at this is that if you start from zero, you can progress indefinitely (infinity) to larger and larger functions, or you can extend indefinitely in the opposite direction to smaller and smaller values.

Karl F. Gauss, the physicist and mathematician, stated that infinity was not permitted in mathematics and was only a "figure of speech." The concept of endlessness, whether in time, distance (space), or mathematically, is difficult to grasp. Galileo considered the study of infinity as an infinite set of numbers, later defined as Galileo's paradox. However, it was not until Sir Isaac Newton's and Gottfried Leibniz's development of calculus that it became necessary for a mathematical explanation of infinity. John Wallis, who developed the law of conservation of momentum, also proposed the symbol for infinity (∞), which was known as the "lazy eight" or "love knot." It was Georg Cantor, however, who postulated that consecutive numbers could be counted high enough to reach or pass infinity. His development of transfinite numbers, a group of real numbers (both rational and irrational) that represent a higher infinity, led to *set theory*, which permits the use of numbers within an infinite range.

Cantor's Set Theory: *The study of the size (cardinality) of sets of numbers and the makeup or structure (countability) of groups of rational or irrational natural numbers*. Following are two examples of sets of numbers:

$$
\begin{array}{cccccc}
1 & 2 & 3 & 4 & 5, \ldots, n, \ldots, \infty \\
\updownarrow & \updownarrow & \updownarrow & \updownarrow & \updownarrow \\
2 & 4 & 6 & 8 & 10, \ldots, 2n, \ldots, \infty
\end{array}
$$

$$
\begin{array}{cccccc}
1 & 2 & 3 & 4 & 5, \ldots, n, \ldots \infty \\
\updownarrow & \updownarrow & \updownarrow & \updownarrow & \updownarrow \\
1 & 4 & 9 & 16 & 25, \ldots, n^2, \ldots, \infty
\end{array}
$$

where n is a continuation of the sequence or set. By finding a one-to-one match in a set (in the example, a number below the number above it), even if the set is infinite, it is possible to determine the size of the number if the entire structure is unknown. If the set is infinite, some of the numbers can be eliminated without reducing the size or eliminating the structure of the set. *See also* Galileo; Gauss; Leibniz; Newton.

Cardano's Cubic Equation: Mathematics: *Gerolamo Cardano* (1501–1576), Italy.

Definition of Cubic Equation: *A polynomial equation with no exponent larger than three—Specifically: $x^3 + 2x^2 - x - 2 = 0$ $(x - 1)(x + 1)(x + 2)$, which can be further treated by algebraic functions.*

Earlier mathematicians solved equations for x and x^2 but were unable to solve x^3 (cubic) equations. (First-degree equations are linear [straight line], or one-dimensional, involving x; second-degree equations are quadratic [plane surface], or two-dimensional, involving x^2; third-degree equations are cubic [solid figures], or three-dimensional, involving x^3.) The graphic depiction for the solution of a cube is usually easier to understand than is the algebraic representation. When a graph is used to solve cubic equations, the x-axis is eliminated and reoriented. Cardano was not the first to come up with a solution to cubic equations, having been given the explanation for cubic and biquadratic equations by Niccolo Tartaglia (1500–1557), who made Cardano promise not to reveal the secret. After Cardano found a partial solution had been achieved by someone previous to Tartaglia, he published the results as his own. As the first to publish, Cardano has been credited with the discovery. As a result of this controversy, a new "policy" stated that the first person to publish the results of an experiment or a discovery, and *not* necessarily the first person who actually conducted the experiment or made the discovery, is the one given credit. This is based on the belief that science should be open and available to all rather than kept secret.

Carnot's Theories of Thermodynamics: Physics: *Nicholas Leonard Sadi Carnot* (1796–1832), France.

Carnot Cycle: *The maximum efficiency of a steam engine is dependent on the difference in temperature between the steam at its hottest and the water at its coldest. It is the temperature differential that represents the energy available to produce work. $T_2 - T_1 = E$, where T_2 is the higher temperature, T_1 is the lower temperature, and E is the energy available to do work. The efficiency is unity only if $T_1 = 0$ kelvin.*

The concept of a temperature differential is analogous to the potential energy of water flowing over a water wheel to produce work. This concept enabled inventors to develop more efficient steam engines and locomotives and became known as the Carnot cycle, based on the difference in the temperature of the steam at its highest temperature and the water at its coldest temperature—not on the total amount of internal heat. Carnot pointed out that energy (heat) is always available to do work, but it cannot completely be turned into useful work. Even so, it is always conserved; none is ever "lost." For example, within the steam engine, which is not 100 percent efficient, some energy is not available to accomplish work due to "lost" heat. The first law of thermodynamics (which means "flow of heat" in Greek) states that energy is always conserved, while the second law states there is a limit to how much of the energy can be converted into work. A system cannot produce more energy (work) than the amount of energy that is expended. Perpetual motion is impossible, as some energy is always lost to heat due to friction. *See also* Clausius; Fourier; Helmholtz; Kelvin; Maxwell.

Caspersson's Theory of Protein Synthesis: Chemistry: *Torbjorn Oskar Caspersson* (1910–   ), Sweden.

Proteins are synthesized in cells by large RNA (ribonucleic acid) molecules.

Torbjorn Caspersson invented a new type of spectrophotometer, enabling him to trace the movement of RNA. He concluded that RNA is involved in the synthesis of amino acids in the production of proteins. RNA is a type of single, long, unbranched, organic macromolecule responsible for transmitting genetic information to deoxyribonucleic acid (DNA) molecules. (See Figure C2 under Crick.) All organisms, except viruses, depend on RNA messengers to carry inherited characteristics to the DNA molecules, which can then be duplicated to pass genetic information to offspring. In RNA-type viruses, the RNA itself acts as the DNA, since it contains all the genetic information required for the virus to replicate. Caspersson was the first to determine that DNA had a molecular weight of 500,000. In addition, he discovered a way to dye specimens of DNA so that the nucleotides would appear in dark bands.

Tests comparing one person's DNA with another person's or their offspring can determine who is genetically related to whom, often years after death. Many police departments and laboratories use this procedure to test and compare hu-

man DNA when investigating the crimes of rape and murder. *See also* Chargaff; Crick.

Cassini's Hypothesis for Size of the Solar System: Astronomy: *Giovanni Domenico Cassini* (1625–1712), France.

The mean distance between the Earth and the sun is 87 million miles.

Giovanni Cassini developed this figure by working out the parallax for the distance of Mars from Earth, which enabled him to calculate the astronomical unit (AU) for the distance between the Earth and sun. He accomplished this by using calculations attained by other astronomers, as well as his own observations. The figure greatly increased the estimations at this time in history for the size of the solar system. Tycho Brahe calculated the distance between the sun and earth at just 5 million miles. Kepler's estimation of 15 million miles was better, but still greatly underestimated, while in 1824 Johann F. Encke (1791–1865) used Venus's transit with the sun to overestimate the distance as 95.3 million miles. (The correct mean distance from the sun's center is approximately 92.95 million miles = 1 AU.)

Cassini was the first to distinguish the major gap separating the two major rings of Saturn, now called the Cassini division. He also discovered four new moons of Saturn, determined the period of rotation for Jupiter as 9 hours, 56 minutes, and observed that Mars rotates on its axis once every 24 hours, 40 minutes. Cassini measured earth's shape and size but incorrectly identified it as a perfect sphere. His work is convincing, inasmuch as the Earth is not unique in the solar system and is similar to the other inner planets, with the obvious exception of its human habitation. *See also* Brahe; Spencer-Jones.

Cavendish's Theories and Hypothesis: Chemistry: *Henry Cavendish* (1731–1810), England.

Cavendish's Theory of Flammable Air: *When acids act on some metals, a highly flammable gas is produced, called "fire air."*

Henry Cavendish was one of the last scientists to believe in the **phlogiston** theory of matter, which states that when matter containing phlogiston burns, the phlogiston is released. Since his "fire air" would burn, he called it *phlogisticated air,* but it was Antoine Lavoisier who named the new gas *hydrogen,* meaning "water former" in Greek. Cavendish was the first to determine the accurate weights and volumes of several gases (e.g., hydrogen is 1/14 the density of air). Until this time, no one considered that matter of any type could be lighter than air. *See also* Lavoisier; Stahl.

Cavendish's Theory of the Composition of Water: *When "fire air" and oxygen are mixed two to one by weight and are burned in a closed, cold, glass container, the water formed is equal to the weight of the two gases. Thus, water is a compound.*

Although similar experiments and claims that water is a compound, and not an element (as believed for hundreds of years) were made by other scientists, Cavendish was credited with his discovery of hydrogen, as well as the concept of water as a compound, although he delayed publishing the results of his experiments.

Cavendish developed many concepts but had difficulty making generalizations from his experimental results. Two examples of his "new" discoveries are (1) the distinction between electrical current, voltage, and capacitance, which later led to Ohm's law, and (2) the anticipation of the gas laws dealing with pressure, temperature, and water vapor. He also foresaw the chemical concepts of multiple proportions and equivalent weights for which John Dalton and others were given credit. Cavendish also determined that oxygen gas has the same molecular weight regardless of where it is found and that the percentage of oxygen in ordinary air is approximately the same wherever found on the earth.

Cavendish's Hypothesis for the Mass of Earth: *Based on the determination of the gravitational constant and the estimated volume of Earth, its mass should be 6.6 × 10^{36} tons, with a density about twice that of surface rocks.*

Cavendish's procedure for determining the mass of Earth is known as the Cavendish experiment. It involved two small lead balls, one on each end of a single rod suspended at its midpoint by a long, fine wire. As the two large lead balls were brought close to the smaller ones from opposite directions, the force of gravity between the balls produced a minute twist in the wire that could be measured. From this twist, Cavendish calculated the gravitational force between the balls, which he used as a gravitational constant for calculating the mass of Earth. His figure was approximately 6,600,000,000,000,000,000,000 tons, very close to today's estimation of 5.97 × 10^{24} kilograms.

Chadwick's Neutron Hypothesis: Physics: *Sir James Chadwick* (1891–1974), England. Chadwick was awarded the 1935 Noble Prize for physics.

Physicists believed there were only two elementary particles in atoms: the positive proton and the negative electron. However, this concept was not adequate to explain many physical phenomena. One idea was that the internal nucleus of a helium atom has four protons, as well as two internal neutralizing electrons, which could explain the 2$^+$ charges for helium. But scientists could not identify any electrons originating from the helium nuclei when it was bombarded with radiation. Another dilemma was that nitrogen has a mass of 14 but an electrical charge of 7$^+$. This was confusing because if the nitrogen nucleus has 14 protons, it would have a charge of 14$^+$. One solution was to assign 7 negative electrons to neutralize 7 of the positive protons, meaning the nucleus of nitrogen now has 21 particles. Since this did not make sense to many scientists, it was determined that the measured "spin" of particles in the nitrogen nucleus could be only a whole number; thus there could not be 21 particles in the nucleus. Further work demonstrated that gamma rays did not eject the electrons from positive protons, nor did the radiation eject the protons. Therefore, it was hypothesized that some new particle(s) with no electrical charge and very little mass must exist in the nuclei of atoms. Some years later this "ghost particle" was found and named the *neutrino. See also* Fermi; Pauli; Steinberger.

Chadwick's Hypothesis: *Particles with the same mass as protons, but with no electrical charge, were one of the particles ejected from helium nuclei by radiation.*

Walther Bothe (1891–1957) bombarded beryllium nuclei with alpha particles (helium nuclei) and detected some unidentified **particle** radiation. Chadwick continued this work and called these nuclear particles named *neutrons*, because they carried no electrical charge but had a similar mass as protons. This discovery assisted in understanding both the atomic number and the atomic mass of elements and provided an excellent tool for further investigations of the structure of atoms. Neutron bombardment is used today to produce **radioisotopes**. Chadwick also was the developer of the first **cyclotron**, or "atom smasher," in England. Some physicists consider Chadwick's neutron discovery the beginning of the nuclear age, because it led to understanding the physics necessary to develop practical uses for nuclear fission and **fusion**, such as nuclear electric power plants and the atomic and hydrogen bombs. *See also* Anderson; Fermi; Heisenberg; Pauli; Soddy.

Chambers' Theory for the Origin of Life: Biology: *Robert Chambers* (1802–1871), Scotland.

If the solar system can be formed by inorganic physical processes without the assistance of a supreme being, then it follows that organic plants and animals can develop by a similar physical system.

Robert Chambers based his theory for the origin of life on his acceptance of Laplace's nebula hypothesis for the origin of the solar system. His concept that organisms could be created by the same laws of physics and chemistry that created the universe was accepted as a logical explanation by many scientists. But it was never accepted by theologians and the general public, since the theory denied credit for the origin of life to a supreme being (deism) or personal god (theism). Chambers proposed several other theories related to the embryonic development of species, which engendered tremendous negative reactions from the public. *See also* Baer; Haeckel; Laplace; Ponnamperuma; Swammerdam.

Chandrasekhar Limit: Astronomy: *Subrahmanyan Chandrasekhar* (1910–1995), India.

The Chandrasekhar limit is a physical constant that states, *If white dwarf stars exceed a mass of 1.4 greater than the sun's mass, they will no longer be self-supporting. As their mass increases past this limit, internal pressure will not be balanced by the outward release of pressure generated by atoms losing electrons to form ions. Thus, white dwarfs exceeding this limit will "explode."*

Interestingly, no white dwarfs have been found with a mass greater than the 1.4 limit. To date, most white dwarfs discovered average about 0.6 the mass of our sun. It is speculated that the **supernova** observed and reported by Tycho

Brahe in 1572 and another in 1604 by Johannes Kepler may each have been the collapse of a white dwarf star that pulled off mass from nearby red supergiants at a rate greater than could be eliminated by the white. This new mass was attracted by the gravitational pull of the white dwarf, causing an increase in its mass exceeding the 1.4 limit. Thus, a giant nuclear explosion occurred, creating a supernova.

Chandrasekhar also believed that as stars exhaust their nuclear fuel, they begin to collapse by internal gravitational attraction, which ceases when a balance is established by the outward pressure of ionized gases. This gas is a high concentration of electrons, leaving behind ions (charged nuclei), which become very dense. One thimblefull would weigh several tons on the earth.

It is estimated that most white dwarfs are "leftovers" of mass pulled from more massive stars. It is also speculated that in the long-distant future of the universe, all stars will become white dwarfs, and then we may have a static, or unchanging, universe, since no other bodies would be available from which they could gather in extra mass and exceed the Chandrasekhar limit.

Chang's Theories and Concepts: Mathematics: *Heng Chang* (78–142), China.

Chang's Concept of Pi: *Pi is the square root of 10, which equals 3.1622.*

This was one of the most original and most accurate methods of determining the value of pi up to this time. Many early mathematicians were intrigued by the relationship of the diameter of a circle to its circumference. This led to the idea of forming a series of geometric squares within the bounds of the circumference of a circle and then calculating the known parameters of the squares. *See also* Archimedes.

Chang's Concept of the Universe: *The universe is not a hemisphere rising over Earth, but rather the universe consists of a large sphere with Earth at the center, similar to the yolk in the center of an egg.*

The early Greeks, including Aristotle, believed the universe consisted of a series of crystal hemispheres suspended above earth. Chang's idea was the first to consider a spherical universe. Further, he developed an instrument to measure the major circles of the celestial bodies and demonstrated how they intersected at various points on earth. He also developed an instrument that used flowing water to measure the movement of the stars. *See also* Aristotle.

Chang's Concept of Earthquakes: *Earthquakes are caused by dragons fighting in the center of earth.*

From this idea Chang developed an instrument shaped in the likeness of several dragon heads; the inside of each held a ball. When an earthquake occurred, a pendulum device expelled the ball from one of the dragon's mouths, which would then determine the direction of the earthquake. Although Chang's "seismograph" was inaccurate, it was developed and used in the Far East for over seventeen hundred years before a more accurate one was designed in the West that used the Richter scale.

Chang's Theory of Capacitation: Biology: *Min Chueh Chang* (1909–1991), United States.

The male sperm must spend some time traveling in the reproductive organs of the female before it can fertilize the egg.

Chang called this time factor, whereby sperm become more potent while in the female reproductive tract, the *capacitation factor*. He also believed that male seminal fluid has a "decapacitation" substance that prevents sperm from fertilizing the egg. Although no such factor or substance has yet been identified, it is assumed it is this decapacitation factor that prevents other sperm from uniting with an egg once one sperm has penetrated the ovum.

Using rabbits, Chang conducted much of the research with in vitro fertilization, where an ova (egg) is fertilized by a male sperm outside the body and then transplanted into the female rabbit's uterus. He also used rabbits to demonstrate that progesterone would act as a contraceptive. His research pioneered the way for others to perfect human in vitro fertilization.

Chapman-Enskog Kinetic Theory of Gases: Mathematician: *Sydney Chapman* (1888–1970), England.

In the nineteenth century James Clerk Maxwell and Ludwig Boltzmann developed the general mathematical concept that describes the properties of gases as partially determined by the molecular motion of the gas particles, assuming the gas molecules follow classical mechanics. This is referred to as the *Maxwell-Boltzmann distribution* which gives gas particles a specific momentum.

The Chapman-Enskog theory provides a complete treatment and solution to the mathematics of the Maxwell-Boltzmann equation by using approximations to determine the average path of gas particles.

This joint theory is also known as the *Enskog theory*. (The mathematics for these kinetic energy equations is beyond the scope of this book.)

Chapman also used mathematics to predict the thermal diffusion of gases and the electron density at different levels of the upper atmosphere. In addition, he determined the detailed variations in the earth's magnetic field and related this to the length of the moon's day. (Since the moon keeps the same side pointed toward the earth, it rotates once every 27.3 days.) He demonstrated that the moon not only causes tidal effects on the earth's water and land, but there is a much weaker tidal effect on the earth's atmosphere. Chapman's work enabled other scientists to measure more accurately the kinetic motion of molecules as related to heat, and to understand better the ideal gas law, thermodynamics, and **geomagnetism**. *See also* Boltzmann; Boyle; Gilbert; Ideal Gas Law; Maxwell.

Chargaff's Hypothesis for the Composition of DNA: Biology: *Erwin Chargaff* (1905–   ), United States.

In the DNA molecule, the number of adenine (A) nucleotide units always equals the number of thymine (T) nucleotides, and the number of guanine (G) units always equals the number of cytosine (C) units.

Chargaff's earlier work used paper chromatography and spectroscopy to study the composition of DNA in different species. He found that within one species, the DNA was always the same, but there was a difference in DNA composition between species. He believed there must be as many forms of DNA as there are species, even though much of the DNA was similar for all species. At this point, he realized that a pattern of consistency of AT pairs and GC pairs appeared in the four nucleotides found in nucleic acid molecules. Although he did not follow up on this discovery, it enabled Crick and Watson to arrive at the placement of the AT and GC pairs inside the helix structure of the DNA molecule. (See Figure C2 under Crick.) Scientists are continuing to explore the many possible benefits of this discovery for the betterment of humankind. *See also* Crick; R. Franklin; Watson.

Charles' Law: Chemistry: *Jacques Alexandre Cesar Charles* (1746–1823), France.

When the pressure remains constant, the volume of a gas is directly proportional to its temperature. V/T = constant = n/P.

Jacques Charles, a French physics professor, established the direct relationship between the temperature and the volume of gases. Aware of the flights of hot air balloons, he realized that as the air in the balloon became hotter, it expanded and thus became lighter due to a decrease in density and an increase in its volume. As long as the fire at the bottom opening of the balloon heated the air, the air would expand and become lighter. When the air cooled, it decreased in volume and became heavier, causing the balloon to descend. He also knew that Henry Cavendish, a British chemist, produced hydrogen that was much lighter than hot air and thus more buoyant. Even better, hydrogen did not lose its buoyancy as it cooled. On August 27, 1783, Jacques Charles, and his brother, Robert, made the first flight in a hydrogen-filled balloon. On a later flight, they reached an unprecedented altitude of 2 miles. The danger of using hydrogen for lighter-than-air ships was recognized even in these early days, and its use for this type of airship was halted after the 1937 explosion of the Hindenberg zeppelin while mooring in New Jersey.

Charles' gas law led to further experimentation by Gay-Lussac, Dalton, and other scientists interested in the nature of matter and resulted in what is known as the "ideal gas law," a combination of several gas laws. *See also* Avogadro; Boyle; Galileo; Gay-Lussac.

Charpentier's Glacier Theory: Geology: *Jean de Charpentier* (1786–1885), Switzerland.

Glaciation is the agent responsible for the movement of boulders of one composition to an area where the boulders and rocks of a different composition are found.

Early geologists were puzzled for many years by the presence of large boulders that did not seem to belong in the areas in which they were found. Some thought that they were carried to their new locations by icebergs, but there was no evidence that icebergs ever existed in these areas. Charles Lyell contended they were brought to their locations by enormous floods; however, there existed no other evidence of such giant water flows needed to move these huge rocks. Charpentier's **glaciation** theory was correct but not well accepted by most scientists, except Louis Agassiz, who believed the idea viable and published his glacier theory before Charpentier published his. *See* Agassiz; Lyell.

Chevreul's Theory of Fatty Acids: Chemistry: *Michel Eugene Chevreul* (1786– 1889), France.

When treated with acids or alkali, animal fats break down to produce glycerol and fatty acids.

Chevreul experimented with saponification, the process of producing soaps by treating animal fats with alkalis. By using alcohols to crystallize the product, he identified several fatty acids, including oleic acid, stearic acid, butyric acid, capric acid, and valeric acid, which are used in organic chemistry. He and Joseph-Louis Gay-Lussac patented the process to make candles from the stearic fatty acid. Until this time tallow candles rendered from animal fat were used. They had an unpleasant odor, burned poorly, and were messy. The new fatty acid candles were harder, burned brighter, and were less odiferous. They were a better product and made a fortune for Chevreul and Gay-Lussac, the patent holders.

Chu's Hypothesis for "High Temperature" Superconductivity: Physics: *Paul Ching-wu Chu* (1941–), United States.

The combination of the proper amounts of yttrium, barium, and copper can be used under pressure to reach a critical temperature above that of liquid nitrogen.

Karl Alex Muller (1927–) developed material that achieved **superconductivity**, at an unprecedented high temperature (35 K or $-238°C$). This temperature is very low compared to ordinary temperatures, and liquid helium is required to maintain a system at this temperature. Using Muller's results, Chu's goal was to make superconductivity practical. In order to achieve this, it was necessary to make a material that would be superconductive at a temperature to which a material could be cooled by liquid nitrogen. Liquid helium has a lower boiling temperature than does liquid nitrogen but is too expensive to use on a regular basis. Nitrogen becomes a liquid at about 77° kelvin ($-195.5°C$). Chu tried several combinations of metals as superconductors and finally developed a mixture that was stable and would become superconductive of electricity at about 93° kelvin. Since there is no resistance to the flow of electricity through superconductors at these low temperatures, the potential for research possibilities is manifold. Superconductivity at even higher temperatures may lead to superefficient magnets, improved electromagnetic devices and electrical equipment such as super-fast trains and less expensive transmission of electricity. *See also* Nernst; Simon.

Clarke's Supergene Theory: Biology: *Sir Cyril Astley Clarke* (1907–), England.

Supergenes are groups of closely linked genes that act as an individual unit and carry a single controlling characteristic.

Clarke was interested in the phenomenon of butterflies' inheriting particular wing patterns referred to as *mimicry*—that is, one species (mostly insects) exhibits a color, body structure, or behavior of another species that acts as camouflage. For example, the "eye" spots on the wings of butterflies that mimic another species protect the butterfly from predators. He found that male swallowtail butterflies carry supergenes as recessive. However, the characteristics are expressed as patterns for the females. It is now known that many inherited human traits and characteristics are controlled by supergenes. Clarke recognized a similarity between the supergenes that resulted in butterfly wing patterns and the blood antigen of rhesus monkeys, referred to as the Rh factor. An Rh-negative mother and Rh-positive father may produce an Rh-positive child, which can lead to the development of Rh antibodies in the mother if the child's blood leaks through the placenta. This can cause the destruction of red blood cells in future Rh-positive children born to the mother. The proposed solution to this problem was testing prospective mothers and injecting Rh-negative mothers with Rh antibodies, thus counteracting the effects on future Rh-positive children.

Claude's Concept for Producing Liquid Air: Chemistry: *Georges Claude* (1870–1960), France.

When air is compressed and the heat generated by the increase in molecular activity is removed, a point will be reached at which the major gases of air will liquify.

Claude was successful in applying this principle on an industrial scale, forming the worldwide Air Liquid Company. The gases that make up air can be fractionally separated by this process, producing oxygen, nitrogen, carbon dioxide, argon, neon, and other noble gases. Since each of these gases becomes liquid at a specific temperature, the reverse also occurs. As the temperature of liquid air containing all of these gases in liquid form increases, each specific gas "boils" off at its specific temperature, where it can be isolated and collected as a pure gas. Claude was the inventor of neon lighting (glass tubes containing neon gas at less than normal pressure). When an electric discharge is sent though the gas, it glows with the familiar red light.

Clausius' Laws and Theory of Thermodynamics: Physics: *Rudolf Julius Emmanuel Clausius (1822–1888), Germany.*

Clausius' Law: *Heat does not flow spontaneously from a colder body to a hotter body.* In an early form, the law was stated as: *It is impossible by a cyclic process to transfer heat from a colder to a warmer reservoir without changes in other close-by bodies.*

The conservation of energy is a fundamental law of physics and is often called the first law of thermodynamics, which states that the total energy of a closed system is conserved. Clausius' is an early statement of the second law of thermodynamics, and it led to the concept of entropy.

Clausius' Theory of Entropy: *In an isolated system, the increase in entropy exceeds the ratio of heat imput to its absolute temperature for any irreversible process.*

There are three classifications assigned to thermodynamics processes; natural, unnatural, and reversible. The natural process proceeds only in a direction of equilibrium. It is a reversible process *only* if additional energy (heat) enters the isolated system. A hot body transfers some of its heat to a cold body until both are at the same temperature as their surroundings—equilibrium. The unnatural process never occurs over extended time since the unnatural process would require moving away from equilibrium (reverse the arrow of time). And reversible systems are idealized in the sense they are always arriving at different states or stages of equilibrium (e.g., growing living organisms) until their death. In a perfect closed system, Clausius' ratio (entropy) would always remain constant. But in real life, every process occurring in nature is irreversible and directional (the arrow of time is pointed in only one direction) toward disorder. Thus, there is always an increase in entropy, which sooner or later leads to complete disorder and randomness and a static (unchanging) universe.

More recently, concepts of entropy have been expanded to the analysis of information theory.

Consequence of Clausius' Second Law of Thermodynamics: *The amount of energy in the universe is constant while the entropy of the universe is always increasing toward a maximum. At some future point, this maximum disorder will result in the unavailability of useful energy.*

For any irreversible process, entropy will be increased. In other words, a greater state of disorganization or randomness will exist until equilibrium is reached. The only way entropy of an entire system can be decreased is by extracting energy from the system. The entropy of a part of the system may decrease if the entropy of the rest of the system increases enough. For example, the sun continually supplies energy to the earth. Otherwise, total entropy (disorder) would soon occur on the earth. This also applies to small temporary systems, such as some chemical reactions, crystallization, and growth of living organisms. The sun provides all the energy used on earth, with the exception of radioactive minerals. Energy is pumped into all living systems through plants as food, resulting in highly ordered complex molecules and growth in living organisms. Thus, the entropy of some parts of the universe is decreased as organisms grow and become more organized through the input of energy. However, the entropy of the universe is increased by the radiation of the sun and other stars. Upon death, the second law of thermodynamics proceeds, resulting in the disorganization of the organism and its complex molecules. Without earth receiving the sun's energy, entropy will increase and becomes all encompassing leading to death of all living organisms. *See also* Carnot; Fourier; Heisenberg; Kelvin; Rumford; Simon.

Cockcroft-Walton Artificial Nuclear Reaction: Physics: *Sir John Douglas Cockcroft* (1897–1967) and *Ernest Thomas Sinton Walton* (1903–1995), England. They were joint recipients of the 1951 Nobel Prize for physics.

The transmutation reaction progressed as follows: $_3Li\text{-}7 + {}_1H\text{-}1 \rightarrow {}_2He\text{-}4 + {}_2He\text{-}4 + 17.2\ MeV;$ *where* $_3Li\text{-}7$ *is a heavy isotope of lithium,* $_1H\text{-}1$ *is a proton (hydrogen nuclei),* $_2He\text{-}4$ *are alpha particles (helium nuclei), and 17.2 MeV is in millions of electron volts of energy.*

This first artificial nuclear reaction was the lithium reaction that occurred in 1932 and was made possible by Cockroft's development of a proton accelerator and E. T. S. Walton's invention of a voltage multiplier that increased the speed of the proton "bullets." The lithium nuclei (the target) were bombarded by high-energy hydrogen nuclei (protons), resulting in the production of two alpha particles (helium nuclei), plus a small amount of energy. This provided the information and knowledge needed to continue the development of nuclear fission reactions, giving rise to the use of nuclear energy for the production of electricity, radioisotopes, and the atomic (fission) bomb. *See also* Chadwick; Fermi; Hahn; Meitner; Szilard.

Cohn's Bacteria and Cell Theories: Biology: *Ferdinand Cohn* (1828–1898), Germany.

Cohn's Infectious Disease Theory: *Microscopic bacteria are simple organisms that can cause diseases in other plants as well as animals.*

Cohn's study of microscopic organisms led him to develop the first classification of bacteria, which basically is still used today. Although he experimented with boiled solutions of bacteria and suggested that some bacteria can develop resistance to external influences, including heat-resistant spores, Cohn still believed in spontaneous generation, a theory that life can start in rotting garbage. It is no longer accepted as viable. *See also* Redi.

Cohn's Theory of Protoplasm: *The protoplasm found in plant and animal cells is essentially the same.*

Protoplasm is the colloidal substance composed of mostly complex protein molecules found in all living cells. In green plants the protoplasm contains chlorophyll, which in the presence of sunlight manufactures complex organic compounds (mostly carbohydrates from carbon dioxide and the hydrogen from water), while at the same time liberating oxygen as a waste product. All animal and plant cells require the food produced by this process, called *photosynthesis.* Cohn's research and publications were used by others, including Louis Pasteur. *See also* Pasteur; Redi.

Compton's Wave/Particle Hypotheses: Physics: *Arthur Holly Compton* (1892–1962), United States. Arthur Compton and Charles T. R. Wilson jointly received the 1927 Nobel Prize for physics.

Compton Effect: *When x rays bombard elements, such as carbon, the resulting radiation is scattered (reflected) with a wavelength that increases with the angle of scattering.*

In order for this effect to occur, the x rays must behave as particles (photons) that during the collision transfer their energy to the electrons of the carbon. This reaction indicates that x rays behave as particles as well as waves. Compton used Charles Wilson's "cloud chamber" to assist in detecting, tracking, and identifying these particles. The Compton effect was the first experimental evidence that electromagnetic radiation, such as radio waves, light, ultraviolet radiation, microwaves, x rays, and gamma rays, exhibits characteristics as both waves and particles.

Compton's Hypothesis for Cosmic Rays as Particles: *If cosmic rays have characteristics similar to charged particles, then there should be a variation in their distribution by latitude caused by the earth's magnetic field.*

If such an effect by the magnetic field could be detected, then it could be concluded that cosmic radiation consists of charged particles and are not pure **electromagnetic** radiation. This hypothesis was later proved correct by measuring the slight distribution of cosmic rays from outer space in relation to different latitudes of the earth's surface. *See also* Anderson; Hess; Rutherford; C.T.R. Wilson.

Copernicus' Cosmology Theories: Astronomy: *Nicolaus Copernicus* (1473–1543), Poland.

Copernicus' Heliocentric Theory of the Universe: *All the spheres (planets and moons) revolve about the sun as their midpoint; therefore, the sun is the center of the universe.*

At the time of Copernicus' pronouncement, the concept of a sun-centered universe was not really new. Aristarchus of Samos (c.320–c.250 B.C.) reportedly stated that the sun and stars were motionless and the planets, including earth, revolved in perfect circles around the sun. However, no attention was paid to his theory for almost 2000 years because the Earth-centered universe, as postulated by Aristotle and Ptolemy, was the accepted truth—that is, until the Copernican **heliocentric** model was proposed. Copernicus was influenced by the Pythagorean Philolaus (c.480–c.400 B.C.) who theorized that the planets, including Earth, moved around a central fire. Philolaus said we could not see this fire since we lived on the side of Earth that was always turned away from it. Copernicus' model engendered much controversy among the clergy, as well as most scientists, until others were able to study and understand his new model, which also explained planetary motion. (See Figure C1.)

Copernicus' Theory of Planetary Positions: *Superior planets are those whose orbits are larger than the earth's and are farther from the Sun than the earth.*

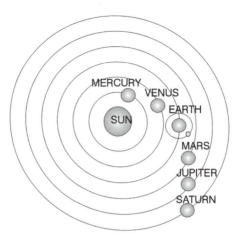

Figure C1: Artist's conception of Copernicus' heliocentric universe where all planets, including earth, revolve around the sun. His rationale was that if the sun is the midpoint of the planets' orbits, then the sun must be at the center.

Inferior planets are those whose orbits are smaller than the earth's and are closer to the Sun.

As a result of this theory, several other concepts to explain planetary motion were proposed. When an inferior planet is between earth and the sun, it is in line of sight with the sun and is said to be in "inferior conjunction." When the planet passes on the far side of the sun, away from earth, the planet is in opposition and is said to be in "superior conjunction." Further observations of planetary motion led Copernicus to distinguish between the planet's *sidereal period*—its actual period of revolution around the sun—and the *synodic period*—the time period of two successive conjunctions of a planet as seen from Earth. Therefore, the synodic period is what is directly viewed from Earth, as it is the time required for the planet to progress from one 180-degree opposition to the opposition of the next cycle, as related to earth.

Copernicus was also aware his model required the inferior planets to move faster in their smaller orbits, while the speed of the superior planets was slower as they revolved around the sun. This information later became important for Kepler in the development of his laws of planetary motion.

Copernicus' Theory of Planetary Distance from the Sun: *Since the planets travel in perfect circles, they can be viewed as a direct line of sight to the sun as well as at a 90-degree right angle, which forms a right triangle with the sun. The distance from the sun to the planets can be determined by geometry and trigonometry.*

At this time in history it was believed that all celestial motion, including planetary motion, progressed in perfect circles. Also, there were only six known planets. Copernicus' calculations for the distance of planets to the sun were excellent, considering that he believed their orbits to be perfect circles. His figures are close to what we now know about elliptical orbits of planets and their distances from the sun. Current figures are based on earth's distance to the sun as being 1 unit. This distance is the standard astronomical unit (AU), which equals the mean distance of 92,956,000 miles from earth to the sun. AUs are used to calculate the distances for the other planets to the sun.

Planet	Copernicus's Data	Modern Distance
Mercury	0.38	0.39
Venus	0.72	0.72
Earth	1.00	1.00
Mars	1.52	1.52
Jupiter	5.22	5.21
Saturn	9.18	9.55

Copernicus' Theory of Planetary Brightness: *All the planets travel in circles around the sun; thus they are at different distances from the earth. Therefore, as their distances from earth differ, so does their brightness.*

These theories raised another question: If earth actually revolved around the sun, why didn't the position of the stars change every six months of earth's yearly orbit of the sun? This question is based on **parallax**—the apparent change in position of an object when viewed from two different positions.

Copernicus' Conclusion about the Size of the Universe: *Stars seem not to move and therefore must be located at tremendous distances in space. Thus, the universe is much larger than formerly believed.*

Copernicus based this theory on the concept of parallax, which is dependent on both the distance between two geographic sites at which the stars are viewed and the distance separating the observer and the actual object. In other words, even when viewed from opposite positions every six months in earth's orbit, the stars are so distant they seem to stay in one position. About two hundred years later, with the development of improved instruments, a slight parallax of the stars was measured. But for the ordinary viewer on Earth, this slight parallax displacement is not noticeable. Copernicus was one of the first to conceptualize a vast universe that for the next hundred years remained incomprehensible.

Copernicus' Theory of Epicycle Motion of the Planets: *The minor irregularities in the motion of planets revolving around the sun can be explained by the epicycle each planet traces as it progresses in its own orbit.*

Ptolemy's earth-centered model of the universe required the extensive use of epicycles to explain planetary motion. Since Copernicus maintained that planets revolved around the sun in perfect circles, he too required the use of epicycles to explain the irregularities in their motion. Several centuries later, it was determined that not all celestial motion is in circles, and planets, moons, and comets travel in ellipses of one type or another. An epicycle may be thought of as the planet moving in its own series of small circles as it progresses around the circumference of its orbit.

Copernicus' Theory of the Earth Spinning on its Axis: *The motions of earth consist of two or more component motions. One motion is revolving, the other is rotating.*

The concept of earth's revolving around the sun was not accepted by most philosophers and scientists in Copernicus' day. The second contention—that earth spins on its axis as it revolves around the sun—was even more difficult to comprehend. Most people claimed that their common sense dictated it was not possible and offered these arguments against a rotating earth: (1) If the earth spun on an axis, why didn't objects fly off into space? (2) It would be impossible for anything moveable to be firmly affixed to the Earth. (3) Birds would have to fly faster in the direction of the earth's rotation just to stay in the same place. (4) If a person jumped up, he would come down in a different spot because the Earth would have moved. *See also* Brahe; Coriolis; Galileo; Kepler; Ptolemy.

Corey's Theory of Retrosynthetic Analysis: Chemistry: *Elias James Corey* (1928–   ), United States.

Retrosynthetic analysis occurs when the whole target compound (the compound to be studied) is divided into subunits for analysis and then synthetically recombined.

This was a new approach to analyzing chemical substances, particularly complex organic molecules. Corey and his research teams reduced complex molecules to smaller and smaller "pieces" and then recombined them to arrive at the original or an altered molecule. In this manner, they determined how to synthesize many organic compounds for medical use. An example is the synthesis of prostaglandin hormones useful in treating infertility and inducing labor. Using his retrosynthetic concept, Corey developed a new computer program that greatly assisted chemists in their analysis and synthesis of organic compounds.

Coriolis' Theory of Forces Acting on Rotating Surfaces: Physics: *Gustave-Gaspard Coriolis* (1792–1843), France.

An inertial force acts on rotating surfaces at right angles to the rotating earth, causing a body to follow a curved path opposite the direction of the rotating earth.

The Coriolis effect is greatest if an object is moving longitudinally on earth— from either pole to the equator along longitudinal lines. In the Northern Hemisphere the apparent motion, when viewed from the North Pole, is to the right; for the Southern Hemisphere, when viewed from the South Pole, it is to the left. The Coriolis effect is more apparent near the equator than at the poles, since the speed of rotation is greater at the larger circumference. It affects both the oceans and atmosphere on earth but is a much weaker force than gravity. Even so, over great distances it causes cyclones, which are low-pressure areas that can develop into hurricanes, and water whirlpools to circle clockwise in the

Northern Hemisphere. Conversely, anticyclones may develop into typhoons, which rotate counterclockwise in the Southern Hemisphere. The Coriolis effect influences ocean currents, including El-Niño, as well as other local and world-wide weather and climate phenomena.

The magnitude of the Coriolis effect is the velocity of the object compared to Earth's angular velocity for a given latitude and is the reason rockets and spacecrafts are launched to the east. The Coriolis effect gives rockets an extra boost as Earth spins eastward. Also, missile launching sites are usually located on coastal areas so that any defective rockets or missiles will fall on water rather than land. Although other scientists recognized the effect caused by Earth's rotation, it was Coriolis who worked out the mathematics for it and was first to publish his results. This complex force resulting from the rotation of Earth on its axis was not apparent in the days of Copernicus and Galileo because it is a force much too weak to have been recognized or measured in their times.

Cori Theory of Catalytic Conversion of Glycogen: Chemistry: *Carl Ferdinand Cori* (1896–1984) and *Gerty Theresa Radnitz Cori* (1896–1957), both from the United States, jointly received one-half of the 1947 Nobel Prize for physiology or medicine. The other half of the 1947 prize was awarded to Bernardo Houssay (1887–1971) for his work with the pituitary gland and sugar metabolism.

Cori Hypothesis for Glucose Conversion: *The complex carbohydrate glycogen stored in the liver and muscles is converted, as needed, to energy, in the form of glucose-6-phosphate by the enzyme phosphoglucomutase. The process is reversible.*

Carl and Gerty Cori collaborated on biochemical research projects dealing with the analysis of enzymes and glycogen (sugars and starch). They isolated a chemical from dissected frog muscle identified as glucose-1-phosphate, where this complex molecule was joined by a molecular ring containing 6 carbon atoms. This new compound was named Coli ester and was shown to convert to the more complex sugar form after it was injected into animal muscles. Their research led to a more complete understanding of the role of high blood sugars, diabetes, insulin, hormones, and the pituitary gland.

Coulomb's Laws: Physics: *Charles Augustin de Coulomb* (1736–1806), France.

Coulomb's Fundamental Law of Electricity: *Two bodies charged with the same sort of electricity will repel each other in the inverse ratio of the square of the distance between the centers of the two bodies.*

Coulomb devised this law after developing an extremely sensitive torsion balance, which consisted of a thin silk thread supporting a wax-covered straw (thin, very light reed or grass) with a small pith ball suspended on one end. The straw was balanced and suspended in an enclosed jar to prevent air drafts from affecting the results. His torsion balance could measure a force of only 1/100,000 of a gram by gauging the twist of the thread. There were markings around the circumference of the jar so Coulomb could measure, in degrees, any changes in the ball's position. Coulomb then used static electricity to charge the pith ball on the straw and another pith ball outside the jar. He brought the outside charged ball close to the jar. The ball inside the jar was repelled. He measured the distance of its movement on the degree markings on the jar and compared it with the distance between the centers of the inner and outer balls. He discovered that one of the basic laws of science applied (the inverse square law). When moving the outer ball twice the distance away from the inner ball, the effect on the inner ball was not one-half its movement but one-fourth. In other words, the effect of the electrical charge decreased as to the square of the distance between the centers of the charged balls.

Coulomb's Theory of the Relationship between Electricity and Magnetism: *Both electricity and magnetism follow the same physical laws, including the inverse square law.*

Coulomb demonstrated the similarity of the attractive and repulsive forces for both electricity and magnetism and concluded that both were similar physical phenomena that followed the same physical laws. (See Figure A1 under Ampere.) Even so, many scientists of his day rejected this theory. When the theory of electromagnetism was later refined and better understood, it became important for the development and use of electromagnetic devices, such as motors and generators. *See also* Faraday; Maxwell.

Coulomb's Law of Electrical Charge: *The attractive and repulsive forces for electricity are proportional to the products of the charge.*

This famous theory is now known as Coulomb's law, which states that a Coulomb is the unit quantity of electricity carried by an electric current of 1 ampere in 1 second. As a result, a much better understanding of electricity as a measurable current forced through a conductor by a voltage differential was achieved.

Couper's Theory for the Structure of Carbon Compounds: Chemistry: *Archibald Scott Couper* (1831–1892), Scotland.

Carbon atoms have the unique ability to bond their valence electrons together to structure both chains and branches of chains to form carbon (organic) compounds.

Couper knew carbon must have four valence electrons since it could form inorganic compounds such as carbon dioxide, CO_2, where oxygen has a valence of 2. Thus, it takes only one carbon atom with a valence of 4 to join with two oxygen atoms to form the molecule $O=C=O$. Further research indicated that carbon atoms have the unique ability to bond with other carbon atoms to form chains and branches. (See Figure V3.) Couper was the first to use notations such as $=C=C=C=$ for carbon atoms to illustrate his theory, advancing the concept of isomers, which are compounds with the same molecular formulas but are structured differently. Isomers have different chemical and physical characteristics due to different arrangements of the elements making up the molecules. The greatest variety of isomers are found as the varied hydrocarbons and the many complex organic compounds. A typical example is the hydrocarbon isomer of C_4H_{10}, which can be structured as a chain of 4 carbon atoms with 10 attached hydrogen atoms, or a different structure of a branched group of 4 carbons atoms, with 10 hydrogen atoms attached to the carbon atoms. Their characteristics are quite different. The whole chemistry of carbon isomers, while complicated, has pioneered many new medicines and useful products. *See also* Kekule; Van't Hoff.

Crick-Watson Theory of DNA: Biology: *Francis Harry Compton Crick* (1916–   ), England. Francis Crick shared the 1962 Nobel Prize in physiology or medicine with James Watson and Maurice Wilkins.

DNA is a double helix joined by pairs of nucleotides of adenine + thymine (A+T) and guanine + cytosine (G+C), with the sugar-phosphate structure attached to the outer sides of the helix strands.

Early in Crick's career in England, he was joined by an American, James Watson, who suggested that the first step in determining the structure of the basic molecule of life would be to learn more about its chemical nature, which they then pursued together. Crick and Watson were not the only scientists searching for the holy grail of life. Others, such as Maurice Wilkins (1916–   ) and his assistant, Rosalind Franklin, as well as Alexander Todd (1907–   ) in England, and Linus Pauling, Erwin Chargraff, and Phoebus Levene (1869–1940) in the United States, were conducting similar research.

Nationalism, competition, jealousy, and secrecy, all of which are the antithesis of scientific investigations, were part of the search for the structure of DNA, Crick's work depended on information he obtained from crystallographic x rays (x-ray photos) of DNA crystals to determine its structure. (See Figure C2.) Franklin was very secretive about her work and refused to share her photographs or her crystallography techniques. Her supervisor, Wilkins, considered the study of DNA a joint project. Franklin also withheld information about the placement of the sugar-phosphate backbone for the DNA molecule. Crick was acquainted previously with Wilkins, and as the story goes, there may have been a break in trust when Wilkins provided Crick with some of Franklin's vital information that enabled Crick to succeed with his project. In 1953 Crick and Watson completed their model based on information known at that time. They shared the 1962 Nobel Prize for their discovery along with Wilkins, but Rosalind Franklin, being deceased, was not included. (Some say that Franklin was not given adequate credit for her contributions to the final outcome.) *See also* Chargaff; R. Franklin; Pauling; Watson.

Crookes' Radiation Theories: Physics: *Sir William Crookes* (1832–1919), England.

Crookes' Radiometer: *A closed glass container in which most of the air has been evacuated will continue to radiate heat.*

From his spectrographic work in identifying the element thallium, Crookes

Figure C2: Depiction of the Crick and Watson double-helix model of deoxyribonucleic acid (DNA). A number of other scientists contributed to the background information that led to their model.

noticed that heat radiation caused unusual effects on the thallium gas while in the sealed glass container. He designed a device with four vanes. One side of each of the four vanes was painted black, and the other sides of the vanes were polished like a mirror. The vanes were balanced on a vertical pivot in a closed glass in which most of the air had been removed. When heat radiation (sunlight) struck the vanes, molecules on the dark hot side had greater momentum, and thus pushed the vanes backward to a greater extent than did the molecules from the cooler shiny side. This led to further investigation of the effects of electricity on low-pressure gases. At the time, the radiometer demonstrated the essence of kinetic energy of gas molecules. Today, Crookes' radiometer is more like a toy used to demonstrate the effects of heat on dark and shiny surfaces.

Crookes' Cathode Ray Tube: *The air will glow when an electric current is passed through a closed glass tube containing low air pressure.*

To demonstrate that the glow in the Crookes tube and the slight fluorescence on the inner walls of the tube were due to electricity, Crookes placed a Maltese cross in the path of the rays. The form of the Maltese cross was used because its symmetrical design would produce a recognizable image as it interrupts the flow of cathode rays. Where the rays were blocked by the cross, a distinct shadow-like pattern appeared on the end of the glass tube. (See Figure C3.) Crookes also demonstrated that a magnet brought near the glass tube would deflect the cathode rays in a curved pattern that suggested the rays were com-

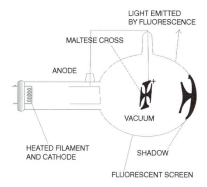

LIGHT EMITTED
BY FLUORESCENCE

MALTESE CROSS

ANODE

VACUUM

SHADOW

HEATED FILAMENT
AND CATHODE

FLUORESCENT SCREEN

Figure C3: The Maltese cross design was used to demonstrate that a stream of electrons sent to a target fluorescent screen were blocked by the metal cross. The shadow of the cross on the screen prevented the electrons from producing light on the fluorescent material on the screen. A similar device was used to demonstrate the nature of the electron: it has a negative charge, a magnet can deflect its path, and it has the mass of just 1/1840 of the mass of the hydrogen atom.

posed of particles with an electric charge. He concluded it was impossible for electromagnetic radiation, such as light, to carry an electric charge and be deflected by a magnet. Therefore, the cathode rays must be charged particles. J. J. Thomson later demonstrated the cathode rays were really electrons. The shadow of the Maltese cross in Crookes' cathode ray tube might be considered the first TV picture, since a similar process is used in modern television receivers. *See also* Thomson.

Curies' Radiation Theories and Hypotheses: Chemistry: *Marie Sklodowska Curie* (1867–1934), France. Marie and Pierre Curie shared the 1903 Nobel Prize for physics with Antoine Henri Becquerel, who discovered spontaneous radioactivity. Madame Curie was also awarded the 1911 Nobel Prize for chemistry for her discovery of radium and polonium. She is one of only four people ever to receive two Nobel Prizes. (The other three are L. Pauling, J. Bardeen, and F. Sanger).

Curie's Radiation Hypothesis: *Chemical reactions and mixtures of uranium with other substances do not affect the level of radiation. Only the quantity of uranium determines the level of radiation. Therefore, radioactivity must be a basic property of uranium.*

Madame Curie separated chemicals from uranium minerals and found the ore pitchblende was more radioactive than uranium earth itself. Pitchblende is a heavy black ore containing a yellow compound that Martin Heinrich Klaproth (1743–1817) thought was a new element. He named it *uranium* after the planet Uranus. The Curies brought many cartloads of pitchblende from northern Europe to her shed laboratory in France. (Pitchblende is also found in Colorado, Canada, and Zaire.) Over a period of months, Curie and her assistant chemically extracted this new element. She also theorized there must be more than one type of radioactive element in the ore, leading to a new hypothesis.

Curie's Hypothesis for New Radioactive Elements: *Since the ore contained substances with greater radioactivity than uranium, pitchblende must contain new radioactive elements.*

Curie continued to separate and test these new substances, which proved to be new elements. She named one polonium after her native country (Poland), and the other radium, for its high radioactivity. She discovered that the heavy metal thorium also produced radiation. Curie is credited with coining the word *radioactivity*. She and her assistant used several chemical processes to separate the radium, which exists in very small amounts in pitchblende. After many months, she had produced only about 0.1 gram of radium chloride.

Pierre Curie (1859–1906), Marie Curie's husband, was a well-known physicist who assisted his wife in her research. With his brother, he developed several techniques for detecting and measuring the strength of radiation. Their instrument, the electrometer, was sensitive enough to produce an electric current between two metal plates separated by the radioactive sample. They also discovered piezoelectricity, from the Greek word *piezo* meaning "to press." The piezo effect occurs when certain types of crystals are put under pressure. The slight deformation caused by the squeezing will produce opposite electric charges on opposite faces of a crystal. Pierre Curie and his brother also discovered it would work in the opposite manner: applying an electric charge to a crystal will produce a change in the crystal's structure. This discovery was incorporated in their electrometer used to measure minute electric currents, as well as radiation. Pierre Curie also measured the amount of heat given off by radium. Each gram of radium gives off 140 calories of heat per hour, with a **half-life** of about 1600 years. They realized this amount of energy was beyond normal chemical reactions and must be from some other part of the atom. Thus began the age of nuclear energy, even though the nuclei of atoms had yet to be discovered.

At the time the Curies worked with radiation, particularly radium, the extent of the dangers of radiation was unknown. It is assumed the Curies may have been the first humans to suffer from radiation sickness, but Pierre died after an accident with a horse and carriage. The *curie*, the unit measurement for radioactivity, was named for Pierre Curie. Marie Curie's notebooks are still considered extremely radioactive. *See also* Becquerel; Rutherford.

Curl's Hypothesis for a New Form of Carbon: Chemistry: *Robert F. Curl, Jr.* (1933–   ), United States. Robert F. Curl, Richard E. Smalley, and Sir Harold W. Kroto jointly received the 1996 Nobel Prize in Chemistry.

Curl's C_{60} Hypothesis: *Vaporized carbon atoms in a vacuum can form single and double bonds, similar to aromatic compounds, to produce a symmetrically closed shell with a surface consisting of multiple polygons.*

The new complex carbon molecules are called *fullerenes*. The most common

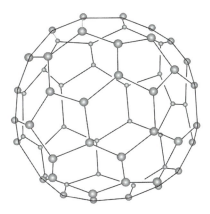

Figure C4: Artist's conception for the structure of the C_{60} atom discovered by Robert Curl and named *buckminsterfullerene* (nickname *Buckyballs*) by Harold Kroto. These compact masses of carbon are a third form of pure carbon, each composed of 60 carbon atoms formed into a soccer ball shape by 20 hexagons and 12 pentagons of bonded carbon atoms.

is a group of 60 atoms shaped similar to a soccer ball, which is formed by 20 hexagons and 12 pentagons of bonded carbon atoms. (See Figure C4.) Due to the geodesic shape of the surfaces it was named *buckminsterfullerene*, after the architect R. Buckminster Fuller, who designed geodesic dome structures. However, it is usually referred to by its nickname, *Buckyballs*. Additional complex ball-shaped molecules of bonded carbon have been developed in addition to C_{60}. They include C_{69}, C_{70}, C_{76}, C_{78}, and C_{80}.

The discovery of fullerene Buckyball molecules has opened up new research vistas in the areas of superconductive materials, plastics, polymers, and medicines, as well as new theories to explain the beginning of the universe and the structure of stars. *See also* Kroto.

Cuvier's Theories of Anatomy and Taxonomy: Biology: *Baron Georges Leopold Chrétien Frédéric Dagobert Cuvier* (1769–1832), France.

Cuvier's Classification of Animals: *Based on functional integration of animals there are four classes or "branches": (1) vertebrata (with backbones; e.g., mammals), (2) articulata (arthropods; e.g., insects), (3) mollusca (bilateral symmetrical invertebrates; e.g., clams), and (4) radiata (radially symmetrical sea animals; e.g., echinoderms).*

Cuvier demonstrated how the anatomy of different animals compared with one another and proposed these four **phyla** of animals in his classification scheme. Cuvier is often referred to as the founder of the science of comparative anatomy.

Cuvier's Concept of Form and Function: *All organisms are integrated wholes into which parts are formed according to their functions.*

Cuvier was a firm believer that form follows function, not the reverse. Generally this means the structure of tissue or an organ is based on what function it is required to execute. He believed that if a part of an animal changed, it would change the entire animal's form. He stated that all the parts of each animal are arranged to make it possible for the animal to be complete. Thus he rejected organic evolution.

Cuvier's Theory of Evolution: *Organic evolution cannot exist because any change in an organism's structure would upset the balance of the whole organism, and thus it would be unable to survive in its environment.*

Cuvier rejected Darwin's theory of organic evolution based on natural selection that results in changes in species and the emergence of new species over long periods of time. Cuvier believed similarities between and among different organisms were due to common functions of their parts, not evolutionary changes. However, he did believe that catastrophic events occurred on earth (he preferred the term *periodic revolutions*). Natural disasters such as floods, fire, and earthquakes caused massive extinction of animals and provide situations for the arrival of new species. The catastrophic theory was recently revived as *punctuated evolution* by Niles Eldredge and Stephen Jay Gould.

Cuvier's Theory of Fossils: *Fossils represent ancient species that became extinct due to periodic revolutions in their environment.*

In his study of fossils, Cuvier recognized some fossils were found deeper in the strata of rocks and earth and that their age could be determined by the depth of the strata. He used a similar classification system for fossils as his four phyla for living animals. Cuvier also recognized the detailed structure in some fossils, particularly the structure of wings. He was the first to identify the fossil of a flying reptile he named *pterodactyl*. Cuvier has also been referred to as the founder of paleontology. *See also* Buffon; Darwin; Eldredge; Gould; Wallace.

D

Dale's Theory of Vagus Nerve Stimuli: Biology: *Sir Henry Hallet Dale* (1875–1968), England. Sir Henry Hallet Dale and Otto Loewi shared the 1936 Nobel Prize for physiology or medicine.

Both chemical and electrical stimuli are responsible for affecting nerve action.

Henry Dale discovered that the dangerous ergot fungus contained the chemical acetylcholine, a neurotransmitter, which was later demonstrated to affect the parasympathetic nervous system, which controls various organs. Acetylcholine is an **alkaloid** that poisons animal tissue. Eating spoiled grain containing this fungus can result in a serious disease called *ergotism*. The symptoms are a burning sensation in the limbs that may lead to gangrene, hallucinations, and convulsions. It has been known to cause epidemics in poor people who eat rotten rye grain. Outbreaks of ergotism in the Middle Ages were called St. Anthony's fire. Along with the plague and scurvy, it caused psychic epidemics, with symptoms of dancing manias and mass madness where people claimed to be possessed by the devil, often ending in the killing of Jews, children, and witches.

Otto Loewi (1873–1961) identified a chemical substance he extracted from the vagus nerves of frogs and called it *vagusstoffe*. Dale recognized it was similar to acetylcholine produced by ergot, which he associated as the same chemical resulting from the electrical discharge that stimulates the nervous system. Dale hypothesized that both the electrical stimulation and acetylcholine were involved in controlling the heartbeat rate of humans as well as the nerve responses for other organs. This discovery that acetylcholine is a chemical released from **autonomic** nerve endings led to a better understanding of the electrochemical nature of the nervous system and the development of drugs similar to acetylcholine to control heart abnormalities.

Dalton's Laws and Theories: Chemistry: *John Dalton* (1766–1844), England.

Dalton's Law of Partial Pressure: *At an initial temperature, the individual gases in a mixture of gases expand equally as they approach a higher temperature.*

Another way to say this is that all gases expand equally when subjected to equal heat. Since this relationship cannot be observed directly, it was established as a viable law by Dalton's observations and calculations dealing with his study of the atmosphere, humidity, dew point, and vapor pressure. This concept that all gases behave in a similar manner under similar temperatures led to other gas laws and Dalton's theories of the atom.

Dalton's Atomic Theory for Elements: *(1) The smallest particles of all matter are atoms; (2) Atoms are indivisible particles that cannot be either created or destroyed; (3) Atoms of the same element are the same; (4) Each element has its own type of atoms; (5) Atoms of one element cannot ever be changed into atoms of another element.*

Dalton's atomic theory was based on Democritus' philosophical concepts. A main difference was that Dalton was more empirical and documented his observations. He based his ideas about the atom on concepts developed by the "gas chemists," such as Avogadro, Boyle, Charles, and Gay-Lussac.

Dalton's Theory for Compounds: *(1) Chemical reactions occur when atoms of different elements are separated or arranged in exact whole number combinations, and (2) compounded atoms (molecules) are formed by the joined atoms of the elements that make up the compound.*

Dalton used his observations and measurements to assert his theory of compounds. Although molecules were not yet identified, his concepts of atoms' combining by weight and whole numbers remain essentially correct.

Dalton's Law of Definite Proportions: *A specific chemical compound always contains the same elements at the same fixed proportion by weight.*

Dalton rationalized these laws based on his theories for elements and compounds and on what was known about atomic weights at the time. The law of definite proportions led to his law of multiple proportions.

Dalton's Law of Multiple Proportions: *When two elements form more than one compound by combining in more than one proportion by weight, the weight of one element will be in simple, integer ratios to its weight when combined in a second compound.*

This means that atoms of one element can combine in different ratios, by weight, with atoms of another element. Dalton's laws were in essence correct. The problem he had at the time he formulated them was that accurate atomic weights of elements were not known, nor was the concept of **valence** for atoms forming molecules. Regardless, his insight enabled him to formulate two major laws of chemistry: the laws of definite and multiple proportions.

Dalton conceived this law from his knowledge that oxygen and carbon can form two different compounds with different proportions of oxygen and carbon. For example, CO_2 (carbon dioxide, with a 2:1 ratio of oxygen) contains twice as much oxygen as does CO (carbon monoxide). Dalton assumed the composition and ratio of elements in all compounds would be the simplest possible.

This led to a mistake when he tried to apply his law to the compound water molecule. He assumed the ratio was 1:1 for hydrogen to oxygen (HO). This error occurred because at this time in history, oxygen was given the atomic weight of 7, while hydrogen was given the arbitrary weight of 1, since it was the lightest of the elements. Once water molecules were separated by electrolysis, it became obvious there was twice as much hydrogen gas (by volume) derived than oxygen gas. Therefore, the water molecule had to be composed of two molecules of hydrogen (by volume not weight) to one molecule of oxygen (2H + O) or (H_2O).

Dalton's laws were not well received until other chemists rediscovered Avogadro's theories dealing with particle relationships of gases. Dalton's laws have been refined and improved over the years, but his work formed the central basis for modern chemistry. He is considered one of the fathers of modern chemistry. *See also* Atomism Theories; Avogadro; Cannizzaro; Dumas; Lavoisier; Thomson.

Darwin's Theory of Evolution by Natural Selection: Biology: *Charles Robert Darwin* (1809–1882), England.

Environmental pressures on organisms, such as climate and availability of food, act to select, by natural processes, those individuals better adapted to survive and who thus will pass viable traits related to survival to subsequent generations.

Charles Darwin, familiar with Thomas Malthus' (1766–1834) principles of population, also recognized these theories as applicable to humans. Darwin based his new theory of natural selection on his years of collecting plants and animals and his study aboard the *H.M.S. Beagle* as it visited islands of the Southern Hemisphere. He recognized the selection of individuals within species can, over long periods of time, alter the species, including the appearance of new characteristics and species. Darwin used the term *descent with modifications* instead of *biological evolution* as we think of it today.

Darwin received a letter from Alfred Russel Wallace that outlined the same theory of natural selection as Darwin was developing. Since Darwin formulated the theory first, he is credited, along with Wallace, with the concepts of organic evolution. Darwin's book *On the Origin of Species by Means of Natural Selection* (1859) caused much debate in both the general public and religious groups. Although Darwin did not emphasize the evolution of humans from lower forms of animals, it was the natural conclusion to be drawn from the theory. Since the concept of **genetics** and heredity was unknown to Darwin, he relied on the mistaken Lamarckian idea of inheriting acquired characteristics to explain the transfer of characteristics from parents to offspring. Many years before Darwin's time, Gregor Mendel had proposed the general concept of inherited characteristics. It was only after Darwin's death that Mendel's work was rediscovered and applied to organic evolution. Since then, the role of random genetic mutations (in RNA and DNA molecules) is better understood as to how these inheritable changes can equip living organisms to survive in their environments and thus produce more offspring with similar traits. *See also* Dawkins; DeVries; Dobzhansky; Lamarck; Lysenko; Mendel; Wallace.

Dawkins' Theory of Evolution: Biology: *Richard Dawkins* (1941–   ), England.
Hierarchical reductionalism occurs in genes and the DNA molecules, which are the basic units of natural selection responsible for the evolution of organisms.

Richard Dawkins applied knowledge of genes and heredity to Darwin's theory of organic evolution. The genetic and molecular materials in the DNA base pairs of **nucleotides** are the fundamental units of natural selection. Dawkins refers to them as the "replicators," while the entire organism is the "vehicle" containing the genetic DNA "replicators." In his book *The Selfish Gene* (1976), Dawkins described his theory by stating that only the genes and molecules of DNA are important for natural selection to maintain the species. The individual organism is just a means of maintaining and replicating the DNA. How successful the species is depends on how well the replicators build the vehicles (bodies of plants and animals) that "store" and "reproduce" the DNA genetic material through natural selection. Dawkins expands his theory to include a form of sociobiology or "social Darwinism," and he coined the word *meme* as the unit for cultural or social inheritance, with memes responsible for the evolution of ideas through natural (human) selection. Memes are also regulated by evolutionary processes in the sense that families, tribes, social, and cultural groups create human environments that evolve with their culture. Memes, as units of cultural inheritance, evolve just as does genetic material, through the process of natural selection. Dawkins also contends that current living organisms, including humans, are random "accidents." Dawkins' basic idea of evolution states that by following a few rules of physics and starting at very simple points (energy, **amino acids**, self-replicating organic molecules, etc.), life can evolve. Thus, under natural conditions, a variety of complex organisms and their cultures can evolve but not necessarily in any one given direction. Dawkins believes no supreme being is required to start or direct the process.

There has been, and still is, much controversy over Dawkin's concept of hierarchical reductionalism as applied to evolution. It is usually used by physicists to explain the structure and behavior of atoms in terms of elementary particles, molecules in terms of atoms, and so on, up the ladder to the structure and behavior of living cells as related to their component atoms and molecules. Sociobiology continues the hierarchical model to include not only the structure but also the behavior of humans and what species might follow humans based on the most elementary of quanta of energy and matter. Note that it is system-involving feedback. In other words, hierarchical reductionalism also states that small, individual parts made up of differentiated cells and tissues evolve into an entire organism whose structure as well as behavior (culture, society, psychology) are expressed in terms of the most basic particles such as **quarks** and leptons. *See also* Darwin; De Vries; Dobzhansky; Wallace.

De Beer's Germ-Layer Theory: Biology (zoology): *Sir Gavin Rylands De Beer* (1899–1972), England.

The development of animal cartilage and bone cells originates in the ectoderm of animal embryos.

Up to this time, the accepted germ layer theory stated that bone and cartilage cells were formed in the mesoderm (the middle layer of tissue) rather than the ectoderm, the outer layer of embryonic tissue. De Beer's theory contributed to the knowledge of how the vertebrae are developed in reptiles, birds, and mammals. Recent research used genetically engineered cells implanted in chicken embryos to produce a protein that determines the bone structure of a bird's wing. This led to the knowledge that similar genes shape the human arm, as well as the general skeletal structure and organs of all animals, including humans. De Beer also demolished Haeckel's law of recapitulation (also known as the biogenic law), which states that ontogeny (embryo development) recapitulates the phylogeny (evolutionary history) for each individual. In other words, each embryo goes through all the stages of development that resemble all the stages of the ancestral evolution of that organism's species. The law of recapitulation is an oversimplification of embryology as well as evolution and is no longer considered viable. Instead, De Beers framed his argument on the concept of *pedomorphosis*, which is the evolutionary retention of some youthful characteristics by adults. *See also* Darwin; Haeckel; Linnaeus; Wallace.

De Broglie's Wave Theory of Matter: Physics: *Prince Louis Victor Pierre Raymond de Broglie* (1892–1987), France.

A particle of matter with momentum (mass × velocity) behaves like a wave when the wavelength is expressed as $\lambda = \hbar / p$. (λ = wavelength, $\hbar$ = Planck's constant, and p is the particle.)

In the macro world (large masses) when a body, such as an automobile is moving, its momentum is very great. Therefore, the wavelength is so short that the wavelength behavior of the automobile cannot be discerned. However, in the sub-micro world (very small particles of mass), such as electrons and protons, the particle will have little momentum and therefore its wavelength is easily measured.

The evidence for this theory is demonstrated by the effects of interference. (See Figure Y1 under Young.) If a beam of particles is divided into two parts as it passes through two slits in a screen and the number of small particles with mass arriving at different points on a target screen are measured, the results are the same as they are for a similar experiment done with light photons and waves. The characteristics of constructive interference resulting from the split screen for the particles of matter are the same as the characteristics of wave motion. De Broglie's wave theory of matter supported and helped Erwin Schrödinger explain the theories of relativity and quantum mechanics. *See also* Bohr; Einstein; Heisenberg; Schrödinger; Young.

Debye-Huckel Theory of Electrolytes: Chemistry: *Peter Joseph William Debye* (1884–1966), United States, and *Erich Armand Arthur Joseph Huckel* (1896–1980), Germany.

In concentrated solutions, as well as dilute solutions, ions of one charge will attract other ions of opposite charge.

Up to this time, the Arrhenius theory of ionic conductivity was correct only for very dilute solutions. (See Figure A2 under Arrhenius.) This theory initiated the use of **electrolysis** for the separation of ions in very concentrated solutions (e.g., brine), for the extraction of sodium and chlorine, and led to the industrial production of gases, such as bromine, fluorine, and chlorine, as well as the extraction of some metals from their ores. Using x-ray diffraction, Debye determined the degree of the polarity of covalent bonds and the spatial structures of molecules, which disproved earlier theories of conductivity in strong electrolytes. *See also* Arrhenius.

Dehmelt's Electron Trap: Physics: *Hans Georg Dehmelt* (1922–   ), United States. Dehmelt shared the 1989 Nobel Prize for physics with Wolfgang Pauli and Norman Foster Ramsey.

By isolating an electron in an electromagnetic field, it is possible to suspend it, thus providing a means of continuously and accurately measuring its characteristics.

Hans Dehmelt constructed the *penning trap*, a combination of strong magnets in an electric field contained in a vacuum used to isolate and suspend a single electron so it could be studied. He accomplished this by reducing the kinetic energy (motion) of the electron by cooling it, enabling him to measure the single electron accurately. Dehmelt and his colleagues were also the first to isolate and detect individual protons, antiprotons, positrons (positive electrons), and ions of some metals. When light was shone on a suspended single metal ion, it could be seen without the aid of instruments and appeared as a very small, bright, starlike light. Dehmelt and his assistants were the first to view what is known as the *quantum leap*, a very, very small "bit" of energy. This occurred when a single electron of a barium atom jumped to a higher energy state (orbit), became invisible in blue light, and then jumped back to its normal state (orbit), where it became visible again. (See Figure D1.) His work led to confirmation of quantum theory and a better understanding of the physics of atomic particles. *See also* Franck; Heisenberg; Hertz; Pauli; Ramsey; Rutherford; Stern; Thomson.

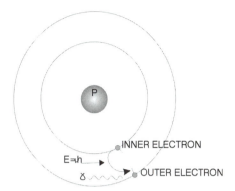

Figure D1: The quantum leap is based on Niels Bohr's idea of electrons being similar to planets orbiting the sun and with specific energy levels. The "leap" occurs when an electron, which cannot revolve around the nucleus in just any orbit, jumps from an inner orbit (higher energy level) to an outer, lower-energy-level orbit, emitting a photon (a tiny packet/quanta of light). The energy level of the photon is equal to the difference in the energy levels of the two orbits of the electron, which is expressed as Planck's hypothesis $E = \hbar v$ where $\hbar$ is Planck's constant and v is the angular momentum of the electron.

Delbruck's and Luria's Phage Theory: Biology: *Max Delbruck* (1906–1981) and Salvador Edward Luria (1912–1991), and Alfred Hershey (1908–1997) all from the United States, shared the 1969 Nobel Prize for physiology or medicine.

Bacteria develop resistance to phages by spontaneously mutating.

Phage, a Greek word meaning "devour" or "eat," is the simplest genetic system known. Phages are simple viruses composed of plain strands of nucleic acid with a more complex "head" that contains DNA material. The phage that infects bacteria is referred to as *bacteriophage*. (See Figure D2.) Delbruck's and Luria's research sought to ascertain how phage could multiply so rapidly in bacteria—up to 100 phage particles are produced in just a few minutes. Delbruck did the mathematical and statistical work on the problem, while Luria conducted the experiments. Along with Alfred Hershey, they investigated and determined the genome of phage virus. The genome is *all* the DNA, including the DNA in the genes, that are contained in the structure of an organism. Delbruck and Luria demonstrated that the phage virus inserts itself into the host bacteria and replaces the bacteria's DNA with the phage's own DNA, in essence cloning itself and resulting in mutation of the bacteria. They also determined some genetically mutated bacteria develop a resistance to the destructive bacteriophage. The three Nobel Prize winners are credited with founding the field of molecular biology. *See also* D'Herelle; Northrop.

Democritus' Atomic Theory of Matter: Chemistry: *Democritus of Abdera* (c.460–370 B.C.), Greece. *See* Atomism Theories.

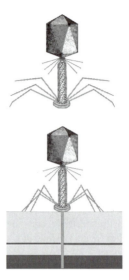

Figure D2: Artist's conception of a typical T4 bacteriophage virus that infects bacteria cells. It consists of two major parts, nucleic acid (RNA or DNA) in the "head" and a coating (capsid) made of protein to protect the nucleic acid. A specific bacteriophage infects only a limited range of bacteria species.

Descartes' Theories and Philosophy: Philosophy: *René du Perron Descartes* (1596–1650), France.

Descartes' Theory of Light and Reflection: *White light is pristine light that can be reflected at an angle equal to the angle at which it strikes a mirrored surface.*

Descartes believed that pristine (white) light produced colored light only when there was a "spinning" sphere of light. He also stated that the angle of incidence is the angle formed between the perpendicular to the surface and the ray of light striking the surface. This was a simple way of explaining the law of reflection, since the angle of incidence is equal to the angle of reflection, which was later developed and became known as Snell's law. Descartes' concept of light and matter was more metaphysical than empirical.

Descartes' Concept of Motion: *The force of motion is the product of mass times velocity.*

Today we refer to his force of motion as *momentum* (mass × velocity). This concept was accepted by Newton but opposed by Leibniz. Leibniz incorrectly assigned the force of a body as the product of mass times the velocity squared, which later developed into the idea of kinetic energy. This dispute led to conflicts in different schools of philosophy.

Descartes' Mathematical Concepts: *(1) A systematic approach to analytic geometry (combining algebraic and geometric functions); (2) the invention of exponential notation; (3) a rule for determining the positive and negative roots for algebraic equations; and (4) the development of equations to describe specific curved lines and curved surface area.* All were important to the development of calculus.

Descartes was the first to relate motion to geometric fields. He saw a curve as if it was described by a moving point. As the moving point describes the curve, its distance from two fixed axes will vary according to that particular curve. This is known as the Cartesian coordinate system, which lends itself to the study of geometry by algebra as a means of interpreting graphs with the x (horizontal) and y (vertical) coordinates that compare two variables. Graphs have become a major geometric and analytic tool for all sciences. All of these concepts contributed to later developments by other mathematicians.

Descartes' Philosophy and System of Knowledge: *(1) Nothing is true until a foundation has been established for believing it to be true. (2) Start with a priori assumptions (first principles), proceeding by mathematics to deductions that use physics and mathematics. One cannot have complete knowledge of nature because there is always doubt; one is never certain of the nature of nature. In addition, reason deceives us. Therefore, a person can be certain only of doubting (not knowing for sure).*

These philosophical concepts led to his famous phrase, which summed up his total philosophy as "I think, therefore I am," which he asserted as justification of self and God. Descartes' greatest contributions were in mathematics and philosophy, and he believed there was great unity in the study of both these sciences. His application of algebraic methods to geometry was a major step in the progress of mathematics as well as other sciences.

De Vries' "Pangenes" Theory of Evolution: Biology: *Hugo De Vries* (1848–1935), Netherlands.

*(1) Organisms consist of large groupings of physical characteristics. (2) Each characteristic is attached to particles inside the nuclei of cells. (3) Although invisible, these hypothetical particles were referred to as "pangenes" by Darwin, which he theorized as units controlling heredity. De Vries equated them to the concept of chemical atoms. (4) Theoretically, when pangenes combine, they cause the appearance of unique characteristics for each species. (5) The more pronounced a particular character or feature of a species, the greater are the number of pangenes. (6) Pangenes may be dominant or latent (recessive and not visible). (7) During **meiosis**, pangenes may split, each causing new features or possibly new species. (8) When splitting or "mutating," pangenes are responsible for the formation of new pangenes, which produce new "mutants."*

Hugo De Vries rejected the idea that natural selection over long periods of time could produce new species and/or variations in existing species, as Darwin's theory of evolution proposed. De Vries believed new characteristics or species could come about only by genetic changes (mutations) in cell nuclei of organisms. His theories were not all correct, but they did explain that mutants have extra sets of **chromosomes** in their nuclei (triploids and tetraploids). Although the term *pangenes* is no longer accurate or used, De Vries' explanations give a reasonable account of how variations, later determined to be due to genetic mutations and natural selection, occur in the evolution of species. *See also* Darwin; Dawkins; Dobzhansky; Lamarck; Mendel.

D'Herelle's Bacteriolytic Theory: Biology: *Felix D'Herelle* (1873–1949), Canada.

When bacteria are infected by viruses and the resulting fluids are filtered, the filtrate contains no live bacteria, but this filtered fluid can still infect bacteria. Since the infection agents are smaller than bacteria, they are thus "filterable viruses" and can still act as parasites on live bacteria.

Felix D'Herelle referred to this filterable virus as a "bacteriolytic agent." The word *virus* is Latin for "poison." His experiments and conclusions were not well accepted by those investigating bacterial-viral problems because it was not understood how, once filtered, the virus could still infect bacteria. (See Figure D2.) The term *bacteriophages* and the importance of the filterable viruses in the field of bacteriology later became significant under Delbruck's phage group of researchers. *See also* Delbruck; Northrop; Pasteur.

Dicke's Theory of the Big Bang: Physics: *Robert Henry Dicke* (1916–1997), United States.

The universe started about 14 to 16 billion years ago as a tiny point of energy exploding in an ever-increasing three-dimensional spread of energy at extremely high temperatures that soon formed elementary matter.

The big bang theory proposes that the formation of the universe originated with an infinitely dense pinpoint of compressed energy that, when first expanding, formed hydrogen and helium and, later, stars and their planets and other objects found in the universe. This "explosion" was the source of all the energy and mass in the universe and was forceful enough to overcome the gravity of the expanding particles that are still expanding after more than 14 billion years.

The big bang theory is explained by considering a number of related physical phenomena:

1. A cataclysmic explosion occurred about 14 to 16 billion years ago.

2. Within a few seconds the temperature reached about 10 billion°C.

3. This high temperature caused intense radiation to spread outward in all directions.

4. Soon after the initial explosion, particles of elementary matter were formed and radiated outward.

5. As the explosion expanded, it lost energy through both heat and radiation.

6. **Cosmology** confirms this expansion is still taking place, and possibly accelerating.

7. Although it took place billions of years ago, the intensity of this radiation should still be detectable.

8. If such residual radiation exists today, it should be detectable as "black body" radiation—radiation, such as x rays from deep space, emitted from a black body at a fixed temperature.

These eight phenomena provide the basis for the origin of the big bang as being a condensed, very small, extremely hot mass of energy. Several different wavelengths of radiation, in addition to visible light, have been detected. For example, radio waves, x rays, and microwaves have been detected as originating from deep space and are used to further our knowledge of the cosmos and support of the big bang theory.

Calculations determined that the residual temperature from the big bang in the universe is about 3 degrees kelvin ($-270°C$), and such radiation was later detected as background microwaves of about 7 cm. These factors led cosmologists to conclude the big bang theory is correct. The question remaining is: What was there before the big bang, and where did it all come from? This is more of a metaphysical question than a part of the theory that assumes infinity in time as well as space in both directions—before and after the event—just as it is possible to calculate infinity as negative and positive numbers. *See also* Gamow; Hoyle; Weinberg.

Dirac's Relativistic Theories: Physics: *Paul Adrien Maurice Dirac* (1902–1984), United States.

Dirac's Quantum Mechanics: *The physics of electromagnetic radiation and matter should interact on the micro subatomic, very small scale of nature, as well as at the macro, very large scale.*

For almost three centuries, fundamental Newtonian mechanistic laws for classical physics were accepted. These laws explained the stability of atoms and molecules but could not explain the small packets of energy emitted by excited atoms. Albert Einstein developed the special and general theories of relativity, and Edwin Schrödinger developed the wave equation representation of quantum mechanics. Enrico Fermi and Paul Dirac revised and improved the concept for quantum mechanics by incorporating Einstein's calculations for relativity into the equation and making some corrections in the energy levels of "spinning" atoms which resulted in the Fermi-Dirac statistics. This statistic was designed to satisfy the Pauli exclusion principle required to account for the behavior of particles with half-integer spins.

Dirac's Theory of Negative Energy: *Energy exists below the ground state of positive energy.*

To explain his theory, Dirac claimed that areas below ground state (normal positive charge) were already filled with negative energy, and if a light photon tried to enter this area, it would become an observable electron, leaving behind a vacant "hole" representing a potential similar to a positive charge. This theory led to a new understanding about the nature of matter. (See Dirac's theory of antimatter, below.) This positive electron was named a positron (^+e) and was confirmed by Carl Anderson in 1932. It proved to be a new way to look at the universe, which now included antimatter somewhat as a mirror image of matter. If there were exactly the same number of electrons (^-e) as positrons (^+e) in the universe, they would collide and destroy each other, resulting in energy with no matter left over. The reason that matter exists in the universe and there are more electrons than positrons is that more electrons (photons/electromagnetic radiation) were produced at the beginning of time. This theory has led to speculation of a "sister" universe based on antimatter. *See also* Anderson

Dirac's Theory of Antimatter: *(1) There are negatively charged particles called electrons, each of which has a counterpart. (2) These counter particles are positively charged electrons called positrons. (3) A positron must always occur in conjunction with an electron (but not the other way around). (4) Their collision destroys both. Thus, (5) electromagnetic waves (radiation) are released, producing more electrons than positrons.*

Dirac's theoretical positron was discovered later by Carl Anderson, which confirmed Dirac's concept of antimatter as applying to all matter. Dirac's theory was an explanation of the duality nature of light where in some instances it behaves as a wave (as indicated by interference and diffraction). In other cases, it resembles particles called **photons**, which are matter similar to electrons with measurable energy, as indicated by frequency and momentum. Thus, the concept of particle-wave duality is dissimilar to the Newtonian classical mechanics concept of matter.

Dirac's Theory of "Large Number Coincidences": *There is a 1 to 10^{40} ratio that is constant in the universe. It is exhibited in various physical phenomena, and represents a model of the universe.*

Dirac's mathematics revealed this phenomenon as a relationship existing between natural physical constants and the quantification of natural properties: (1) The ratio of gravitational attraction and electrostatic attraction between electrons and protons is 10^{40}; (2) the earth's radius is 10^{40} that of the electron's radius; and (3) the square root of the number of particles in the universe is 10^{40}. Dirac considered there was a universal relationship between the ratios of the radii of all objects (e.g., earth) and all forces (e.g., gravity). He also theorized that as the universe continues to expand with age, this force ratio will not change, even when the distance between objects increases. Rather, the gravitational constant will change; thus the effects of gravity become less as time and space increase (expand). The theory that the gravitational constant is decreasing over time is no longer accepted. On the other hand, to explain the increased rate of expansion of the outermost galaxies in the universe, some cosmologists are suggesting that antigravity might exist at the horizon of the universe. This is not the same as Dirac's theory for the change in the gravitational constant, but both concepts end up with the same conclusion about the expanding universe. *See also* Anderson; Born; Einstein; Fermi.

Dobereiner's Law of Triads: Chemistry: *Johann Wolfgang Dobereiner* (1780–1849), Germany.

When considering three elements with similar characteristics and with atomic weights within the same range, the central (middle) element will have an atomic weight that is the average of the total atomic weights of all three elements.

Johann Dobereiner was interested in chemical reactions that involved catalysts, which are chemicals that either speed up or slow down chemical reactions but are not changed or consumed in the reaction. One of his discoveries was that hydrogen gas would ignite spontaneously when passed over powdered platinum. Using this concept, he developed the Dobereiner lamp, which uses metal platinum as the catalyst. (The catalytic converter in the exhaust systems of modern automobiles uses platinum as the catalyst to convert toxic exhaust gases to less harmful compounds.) In his experiments involving catalytic actions, he observed how the atomic weights changed incrementally for elements with similar compositions. This led to his law of triads. An example is the triad for the nonmetals chlorine (atomic weight 35.5), bromine (atomic weight 80), and iodine (atomic weight 127). Calculating the average atomic weight, $242.5 \div 3 = 80.8$, which is the approximate atomic weight of bromine, the middle element of the three examples. A second example is the triad for three metals: calcium, strontium, and barium. This triad's average atomic weight calculates as $265 \div 3 = 88.3$, which is the approximate atomic weight of the middle element, strontium. This relationship of atomic weights to characteristics of similar elements was an important discovery Mendeleev used in designing his **Periodic Table of the Chemical Elements**. (See Figure M4 under Mendeleev.) Being aware of this phenomenon provided Mendeleev a means to leave vacant spaces in his table that could later be filled. He called yet-to-be-discovered elements *eka elements*. He predicted these eka elements would fit into specific blank spaces where the other two elements in the triad, with known atomic weights, were directly above and below the missing eka element. For example, in group IVA (14) yet-to-be-discovered element germanium was named eka-silicon to fill in the vacant space. His eka-predicted elements were very accurate as far as atomic weights were concerned, but they were not always correctly arranged according to specific characteristics of elements in groups. Later, his periodic table was corrected to represent the elements arranged by their atomic numbers (protons) rather than atomic weights. *See also* Dalton; Mendeleev; Newlands.

Dobzhansky's Theory of Genetic Diversity: Biology: *Theodosius Dobzhansky* (1900–1975), United States.

Populations with a high genetic load of debilitating genes confer an advantage to organisms by providing more versatility within changing environments.

Theodosius Dobzhansky believed **species** that have a wide variety of genes, even recessive dormant debilitating genes, will be more successful by providing the entire species with greater genetic diversity. This diversity is related to the evolution of race and species and provides more effective adaptation to changing environments. Historically, this meant that those species that survived over long periods of time were the ones with the greatest pool of genes. When these genes no longer provided an advantage in overcoming environmental conditions, natural selection contributed to their extinction over time. Those that had the greatest genetic variety survived. By all standards, the dinosaurs must have had a very wide and diverse genetic load because there were so many different types of dinosaurs that successfully survived for millions of years. Despite this, they became extinct about 65 million years ago. Humans arrived many millions of years after the dinosaurs' extinction. Prehistoric humans existed for only a few hundred thousand years, and modern humans (*Homo sapiens*) have existed for less than 100,000 years. Over 98 percent of all plant and animal species that ever lived are now extinct due to many natural causes. How long present species of plants and animals survive, including modern humans (*Homo sapiens sapiens*), may depend on the extent of their genetic load, which may mean that humans are at the end of their evolutionary evolvement as a species. Dobzhansky's theories were important in understanding the mathematical relationships of natural selection, as well as Mendelism. *See also* Darwin; Dawkins; De Vries; Mendel.

Domagk's Concept of Dyes as an Antibiotic: Chemistry: *Gerhard Domagk* (1895–1964), Germany. Gerhard Domagk received the 1939 Nobel Prize in physiology or medicine. Because of World War II, he was unable to receive the award until 1947.

By adding sulfonamide compounds to selected dyes, bacterial infections can be controlled.

Gerhard Domagk based his ideas on research by Paul Erhrlich and other scientists who used several coal tar dyes and other chemical compounds to treat diseases. They succeeded in using dyes to treat conditions caused by some large organisms such as protozoa, but were unsuccessful in treating infections caused by small cocci and bacilli bacteria. Domagk added a chemical called 4-sulfonamide-2', 4'-diaminoazobenzene hydrochloride to an orange-red dye. For the first time in history, a chemical was found to combat bacterial infections in humans without poisoning the patient. He named this new chemical compound *prontosil*; it was the first sulfa drug. Sulfonamide compounds soon proved effective in treating streptococcal diseases such as gonorrhea and epidemic meningitis, as well as staphylococcal infections. They were extremely effective in treating erysipelas, urinary tract infections, and undulant fever due to bacilli. They saved many lives during World War II until about 1945 when penicillin, a superior antibiotic, became available. *See also* Ehrlich.

Doppler's Principle: Physics: *Christian Johann Doppler* (1803–1853), Austria.

The movement of a spectrum-yielding body (any source producing electromagnetic frequencies, such as light, radio, and radar frequencies) can be measured by the shifting lines in its spectrum. Another way of stating the principle is: *The observed frequency of a wave depends on the velocity of the source relative to the observer.*

The Doppler effect relates to waves of air particles as well as light. Christian Doppler arrived at his equation about frequency related to velocity of waves based on a unique experiment with sound. He placed a group of trumpet players on an open train car and had them play loudly as the train moved away. As the train moved closer, he noted the change in the tone and pitch (frequency of the sound waves) of the trumpet notes. Almost everyone has experienced the Doppler effect. For example, when a train rapidly approaches, its whistle is shrill or high pitched due to the compressed sound waves, which increase the sound waves' frequency. The reverse takes place as the train recedes, since the sound waves are less close together, and thus at a lower pitch (frequency). Doppler himself did not have much success using his principle with light waves. However, other scientists using his principle demonstrated a color shift of light waves (frequencies) under the same conditions of motion as there was for sound waves. This finding provided astronomers with a valuable tool.

The Doppler principle is much more important in the field of astronomy. It was first used to measure the rotation of the sun on its axis. As the sun rotates, the light spectrum on the side of the sun rotating toward earth is slightly compressed, which makes the light appear bluer. Conversely, the light from the side of the sun rotating away from earth spreads its spectrum, and thus the sun's light looks more reddish. This principle is also used to measure the motion of stars. If the star appears reddish, it may be receding from us, as its light spectrum spreads out to the red area of the electromagnetic spectrum. This is known as the *redshift* which is due to the shift toward the longer infrared frequency of light waves as stars move away from us. Conversely, if stars are approaching us, they appear bluer due to the compression of the electromagnetic spectrum toward the shorter-frequency blue area of the spectrum. (Basically, most stars are receding from us, but at different rates.) The Doppler principle enables astronomers to measure the distances of stars and galaxies and is used as one of the arguments for an ever-expanding universe.

During World War II, British engineers designed radar (*r*adio *d*etection *a*nd *r*anging), which was based on the Doppler effect. It used a specific radio wavelength that could be bounced off a moving object. The returning wave was at a different wavelength as it was picked up by the sending device, which could then determine the object's position, altitude, and the rate it was approaching or receding from the radar operator. It was also used to develop more accurate bomb sights. Since that time, radar has become a valuable scientific tool for navigation, meteorology, and astronomy. *See also* Watson-Watt.

Douglass' Theory of Dendrochronology: Astronomy: *Andrew Ellicott Douglass* (1867–1962), United States.

Climate and environmental history can be determined by the formation of the rings in the cross sections of trees.

Andrew Douglass' first interest was in trying to decipher the eleven-year period of high sunspot activity to the period of low sunspot activity as measured on Earth. (The high to the low in a cycle is just one-half of the cycle. A full cycle is from high to high.) The actual complete cycle is twenty-two years from one crest to the next crest in the complete cycle. Douglass initially tried to relate the sunspots' high-to-low part of the cycle to the distinct rings in trees that represent yearly growth. He thought there might be some correlation between the two, but later determined the rings were a more interesting area for study as a means of dating the past.

Dendrochronology, Greek for "time-telling by trees," is defined as the study of the rings of growth in old trees to verify historic climate, weather, temperature change, rainfall, insect populations, diseases, and so on. For instance, it has been determined there were periods of devastation of plant growth due to insect plagues and volcanic eruptions, as well as extended droughts, long before to-day's pollution problems. The years during which these events occurred are easily ascertained by carefully examining and counting the rings in the cross-sections of old trees. Douglass' goal was not only to learn about prehistoric chronology of climate but to use dendrochronology as a means of predicting future climatic changes, particularly global climate changes. There are some limitations to dendrochronology as a dating tool for historical conditions. Living trees have a definite age; thus you can go back only so far, whereas fossil tree rings can be read back to prehistoric times. Also, using tree rings to correlate global or even hemispheric climate changes has proved to be very inaccurate due to the lack of correlation between the growth rings of trees located in different parts of the same continent and the wide distribution of trees world-wide. More accurate methods for determining past and future global climate changes are now available.

Draper's Ray Theory: Chemistry: *John William Draper* (1811–1882), United States.

Electromagnetic rays, absorbed by some chemical substances, can cause chemical changes in that substance. In addition, the rate of chemical change is proportional to the intensity of the radiation.

The daguerreotype field of photography was based on exposing silver salts to light, which caused the image on a glass plate coated with silver salts to darken at the points where the greatest amount of light occurred. The problem was that the silver salt continued to darken as it became exposed to more light outside the camera. Draper solved this problem when, after exposure, he dissolved the unexposed silver from the plate with a solution of sodium thiosulfate, also known as "hypo," which is still in use today to "fix" the photographic image. Draper determined that electromagnetic rays (visible light, x rays, etc.) cause a chemical reaction when absorbed by some light-sensitive chemicals (e.g., silver nitrate) and that the amount of light and time of exposure are proportional to the chemical changes. Draper applied his theory to the new field of photography. He built cameras out of cigar boxes and perfected the process to the point where he was able to take short exposure photographs of the moon and to take the first pictures through a microscope. He was also the first to record the solar spectrum photographically through a **prism**. (See Figure F4 under Fraunhofer.) In addition, using his sister as a model, he was the first to take a successful, short-exposure portrait. Draper experimented with the size of the **aperture** for the lens, which, if enlarged, would reduce the time of exposure required to expose the image.

Dulbecco's Cancer Cell Theory: Biology: *Renato Dulbecco* (1914–), United States. Renato Dulbecco shared the 1975 Nobel Prize in physiology or medicine with Howard Temin and David Baltimore.

Normal cells, when mixed with cancer-producing viruses in vitro, kill some of the cells, while other cells are changed by the virus, which continues to grow and multiply as cancerous tumors.

Renato Dulbecco refers to this theory as "protective infection." The significance of Dulbecco's concept is that it is possible to grow cells in the laboratory that have been infected by tumor-causing viruses. This simplifies the process of understanding the nature of malignancy and experimenting with possible treatments. It is easier to experiment with different chemical treatments using cancer cells **in vitro** than in the human body. His theory and laboratory techniques advanced cancer research. *See also* Baltimore; Gallo; Temin.

Dumas' Substitution Theory: Chemistry: *Jean Baptiste André Dumas* (1800–1884), France.

An atom or radical can be replaced by another of known quantity that produces the same results.

Jean Dumas believed that organic chemistry was similar to inorganic chemistry as related to the formation of **radicals** of the same types. Organic chemistry involves the element carbon in the construction of large molecules found in living tissues, such as protein compounds. Inorganic chemistry involves the reactions between and among all types of elements that form inorganic (nonliving) compounds. Radicals might be thought of as molecules that contain an electrical charge and can act as units to combine with other elements or compounds, regardless of whether they are "organic" or "inorganic." An example is the hydroxyl radical OH^-, which is part of the water molecule with a negative charge and is thought to have some effect on the aging of cells. Dumas contended that the site of these radicals is where replacements of atoms of one type for another take place. He showed that several different compounds that were composed of the same atoms or radicals exhibited similar characteristics. His famous example demonstrated that trichloroacetic acid was a similar compound to acetic acid. Dumas contributed to the advancement of chemistry with his work with atomic weights based on whole numbers as multiples of hydrogen as 1, and thus carbon as 12. This led him to his theory of types, which today is referred to as *functional groups of elements. See also* Dalton.

Dyson's Theory of Quantum Electrodynamics: Physics: *Freeman John Dyson* (1923–), United States.

Quantum theory can explain the relationships and interactions between minute particles and electromagnetic radiation.

Freeman Dyson combined several related theories into a general theory that described the interactions between waves and particles in terms of quantum concepts. (*See* Planck for a description of *quantum*, which means "how much.") This single theory enabled scientists better to understand quantum electrodynamics, which synthesizes waves, particles, and the interaction of radiation with the electrons of atoms. This theory is also known as the *quantum theory of light or radiation. See also* Dehmelt; Feynman; Planck.

E

Eddington's Theories and Concepts: Astronomy: *Sir Arthur Stanley Eddington* (1882–1944), England.

Eddington's Star Equilibrium Theory: *For a star's equilibrium to be maintained, the inward force of gravity must be balanced by the outward forces of pressure caused by both the star's gas and radiation.*

Sir Arthur Eddington also developed the system indicating that heat generated inside stars was not transmitted outwardly by convection—as heat could be on earth—but rather by a form of radiation. This theory of equilibrium not only explained why stars do not usually explode, but also provided a much better understanding of the internal structure of stars. It was William Higgins who determined that the sun as well as stars, and presumably the entire universe, are composed of the same basic elements as is earth, with carbon being the most important of the first elements formed whose atoms were larger than hydrogen and helium. Our sun consists of several layers that are not sharply divided. The core of the sun is about 250,000 miles in diameter and consists of hydrogen undergoing fusion to form helium, resulting in great quantities of radiation energy. The core's temperature is about 27,000,000°F; its pressure is about 7,000,000,000,000 pounds per square inch. (The air pressure on the earth's surface is less than 15 pounds per square inch.) Next is the convection zone that surrounds the core, which transmits the radiation to the outer layers. The photosphere, which means "sphere of light," is the surface where the radiation is converted to light and heat as we know it on earth. It is about 500 miles thick, with a temperature of almost 10,000°F. Electromagnetic radiation (light) as well as heat is constantly sent outward in three dimensions from the convection zone to the photosphere. The *chromosphere*—the outer "shell" or layer—is about 2000 miles thick with a variable temperature ranging from about 8000 to 90,000°F. The outermost layer is called the *corona*, which is a low-density collection of ionized gases. This layer forms the "spikes" that shoot out from the sun's surface and can be seen during an eclipse. As Eddington's theory of equilibrium states, if this arrangement becomes imbalanced, a star could explode. Exploding stars, which are extremely bright, were recorded in ancient history. Today, a very bright star that lasts only a few days or a week is called a nova. *See also* Higgins.

Eddington's Theory of Star Mass-Luminosity Relationship: *The more massive (very dense) the star, the more luminous it is.*

The brightness of a star is determined almost entirely by its density (mass per unit volume). It is a fundamental principle of astronomy that for stars of constant mass, their luminosity is also constant. The mass of a star is related to its density but not necessarily to its size. Therefore, a small, high-density star is also massive in the sense that it may contain more matter than does a star that is larger but less dense. A small, dense star may be much brighter than a very large star of low density. Up to the time of this concept, only masses of binary stars could be directly calculated—that is, a pair of stars close enough in proximity that their mutual gravity causes them to rotate around a common, invisible center of gravity. This led to the theory that stars, even of different spectral classes, with the same masses also had the same luminosities. This relationship of mass to brightness was of great significance in not only determining the nature of stars but also their distance from earth. Using this mass-luminosity theory, along with the gravity-equilibrium theory, Eddington calculated there was a limit to the size of stars, that is about ten times the mass of our sun. Any star forty to fifty times the mass of the sun would be unstable due to the excessive internal radiation. Eddington's mathematical equations are considered fundamental laws of astronomy and provided a new way to look at the evolution of stars, including our sun.

Eddington-Adams Confirmation of Einstein's General Theory of Relativity: *Einstein predicted light from distant bright stars would be "bent" by the gravity of another star as it passed by that star.*

Eddington reported that during a total eclipse, the light from several bright stars was slightly bent as their light came past the sun. This demonstrated that light as electromagnetic radiation is affected by the gravity of the sun. (See Figure E1 under Einstein.) Eddington's work confirmed Einstein's theory of special relativity. Walter Sydney Adams (1876–1956) further tested the theory by measuring the shift in the wavelength of light from Sirius B, a very dense white dwarf with strong gravity. Einstein predicted the light from a massive star would shift to the red end of the spectrum. Thus, as the light from a massive star was slowed due to that star's gravity, a reddish shift occurs in the star's light (not to be confused with the **Doppler effect** for the "redshift," which is based on the lengthening of the frequency for lightwaves from a star that is rapidly receding from us). *See also* Doppler; Einstein.

Eddington's Constants for Matter: *The total number of protons and electrons in the universe is 1.3×10^{79}, and their total mass is 1.08×10^{22} masses of the sun.*

Eddington arrived at these figures after considering the concept of an ever-expanding universe with "curved" space as theorized by Einstein. This theory was based to some extent on the tremendous velocities of nebulae. Eddington extended this theory and combined it with the theory for the atomic structure of matter in order to calculate his constants by theory alone. These are considered fundamental constants of science, which are important for the concept of an expanding universe.

Eddington's Physical Theory of the Big Bang: *The universe started to expand in all directions when a small, very dense ball exploded with tremendous force.*

Eddington was not the first to come up with a "cosmic egg" concept of the origin of the universe. This "egg," about the size of a marble or less, was assumed to be extremely dense as it contained all the mass-energy in the universe. Eddington's contribution was in developing the mathematical equations to explain the physics of the expanding universe. The concept goes back thousands of years, but it was Willem de Sitter (1872–1934) who first developed a viable cosmological model based on the theory of an expanding universe. Cosmic microwave remnants of a hot primeval fireball as evidence of the big bang were detected in 1964 by Robert Woodrow Wilson and Arno Penzias of the Bell Telephone Laboratories. Although some of the details of the big bang are still elusive, the concept of an inflationary universe is now being considered. It theorizes that at the time of the bang, all the original particles and energy could defy the speed of light and expand at any speed. Several other scientists have contributed to the cosmic egg/big bang/inflationary universe concept. There are still questions as to the origin and state of the universe: What existed before the big bang? Is the universe really expanding? If so, do we really know the rate of expansion? Will it continue to expand forever? Will it reach equilibrium, then contract and start all over again? Will it regenerate or is it always generating new matter? Is it static? *See also* Dicke; Gamow; Hale; Hubble; Lemaitre.

Edison's Theory of Thermionic Effect: Physics: *Thomas Alva Edison* (1847–1931), United States.

Thermions (negatively charged electrons) generated at the hot cathode filament (of the light bulb) will jump to a cooler wire some distance from the filament.

This is commonly referred to as the Edison effect and is the only physical theory Edison developed. All of Edison's other accomplishments were inventions that led to the development of important industries.

In developing his light bulb, Edison followed the lead of Sir Joseph Wilson Swan (1828–1914), who developed the first light bulb and was the first to use a carbonized (charred) thread as a filament. Swan's bulb did not work very well because he could not produce a good vacuum inside the bulb, nor could he develop a battery to produce a strong enough current to cause the carbonized thread to incandesce. Just one year before Edison announced his invention, Swan perfected his carbon filament incandescent bulb and demonstrated it to the public before Edison perfected his bulb. Edison is generally credited with the invention, although sometimes Swan is listed as a co-inventor of the incandescent light bulb.

Edison was the first to explain that electric current flows in only one direction—from the filament to the electrode, and not the other way around. He experimented with several hundred different types of materials to act as filaments as he developed his incandescent light bulb. In 1883 he inserted a metal wire next to the filament, but not connected to it or the source of electricity. His expectation was that such a "cold" piece of metal would reduce the amount of air, thus improving the vacuum and prolonging the life of his filaments. He noticed that some electrons (he called them *thermions* because the filament was hot) flowed across the space gap in the bulb to the cooler metal wire, producing a noticeable glow. He patented his "Edison effect" but did not exploit it. Later, this arrangement of the filament next to a metal grid proved to be a valuable design for the development of the electronic vacuum tubes used in radios and television sets before the days of semiconductors and transistors.

Ehrlich's "Designer" Drug Hypothesis: Chemistry: *Paul Ehrlich* (1854–1915), Germany. Paul Ehrlich, a physician, was awarded the 1908 Nobel Prize for physiology or medicine.

Using the molecular structure of synthetic compounds, specific pharmaceutical drugs can be produced to treat specific diseases.

Aware of the aniline dyes (coal tar dyes) developed by Sir William Henry Perkin (1838–1907) and how different dyes could stain animal fibers (e.g., wool and hair), Paul Ehrlich assumed these dyes could also differentially stain human tissues, cells, and components of cells. Using an aniline dye, scientists saw for the first time the chromosomes of cells, which Ehrlich called *colored bodies*. Other scientists identified specific germs that caused specific diseases. Ehrlich then hypothesized that since specific dyes will stain specific tissues selectively, it might be possible to design a chemical to attack specific types of germs that are also composed of specialized living material. He formulated his side-chain theory of immunity, which led to the development of synthetic chemical compounds designed specifically to attack microorganisms. Although many coal tar dyes can cause disease, including some cancers, the large dye molecules can be manipulated to attack specific types of bacteria, such as those that cause sleeping sickness and syphilis. However, drugs derived from coal tars were not effective for treating other diseases, including streptococci and cancer. Paracelsus, the alchemist, was known as the ancient founder of chemotherapy; Ehrlich is known as the modern founder of chemotherapy. *See also* Koch; Domagk; Paracelsus.

Eigen's Theory of Fast Ionic Reactions: Chemistry: *Manfred Eigen* (1927–), Germany. Manfred Eigen shared the 1967 Nobel Prize for chemistry with George Porter and Ronald Norrish.

Ionic solutions in equilibrium (same temperature and pressure) can be dis-

arranged out of equilibrium by an electrical discharge or sudden change in pressure or temperature resulting, within a short time, in the establishment of a new equilibrium.

Manfred Eigen used the "relaxation technique" along with ultrasound absorption spectroscopy to determine that this reaction occurred in 1 nanosecond (one-billionth of a second). Using this information and his techniques for measuring fast reactions, he ascertained how water molecules are formed from the H^+ (hydrogen ion) and the OH^- (hydroxide ion) to form H_2O. He continued to use his "fast reaction" theory to explain complex reactions and characteristics of metal ions and, later, more complex organic biochemical reactions and nucleic acids. The theory of fast ion reaction is important to the understanding of chemical reactions in all living organisms.

Einstein's Theories, Hypotheses, and Concepts: Physics: *Albert Einstein* (1879–1955), United States. Albert Einstein was awarded the 1921 Nobel Prize in physics.

Einstein's Theory for Brownian Motion: *The motion of tiny particles suspended in liquid is caused by the kinetic energy of the liquid's molecules.*

In 1827 Robert Brown (1773–1858), while using a microscope, observed that pollen grains suspended in water were in constant motion, which he believed was caused by some "life" in the pollen. He added minute particles of nonliving matter to water and observed the same motion. This phenomenon was not explained until the kinetic theory of molecular motion was discovered. Albert Einstein derived the first theoretical formula to explain why these small particles moved in a liquid when the particles themselves were not molecules. His equation was based on the concept that the average displacement of the particles is caused by the motion resulting from the kinetic energy of the molecules in the liquid. This resulted in a better understanding of the atomic and molecular activity of matter, and thus heat.

Einstein's Theory of the Nature of Light: *Electromagnetic radiation propagated through space (vacuum) will act as particles as well as waves since such radiation is affected by electric and magnetic fields, and gravity.*

James Clerk Maxwell developed an equation stipulating that electromagnetic radiation can travel only as waves. This concept disturbed Einstein, as did the experiments by Philipp Lenard who had observed the photoelectric effect of ultraviolet light "kicking" electrons off the surface of some metals. It was determined that the number of electrons emitted from the metal was dependent on the strength (intensity) of the radiation. In addition, the energy of the electrons ejected was dependent on the frequency of the radiation. This did not jibe with classical physics. This dilemma was solved by Einstein's famous suggestion that electromagnetic radiation (light) flows not just in waves but also as discrete particles he called *photons*. Max Planck referred to these as *quanta* (very small bits). Using Plank's equation, $E = \hbar v$, where E stands for the energy of the radiation, Einstein was able to account for the behavior of light as massless particles with momentum (photons) that have some characteristics of mass (e.g., momentum, as well as waves). It resulted in Einstein's being awarded the 1921 Nobel Prize for physics.

Einstein's Concept of Mass: *The at-rest mass of an object will increase as its velocity approaches the speed of light.*

When a body with **mass** is not moving, it is at rest as far as the concept of inertia is concerned, meaning it is resistant to movement by a force. An analogy would be sluggishness, inertness, or languidness in a human being. Once an at-rest mass is in motion (i.e., **velocity**), it attains momentum (mass × velocity). When there is an increase in its velocity, there is also an increase in its mass. Thus, if a mass attained the speed of light, it would not only require all the energy in the universe to accomplish this, but it would equal all the mass in the entire universe. Therefore, it is impossible for anything with mass (except electromagnetic radiation) to attain the speed of light. This is one reason that light must be considered as being both a wave and a particle.

Newton's three laws that relate to mass and motion represent a classical, mechanistic concept of the universe. Newton's laws are deterministic based on the conservation of mass which states that matter cannot be created or destroyed. Although **weight** is proportional to mass, the *weight* of an object varies as to its position in reference to earth and thus gravitational attraction, whereas the *mass* of an object is independent of gravity. The mass of an object (matter) is the same regardless of its location in the universe and is independent of gravity. At the same time, one might say that in deep outer space, mass has zero weight.

Einstein's theories of relativity ultimately changed the Newtonian concepts of mass and motion. In modern physics, the mass of an object changes as its velocity changes, particularly as the velocity approaches the speed of light. This phenomenon is not noticeable on earth, since our everyday velocities are far less than that of light. For instance, the at-rest mass of an object will double when it attains a velocity of 160,000 miles per second. This is approaching the speed of light, which is 186,000 miles per second, and even a very small mass is incapable of attaining the speed of light. When masses with extremely high velocities interact, nuclear reactions can occur, where mass can be converted into energy—thus the famous Einstein equation, $E = mc^2$, where E is the energy, m is the mass of the object, and c^2 is the constant for the velocity of light squared.

Einstein's Theory of Special Relativity: *(1) Physical laws are the same in all inertial reference systems. (2) The speed of light in a vacuum is a universal constant. (3) Measurement of time and space are dependent on two different events occurring at the same time. (4) Space and time are affected by motion.*

An inertial reference system is a system of coordinates (anywhere in space) in which a body with mass moves at a constant velocity as long as no outside force is acting on it. From this concept, other components of the special theory of relativity follow.

Albert Einstein's special theory of relativity provides an accurate and consistent description of events as they take place in different inertial frames of reference in the physical world, with the provision that the changes in space and time can be measured. He developed the special theory of relativity in order to account for problems with the classical mechanistic system of physics. Many people had (and still do have) difficulty understanding his theory. In essence he is *not* describing the nature of matter or radiation, although he recognized their association. His theory describes the world as it might look to two individuals in different frames of reference. For example, in classical earth-bound physics, a person in a car going in one direction at 50 miles per hour meets a car approaching from the opposite direction going 100 miles per hour. This is how these speeds of the two cars are observed and judged (measured) independently by another person standing by the side of the road and not moving. But the person in the car going 50 miles per hour would say that the car approaching him is going 150 miles per hour. This is just common sense and can be proved with classical equations of adding and subtracting velocities, which is known as Galilean transformations. However, this is not how it works with electromagnetic radiation waves such as the velocity of light. Einstein's theory states that the time between the two events (of the cars) is dependent on the motion of the cars. The special theory states that there is no absolute time or space. According to experiments by Albert Michelson and Edward Morley, the speed of light is independent of the motion (velocity) of its source or the observer. For instance, if both cars are traveling at astronomical speeds in space and one car is going twice as fast as the other car and both turned on a spotlight toward the approaching car, the light would travel the same speed in both (either) directions. One driver would not perceive the light as coming toward him at a greater speed than would the other driver since they would judge the combined speeds of 50 and 100 mph of the two cars on earth. In contrast to earth-bound car drivers, Einstein stated that despite how fast you are going, the speed of light will be constant for all frames of reference. The drivers of the two "space" cars, regardless of how fast they are going, will be in two different frames of reference of both time and space, but the speed of light will remain constant. Thus, from their individual frames of reference (points of view), they will not be aware of "earth-bound commonsense" differences in their speeds. No matter how fast you go, the speed of light will always be the same, even if you are speeding in the same direction as the light is being propagated. The theory later included the concept of the three Euclidean coordinates of space—width (x), height (y), and depth (z)—with the addition of the coordinate *time* to arrive at a space–time continuum as developed by Hermann Minkowski.

There is much confusion about the word *relativity*. In science it is used as something "relative" to something else that can be measured mathematically or statistically. Specifically, Einstein's special theory of relativity is related to frames of reference as measured for the four coordinates of space and time. *See also*: Minkowski.

Einstein's Principle of Gravity: *Gravity is the interaction of bodies equivalent to accelerating forces related to their influence on space-time. Gravity meas-urably affects the space-time continuum.*

There are two related concepts of gravity: the Newtonian classical concept and the Einsteinian concept related to his theories of relativity. Newton's law states that the gravitational attraction between two bodies is directly proportional to the product of the masses of the two bodies and inversely proportional to the square of the distance between them, as expressed in $F = G\ m_1m_2 \div d^2$. Following is an example that relates acceleration to the force of gravity on Earth. If you are in a train or car that is traveling on a perfectly smooth surface and cannot see out the windows, you cannot tell if you are going backward, forward, or not moving at all if the vehicle is traveling at a uniform speed. But if the vehicle accelerates or decelerates, your senses will react as if gravity is affecting you. A person also becomes aware of G forces (simulated gravity) when a car or airplane rapidly accelerates. To sum this up, classical physics stated that all observers, regardless of their positions in the universe, moving or stationary, could arrive at the same measurement of space and time.

Einstein's theories of relativity negate this concept because the measurements of space and time are dependent on the observers' *relative* motions regardless of their inertial frames of reference within space coordinates. Einstein combined the ideas of several other physicists and mathematicians that dealt with non-Euclidean geometry, the space-time continuum, and calculus to formulate his gravitational theory. In essence, Einstein's concept of gravity affected space and time, as in his theory of general relativity. Even so, Einstein's concept of gravity was not quite correct because he did not take into account the information developed by quantum theory for very small particles and their interactions, even though these subatomic particles are much too small (or even massless) to be affected by earth's gravity. His concept dealt with the macro (very large) aspects of the universe. As with all other laws of physics, the laws concerning gravity are not exact. There still is room for statements that more precisely interpret the properties of nature. For Einstein, the interactions of bodies are really the influence of these bodies (mass) on the geometry of space-time.

For many decades, scientists have tried to explain gravitational waves in relation to the theories of relativity or some other principle. *How* gravity acts on bodies (mass) can be described, but exactly *what* gravity is or *why* it is has not been discerned. Another hypothesis is based on the particle called the **graviton**, proposed by quantum theory. Gravitons behave as if they have a zero electrical charge and zero mass. Although similar to photons, they do have momentum (energy). The concepts of gravity waves and gravitons are still under investigation.

Einstein's Theory of General Relativity: *The interactions of mass (as related to gravitational force) are really the influence of bodies (masses) on the geometry of space and time. Space and time are affected by gravity.*

This theory is based on two main ideas: that the speed of light is a universal

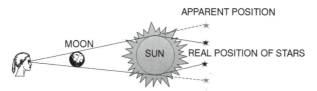

Figure E1: As predicted by Einstein, light from stars located behind the sun can be viewed during a total eclipse. As the star's light passes the sun, it is bent toward earth by the sun's gravity, indicating that light has momentum (mass × velocity) and thus is affected by gravity leading to the theory that light is composed of minute packets (quanta) (e.g., photons). (The actual bending of the light is less than depicted in the diagram.)

constant in all frames of reference and that gravitational fields are equivalent to acceleration for all frames of reference within the space-time continuum.

The first proofs of the general theory of relativity came from astronomy. It explained the previously unknown reason for the variations in the motions of the planets. The theory then was used to predict the bending of star light as it passed massive bodies, such as the sun, and as it was detected during a total eclipse. (See Figure E1.) The theory also predicted that electromagnetic radiation in a strong gravitational field would shift the radiation to longer wavelengths. This was demonstrated by using the Mossbauer effect, which predicts the effects of a strong gravitational field on radiation. An experiment using a strong source of gamma radiation was set up just 75 feet above the earth to measure the gamma rays as they approached the surface of the earth. A minute lengthening of gamma rays (very short wave-length electromagnetic radiation) caused by the gravity of the earth was detected, thus confirming the theory. *See also* Eddington.

Einstein's Unified Field Theory: *A simple general law can be developed combining the four forces of nature (the electromagnetic force, gravitation, the strong force, and the weak force):*

1. An electromagnetic force is exerted between electrical charges and magnetic fields.

2. Gravitational force is related to mass and acceleration and can affect electromagnetic radiation.

*3. The **gluon** of the strong force holds the nuclei of atoms together. The positive protons and quarks in an atom's nucleus would repel each other and fly apart if it was not for the "glueballs" of the strong force that "bind" the quarks and protons.*

4. The weak force is responsible for the slow nuclear processes that produce radiation, such as beta decay of the neutron to generate high-speed electrons.

Albert Einstein spent the last thirty years of his life attempting unsuccessfully to combine these four fundamental forces and the equations incorporating them into a general unified field theory. Einstein did not completely accept the new quantum theory, which did not lend itself to his concept for a unified field theory. Toward the end of his life and later, physicists used **particle accelerators** to separate and identify numerous particles and forces from atoms and their nuclei, which made the unified field theory impossible. But the idea is not completely dead. Today there are several efforts to combine or find symmetry between various theories of matter and energy:

- The grand unification theory (GUT), an attempt to derive an equation to combine the strong and weak forces and explain how the particles of matter were dispersed from each other at the time of the big bang at speeds greater than the speed of light. The GUT theory has led to another concept referred to as the *inflationary universe*.

- The theory of everything (TOE), an attempt to state that there is only one simple force and one ultimate particle in the universe. They have not been found.

- The string theory, a mathematical concept to explain everything in the universe with just one theory, based on the premise that all elementary subatomic particles are really strings that are single-dimensional loops, sometimes described as a doughnut folded over itself several times. Presumably the string theory has as many as twenty-six dimensions (not the three coordinates plus time with which we are familiar). Thus, it is unrelated to the real universe, but is an intriguing concept for mathematicians and physicists.

When various mathematical equations and techniques are used to combine other mathematical equations into one final statement, the results always seem to come out as noise or lead to infinity.

Eldredge-Gould Theory of Punctuated Evolution: Biology: *Niles Eldredge* (1943–) and *Stephen Jay Gould* (1941–), United States.

Evolution of species and individuals by natural selection results from pressure brought about by relatively rapid changes in the environment.

Niles Eldredge and Stephen Jay Gould's theory, also referred to as *punctuated equilibria*, while still accepting many of the tenets of slow organic evolution, such as natural selection, claims the evidence indicates that long periods of slow evolution were "punctuated" by very rapid environmental changes. These periods of rapid change could have been caused by earthquakes, volcanoes, meteors or asteroid bombardments, or other catastrophes that altered the food supply and atmosphere or hastened other rapid changes in the environment. In other words, the abrupt appearance of a new species is the result of *ecological succession* and *dispersion*. One possible example is a 10-kilometer asteroid that impacted about 65 million years ago in Chicxulub, located in Mexico's Yucatan, which exploded dirt and other matter into the atmosphere, blocking sunlight. It is generally accepted as causing the elimination of most plant life on earth. Once plants were gone, so were the dinosaurs and many other animal species. Some primitive forms of life, such as bacteria, minute multicellular organisms, seeds, and small animals, survived and continued to evolve into new species. There are a number of large gaps in the fossil records of early organisms that indicate that evolution may not have been a continuous slow process, but new species were derived from the survivors of catastrophic ecological events. On the other hand, some of these gaps in the fossil records could be that soft-bodied organisms were not fossilized during certain periods of earth's development.

Today most scientists accept the concept that the evolution of species is driven by natural selection. However, there is disagreement concerning the mechanisms of how and why new species appear, and how long it takes for genetic mutations and changes to appear in organisms. *See also* Buffon; Darwin; Gould; Lamarck; Wallace.

Elton's Theory of Animal Ecology: Biology: *Charles Sutherland Elton* (1900–1991), England.

As more species arrive in a given area, space and resources become a limiting factor, restricting the habitat and resulting in the extinction of some species while other species adapt to their limited (changed) environment.

Charles Elton, an early student of the science of **ecology**, performed many of his animal ecology studies on Bear Island off the coast of Norway. Elton named his concept "packing," evidenced by the island's limited number of existing species, which resulted from specialized evolution within the Island's limited environment. Elton was one of the first to advance the concept of an interactive food chain, called a *food cycle* and often referred to as the *food web.* Bear Island's geology was suited to a limited number of plant species, which became food for the island's birds. The birds then became food for a limited number of the island's mammal species, which completed the island's community of species, and food chain. Since the mammals and most other species of organisms on the island could not migrate to escape any limitations of the packing imposed by environmental conditions, they would be subject to different evolutionary pressures than would exist if the animals were located in much larger, diverse environments. Animals living in an environment with a widespread limited number of species can "practice" environmental selection by migrating to change their habitat. Elton's theory of how a limited environment affects the types and distribution of plant and animal species is considered an important contribution to the theory of evolution, but some question its applicability to humans. *See also* Darwin; DeVries; Dobzhansky; Haeckel.

Eratosthenes' Mathematical Concepts: Astronomy: *Eratosthenes of Cyrene* (c.276–c.194 B.C.), Greece.

Eratosthenes' Theory of Prime Numbers: *From a list of ordered numbers (1,2,3,4,5,6,7,8,9, . . .) strike out every second number after 2, every third number after 3, every fourth number after 4, and so on. The remaining numbers in the original list will be prime numbers.*

Eratosthenes of Cyrene, a poet, historian, and mathematician, developed the system of filtering, which became known as the sieve of Eratosthenes. Using his "sieve" procedure, the prime number is a positive integer that has no divisors except the integer itself and the first number selected. For example, select 2, for which the next two numbers are 3 and 4. Strike out 4, which is an integral multiple of the original number 2. A prime number is a positive integer having no divisors except itself and the integer (where the integer is any number, except zero, used for counting).

Eratosthenes' Concept of Measuring the Circumference of Earth: *At the summer solstice (June 21) when the sun is at its zenith in the city of Syene (Aswan), it will be 1/50th of a full circle when measured by the angle of the sun at the city of Alexandra at the same time on the same day.*

Eratosthenes of Cyrene knew that on June 21, the sun cast no shadow in the bottom of a water well in Syene. Therefore, on this date and at this point, the sun was at its zenith. At the same time, he measured the angle of the shadow from a stick placed upright in the ground at the city of Alexandra, which he knew was 5,000 stadia from Syene. (*Stadia* is the plural for the Greek *stadium*, the unit of measurement based on the length of the course in a stadium. It is equal to about 607 feet.) Eratosthenes knew how many stadia a camel can walk in one day, so to estimate the distance between the cities, he multiplied the distance a camel walks in one day by the number of days it took a camel caravan to make the journey. On June 21 at Syene the angle of the stick's shadow was 7°12', which corresponds to about 1/50 of a 360° circle. Multiplying 5,000 stadia by 50 equals 250,000 stadia as the earth's circumference. Eratosthenes' calculation was very close to today's accepted equatorial circumference of 24,902 miles.

Using similar measurements Eratosthenes was able to calculate the tilt of earth to its axis (the ecliptic, which is the inclination of the Earth's equator to its orbital plane) as 23°51'20", which is also close to the modern figure of 23.4 degrees. Modern calculations still use Eratosthenes' geometric and algebraic methodologies to arrive at the current figures.

Ernst's Theory of the Magnetic Moment of Atomic Nuclei: Chemistry: *Richard Robert Ernst* (1933–   ), Switzerland. Richard Ernst received the 1991 Nobel Prize in chemistry.

Atomic nuclei have a magnetic moment that will align with strong magnetic fields that can be altered by submitting the nuclei to specific pulsating frequencies of radio waves.

In the 1940s I. I. Rabi, Felix Bloch (1905–1983) and Edward Purcell (1912–1997) developed the technology of nuclear magnetic resonance (NMR) to probe and study characteristics of nuclei of simple molecules. Nuclei have a natural polarity that aligns themselves with strong magnetic fields. By exposing them to selected frequencies of radio waves, nuclei realign themselves in a new energy state. When the radio waves are removed, they return to their original energy state, giving off specific radiation that can be used to identify the nuclei.

Richard Ernst subjected larger protein (organic molecules) to pulsating high-energy radio waves, which provided a means to produce images of living tissue. The process was originally called nuclear magnetic resonance because it "excited" the nuclei of atoms. However, the name was changed because people mistakenly connected the nuclei of atoms resonated by NMR with nuclear energy. Once the process was improved, with better imaging techniques that could view cross-sections of the human body, a similar process became known as MRI (magnetic resonance imaging). MRI provides a better series of images than does x rays. With no danger of radiation exposure, MRI is safer than x rays because the radio radiation used is of a much longer wavelength and lower frequency than are x rays. The improved images have greatly assisted diagnostic procedures for the medical profession because of its ability to detect various abnormalities in the body more accurately than x rays. *See also* Mansfield.

Euclid's Paradigm for All Bodies of Knowledge: Mathematics: *Euclid* (c.330– 260 B.C.), Greece.
It is postulated that all theorems must be stated as deductions arrived at as self-evident propositions or axioms for which a person can use only propositions already proved by other axioms.
First, some definitions:

- A *paradigm* in geometry is the general plan for the development of the logical statement. In science, it is referred to as a "ruling theory" or a "dominant hypothesis."

- A *postulate* claims something is true or is the basis for an argument, such as in geometry. Euclid set out five postulates: (1) A straight line can be drawn between two points. (2) A straight line can be drawn in either direction to infinity. (3) A circle can be drawn with any given center and radius. (4) All right angles are equal. (5) A unique line parallel to another line can be constructed through any point not on the line (parallel lines never meet).

- A *theorem* in mathematics is a proven proposition.

- *Deduction* is a method of gaining knowledge. A deduction is inferred in the statement "if-then" (from the general to the specific).

- A *proposition* is a statement with logical constraints and fixed values (e.g., if proposition A is true, *then* proposition B must also be true).

- An *axiom* is a self-evident principle that is accepted. Several equivalent synonyms for *axiom* are *primitive proposition, presupposition, assumption, beginning postulate*, and a priori. An example of one of Euclid's axioms is: (A) Things that are equal to the same thing are equal to each other. (B) If equals are added to equals, the wholes are equal. (C) If equals are subtracted from equals, the remainders are equal. (D) Things that coincide with one another are equal to one another. (E) The whole is greater than any one if its parts. These five axioms can be summarized as, "The whole is equal to the sum of its parts."

Euclid's paradigm for knowledge led to his great achievement in the field of plane geometry. He brought together the many statements related to geometry into a logical, systematic form of mathematics. Euclid's Elements, written in about 300 B.C., included thirteen books of what was then known in the field of geometry to which he applied his paradigm. It is still valid today.

Eudoxus' Theory of Planetary Motion: Astronomy: *Eudoxus of Cnidus* (c.400–350 B.C.).

To account for the irregular motion of the planets, earth must be at rest and surrounded by twenty-seven celestial spheres.

Eudoxus of Cnidus was one of the first ancient astronomers to attempt to account mathematically for the irregular motions of the planets and still maintain the earth as the center of the universe. His system required not only a motionless earth, but also twenty-seven crystal-like celestial spheres. The sun and moon each had three spheres, and each of the known planets required four spheres to account for their motions. The outermost twenty-seventh sphere contained all the fixed stars; beyond that were the heavens. (See Figure P4 under Ptolemy.) Eudoxus was able to describe mathematically the rising of the fixed stars and constellations over the period of one year. *See also* Aristotle; Euclid; Ptolemy.

Euler's Contributions in Mathematics: Mathematics: *Leonhard Euler* (1707–1783), Switzerland.

Euler's Three-Body Problem: *The motions of an object moving three ways simultaneously can be predicted using Newton's three laws of motion.*

One example of how a body can move in three different directions at the same time is earth's rotating on its axis, revolving about the sun, and proceeding as part of the solar system toward a distant galaxy. Leonhard Euler, interested in determining the motions of the moon, used analytical techniques that could be applied to this problem to derive a form of mechanics. He also devised a system to analyze how the three Newtonian laws of motion and gravity affected objects that exhibited three-way movements. Euler's equation was based on Newton's dynamics called the *mass point*, for a body containing mass rotating about a point. From this he developed two theories for the motion of the moon that proved to be an asset for sea navigation. These motions of the moon were used before dependable clocks became available to determine longitude. *See also* Newton.

Euler's Theory of Notations: *It is possible to use notations such as sines, tangents, and ratios when the radius of a circle equals 1.*

Euler introduced notations in algebra and calculus that Lagrange, Gauss, Leibniz, Einstein, and others followed. He developed an infinite series of numbers that included notations such as $e^x \sin x$, and $\cos x$ and the relation of $e^{ix} = \cos x + i \sin x$. Euler also wrote the first text on analytical geometry that explains such concepts as prime number theory, differential and integral calculus, and differential equations. The contributions made by Leonhard Euler are numerous and important to the development of mathematical theories by other mathematicians. *See also* Einstein; Eratosthenes; Leibniz.

Everett's Multiple-Universe Theory of Reality: Physics: *Hugh Everett III* (1930–1982), United States.

The wave function of quantum mechanics describes alternate outcomes of events in the same universe.

According to the Copenhagen Interpretation of quantum mechanics, as proposed by Niels Bohr, the quantum mechanical wave function states that only a statistical probability is possible for any explicit event to occur. This traditional interpretation applies only to submicroscopic particles. Hugh Everett proposed a different interpretation. He suggested that every possible outcome that may occur can do so in the same universe, or possibly in multiple universes. Everett's interpretation, also referred to as the "many world interpretation" or the "relative-state model," discounts the Copenhagen interpretation of wave function. His concept relates to as many large and small events and as many outcomes as one could possibly arrive at when measuring the universe. Some scientists discount his theory, but others are attempting to develop a new quantum theory that eliminates the special role of an observer from the process. It seems the observer may account for Heisenberg's uncertainty principle of indeterminacy. Another possible means of justifying Everett's concept of a many-worlds universe is to use probability theory. The main objection to his theory is that it either requires many different outcomes from the same cause in one universe or it requires many parallel universes which do not communicate with each other. *See also* Bohr; Feynman; Hawking; Heisenberg; Schrodinger; Schwarzschild.

Ewing's Hypothesis for Undersea Mountain Ridges: Geology: *William Maurice Ewing* (1906–1974), United States.

The thin crust of the ocean floor enables the sea floor to spread, producing the upward movement of basalt rock and the formation of massive, long, worldwide underwater mountain ridges.

William Ewing employed seismic reflection technology to determine that the crust of the ocean floor is only 3 to 5 miles deep, as compared to the depth of 25–60 miles for the land crust. He was also aware the Mid-Atlantic Ridge had been detected in 1865–1866 when the intercontinental communication cable was laid across the ocean floor. His theory extended the ocean ridge system to over 40,000 miles of underwater mountains worldwide. Ewing and Bruce Charles Heezen (1924–1977) discovered the Great Global Rift, a split in one of the major submerged ranges that created a gap deeper and wider than the Grand Canyon. His theory led to the concept for the movement of continents and the six major tectonic plates, of which five support the Continents while the sixth forms the Pacific Ocean, causing earthquakes and volcanoes to occur along the boundaries where these plates meet.

F

Fahrenheit's Concept of a Thermometer: Physics: *Daniel Gabriel Fahrenheit* (1686–1736), Germany.

The temperature required to reach the boiling point for a liquid varies as to the atmospheric pressure. Thus pressure affects the temperature reading.

Daniel Fahrenheit, glass blower and maker of scientific instruments, knew of Galileo's thermoscope, which used the change in the volume of air (density) to indicate changes in temperature. (See Figure G2 under Galileo.) The thermoscope was inaccurate because it relied on the effects of atmospheric pressure on the water encased in the instrument. In the mid-1600s the first closed alcohol and water thermometer was designed by either Ferdinand II, the grand duke of Tuscany, or Olaus Romer (1644–1710), an astronomer, whose improved designs used wine rather than water, which provided some alcohol. This design responded to temperature changes, but not atmospheric pressure, as did Galileo's air thermometer. Another problem was that the water and alcohol mixture still created internal pressure changes and froze and boiled at temperatures just beyond normal ranges, thus reducing its precision and usefulness.

In 1714 or 1715, Fahrenheit improved the design by enclosing mercury in a glass tube, similar to today's mercury thermometers. He also devised an improved scale by selecting one without fractional units. His design placed the mercury in a vacuum within a sealed glass tube, which eliminated the effects of atmospheric pressure, as well as normal freezing and boiling problems. Fahrenheit then combined ice and salt to determine 0° on his scale, which had each degree divided into four divisions. He then placed it in his mouth to determine human body temperature as 96°, eventually corrected to 98.6°. Later, his thermometer and scale were calibrated to establish 212° as the boiling point and 32° as freezing of water at sea level. The Fahrenheit scale is used only in English-speaking countries, particularly the United States. Scientists worldwide use the more appropriate metric Celsius and kelvin scales. *See also* Kelvin.

Fairbank's Quark Theory: Astronomy: *William Fairbank* (1917–1989), United States.

Quarks originating from cosmic rays exist, and their electrical charge can be detected and measured.

In 1964, Murray Gell-Mann, a particle physicist, postulated the existence of strange, basic, subnuclear particles he called *quarks*. Each had an *antiquark* and needed fractional electrical charges to produce other particles. Also, they were thought not to be producible because they were beyond the energy range of particle accelerators.

In 1977, using the Millikan oil drop experiment, which determined the electric charge of an electron, Fairbank placed a tiny ball of the element niobium (about 0.25 mm in diameter) between two charged metal plates that were kept at a temperature near absolute zero. As a cosmic ray passed through this device, a small electrical charge formed on the ball, which could be measured as a change in the electrical field between the plates. The strength of the charge was extremely small (-0.37) and may have been caused by other sources. There is one theory that says quarks cannot be produced because they are not "free." The question of magnitude of the charge on a quark is still being investigated. Despite recent claims of discoveries of as many as twelve different types of quarks, some physicists believe quarks are "confined" or held together by the "strong force" of **gluons**, which makes their separation inside nuclei, and thus their measurement, impossible. *See also* Gell-Mann.

Faraday's Laws and Principles: Physics: *Michael Faraday* (1791–1867), England.

Faraday's Laws of Electrolysis: *(1) Equal amounts of electricity will produce equal amounts of chemical decomposition. (2) When using an electric current, the quantities of different substances deposited on an electrode are proportional to their equivalent weights.*

Michael Faraday was Sir Humphry Davy's (1778–1829) laboratory assistant who continued Davy's work on the electrolysis of chemical substances by passing electricity through chemical solutions. Davy demonstrated that sodium and potassium metals were deposited on the two electrodes in a solution of sodium chloride (salt) through which an electric current was passed. (See Figure A2 under Arrhenius.) Faraday went one step further and measured the amount of electric current being used and its effect on the deposition of the chemical on one or both electrodes. His hypothesis was that the chemical action of a current is constant for a proportional amount of electricity. The equivalent weight of a chemical is the gram formula weight—the sum of the atomic weights of the elements as expressed in the formula, related to a gain or loss of electricity (electrons). The amount of electricity required to cause a chemical change of one equivalent weight is named a *faraday*. The faraday constant is equivalent to 9.6485309×10^4 coulombs of electricity. Faraday's laws of electrolysis have been used over the past hundred years to produce all kinds of chemicals. For

GALVANOMETER

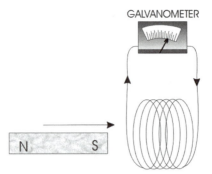

Figure F1: When the magnet is moved into and out of a wire, an electric current is "induced" in the coil by the influence of the moving magnetic field. This is known as the *dynamo effect*.

example, electrolysis can be used to produce hydrogen and oxygen gases by breaking down water molecules in a weak electrolytic solution. No relationship exists between Faraday's laws of electrolysis and the cosmetic process of electrolysis for hair removal.

Faraday's Principle of Induction: *An electric current can produce a magnetic field; conversely, a magnetic field can produce an electric current.*

Both Oersted and Ampere had previously demonstrated that when an electric current flows through a wire placed over a compass, the magnetic needle of a compass is deflected. Faraday rejected the then current belief that electricity was a fluid, and with great insight, he saw electricity as one of several "uniting forces of nature," which he included with magnetism, heat, light, and chemical reactions.

Faraday recognized a connection between the actions of electrical lines of force and the magnetic lines of force. He devised an iron ring with a few turns of wire wrapped around opposite sides of the ring. First, he connected a battery to the two ends of the wire on one side and a **galvanometer** to the two ends of the wire on the other side. When the electrical connection was made on the side with the battery, the needle on the galvanometer on the other side of the iron ring moved. Next, he tried the same experiment without the electric battery by passing a bar magnet through a ring that had a coil of wire wrapped around it. Again, the needle of the galvanometer moved when attached to the ends of this coil. His interpretation was that lines of "tension," as he called the lines of force of the magnet, created an electric current as the magnet moved through the ring (coil) of wire. Thus, an electric current was "induced" in the wire by the moving magnet. (See Figure F1.) Faraday was not an accomplished mathematician, and others, particularly James Clerk Maxwell, developed the mathematics required to make Faraday's concept of induction into a viable field theory. By all and any scientific landmarks, recognizing electromagnetic induction was one of the most important human insights. The concept of induction resulted in the development of the dynamo, or electric generator, which produces alternating electricity by mechanical means, and thus led to the modern electrical age. Induction (brushless) motors have many modern applications, including the small induction motors that run the hard drives of personal computers. *See also* Edison; Henry; Tesla.

Faraday's Principle of Dielectrics: *The conductivity of different substances has different specific inductive capacities for the dissipation of electrical power.*

The *inductive capacity* refers to how much permeability and permittivity a substance exhibits; the *dissipation* of electrical power is the rate of heat loss within the system. This principle states that some substances are very poor conductors of electricity and that induction of electricity relates to the "dielectric" nature (the degree of insulating properties) of the substance. The dielectric strength (permittivity) of a substance is related to how much electricity can be passed through it without breaking down the material. Being able to calculate the dielectric nature of a substance is very important for many industrial uses, including the manufacturing of semiconductor computer chips. As an example, materials with high dielectric constants make excellent **capacitors**, which are important in the electronics industry because they can be made very small and still do the job. Knowing the dielectric properties of substances becomes important when looking for material suitable to make insulators, capacitors, and microelectronic components.

Faraday Rotation Effect: *The plane of polarization of polarized light will be rotated when passed through a magnetic field.*

Faraday's work with electricity and magnetism led him to explore relationships between light and magnetism. He demonstrated that polarized light can be altered when influenced by magnetic fields of force. This *Faraday effect* was developed for instruments used to study the molecular structures of many compounds and later was useful in explaining magnetic fields in the other galaxies of the universe. Faraday's other contributions include the discovery of benzene (C_6H_6) and two new chlorides of carbon, as well as the system for liquefying several common gases, including chlorine. *See also* Ampère; Maxwell; Oersted.

Fermat's Principles and Theories: Mathematics: *Pierre de Fermat* (1601–1665), France.

Fermat's Combination Theory: *Combinations of units are based on the concept of probability.*

Both Pierre de Fermat and Blaise Pascal are credited with developing theories of probability. *Probability* is the ratio of how many times an event will occur as related to the total number of trials conducted.

A famous example of Fermat's theory of combinations follows. In a game where there are just two players, player Allen (A) and player Bill (B), player A wants at least 2 A's or more in a combination of four letters to "win" a point, while player B wants at least 3 or more B's in a four-letter combination to gain points. (Since 3 is higher than 2, Bill felt this higher number would win more combinations.) There are 16 possible combinations for these two letters: *AAAA, AAAB, AABA, AABB, ABAA, ABAB, ABBA, ABBB, BAAA, BAAB, BABA, BABB, BBAA, BBAB, BBBA*, and *BBBB*. Every time an *A* appears at least 2 times or more in a combination, Allen will score a point. At the same time, player Bill requires *B* to appear at least 3 times or more within a combination to win a point. Note that there are 11 "wins" based on 2 or more *A* appearances within the 16 combinations for player A, and only 5 cases containing at least 3 or more "B" appearances within the 16 combinations for player B. The odds for Allen (A) winning the game over Bill (B) are 11 to 5. Also, the game most likely (statistically probable) would be won by A after only four random selections of combinations.

Fermat's Last Theorem: *For the algebraic analog of Pythagoras' theorem for a right triangle, there is no whole number solution for the equation $a^n + b^n = c^n$ (e.g., $3^2 + 4^2 = 5^2$; or $9 + 16 = 25$) for a power greater than 2.*

Integers are positive or negative whole numbers that have no fractional or decimal components and can be counted, added, subtracted, and so forth. Pythagoras' theorem states that the square of the length of the **hypotenuse** in a right triangle is equal to the sum of the squares of the lengths of the other two sides of the triangle. (*See also* Pythagoras.)

Although he had an interest in the theory of numbers and made several contributions to this field, Fermat's record keeping was poor. The equation for the "last theorem" was written as: $a^n + b^n = c^n$ (or as $x^n + y^n = z^n$), and if the n is an integer greater than 2, there is no whole number solution. Fermat professed to have solved this problem, but his solution has not yet been found. For almost four hundred years, scholars have attempted to solve this mathematical conundrum. It seems that when n is 2, it is possible to express the value for a, b, or c, but once whole numbers greater than 2 are used for n, it cannot be calculated. Some claim to have arrived at a proof for $n = 3$, $n = 4$, $n = 5$, and $n = 7$, or even $n = 14$, but only when using prime and complex numbers.

Fermat's Least Time Principle for Light: *Electromagnetic waves (light) will always follow the path that requires the least time when traveling between the two points. In addition, light will travel slower through a dense medium than one less dense.*

Fermat related light to mechanics in the sense that light followed mechanical principles, such as expressed in geometry and the physics of his day. For instance, the *principle of least action*, originally postulated by Aristotle as the *economy of nature*, was also used by Fermat to describe the behavior of light under different circumstances. He based his theory on analytical geometry, which showed that the path of light reflected from a flat surface always took the shortest distance, but for an elliptically curved surface, it took the longest path. Fermat's theory was later restated as the *wave theory of light* during the period when the principle of least action was applied to wave mechanics and quantum mechanics. See also De Broglie; Descartes; Hamilton; Schrödinger.

Fermi's Nuclear Theories: Physics: *Enrico Fermi* (1901–1954), United States. Enrico Fermi was awarded the 1938 Nobel Prize for physics.

Fermi's Slow Neutrons: *Since slow neutrons have no charge and less mass than alpha particles, they are capable of overcoming the positive charge on atomic nuclei, thus allowing them to enter (react with) the atomic nuclei to produce isotopes.*

Neutrons are found in the nuclei of atoms and have approximately the same atomic weight as protons but no electrical charges. Neutrons can be stripped from certain types of nuclei and used as particles to combine with other target nuclei, thus increasing the atomic weight and instability of the target nuclei. Alpha particles are the nuclei of helium atoms composed of two positive protons ($^{++}$He). Thus, an alpha particle is approximately twice the weight of a single neutron. An isotope of an element whose atomic nucleus is composed of a specific number of protons is defined by the number of neutrons. Thus, isotopes of an element have the same atomic number (protons) but different numbers of neutrons, thus different atomic weights.

Enrico Fermi considered using the neutron as the "bullet" to bombard atomic nuclei to produce isotopes of target elements, but he could not produce adequate numbers of neutrons. On a hunch, he placed a thin sheet of paraffin (petroleum wax) between the neutron source and the target. To his astonishment, the production of neutrons increased about a hundred-fold because they were being slowed by the hydrocarbon molecules in the paraffin. Since more neutrons were produced and, more importantly, were slowed down, they would not "bounce" off the target nuclei as did high-speed neutrons. Thus, it provided a better opportunity for the neutrons to interact with the nuclei of the target and produce isotopes of that element by adding neutrons to the target element's nuclei.

Fermi used his new technique to produce numerous radioisotopes of several elements. His theory of slow neutrons was of extreme importance to the new field of nuclear science, which soon led to the nuclear bomb, production of electricity by nuclear power plants, and the production of radioisotopes used in industry and medicine.

Fermi's Theory of Beta Decay: *When a neutron decays, it is converted into a positive proton plus a negative electron and an antineutrino.*

The equation for this reaction is: ($n \rightarrow p^+ + e^- + v$) where *n* is the neutron; *p* the proton, *e* an electron, and v the **antineutrino**. (See Figure F2.) This reaction led Fermi to speculate on a new force he named the "weak force" or "weak interaction," which is responsible for the beta decay process of the neutron. Beta decay occurs when the nucleus emits or absorbs an electron or positron, either increasing or decreasing the element's atomic number but not its atomic weight. The beta decay phenomenon resulted in additional research and discoveries related to the structure of the nuclei and radioactivity.

Fermi's Theory for a Self-Sustaining Chain Reaction: *A self-sustaining chain reaction can be produced by stacking uranium oxide (uraninite ore) and graphite blocks (carbon as a moderator) into an "atomic pile" to produce slow neu-*

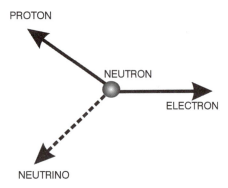

Figure F2: Enrico Fermi predicted that when a neutron disintegrated, the products formed were a proton, an electron, and a very small unknown particle with less than 1 percent the mass of an electron, or possibly no mass at all. This new particle was named the *neutrino* ("the little one"), which became known as the "ghost" particle. In 1956, Frederick Reines and Clyde Cowan confirmed the existence of the neutrino.

trons that will interact with the small amount of fissionable uranium-235 in the refined ore.

After discovery of the neutron in 1932, a number of scientists began to work with nuclei, neutrons, other elementary particles, and nuclear fission reactions. Familiar with this science, Fermi was charged with the supervision of the Manhattan Project. On December 6, 1941 (one day before Pearl Harbor) this ultra-secret project was assigned the task of demonstrating that a self-sustaining chain reaction was possible and could be used to convert a small mass of uranium into tremendous amounts of energy, as theorized by Einstein's famous equation, $E = mc^2$. The atomic pile was constructed beneath the stands at the University of Chicago's football stadium. It was designed to slow the neutrons to the extent they would produce adequate fission of the nuclei, which at some unknown point would enable the reaction to sustain itself. The pile was extra large in order to determine the amount of U-235 needed to reach a **critical mass**, resulting in fission of the U-235. The scientists had one problem. If this critical mass was reached, what would prevent the pile from becoming an exploding bomb? To solve this problem, Fermi and his colleagues devised a means of inserting cadmium rods into the pile to absorb the neutrons before the whole thing became unstable. As the rods were slowly removed, the number of slow neutrons needed to react with the U-235 could be controlled. There were also so-called delayed neutrons produced that kept the pile from going out of control by providing a brief period of safety before the rods needed to be reinserted. This happened on December 2, 1942. Then the neutron-absorbing rods were reinserted, and history was made. Today we know it takes only about 15 pounds of the rare form of U-235 to reach a critical mass, and less than 10 pounds of fissionable plutonium-239 is required to form a self-sustaining chain reaction. Enrico Fermi went on to assist other scientists with the construction of the atomic (nuclear fission) bomb and the H-bomb (thermonuclear fusion bomb). *See also* Hahn; Meitner; Pauli; Oppenheimer; Steinberger; Szilard, Teller.

Feynman's Theory of Quantum Electrodynamics (QED): Physics: *Richard Phillips Feynman* (1918–1988), United States. Richard Feynman, Julian Schwinger and Sin-Itiro Tomonaga shared the 1965 Nobel Prize for physics.

Quantum electrodynamics joins three seemingly unrelated physical phenomena: Einstein's concepts of relativity and gravity, Planck's quantum theory, and Maxwell's electromagnetism.

This theory is also known as the quantum theory of light, which synthesizes the wave and particle nature of light with the interaction of radiation and electrically charged particles.

Richard Feynman's theory grew out of the problem presented with the interactions between electrons and photons of light in which the strength of the charge on the electron mathematically came out to zero, while in actuality the basic unit of electricity is the electron. This did not coincide with known properties of electrons, so a new approach was required. Feynman's concept was based on the probability that if something happens, it can happen in many different ways. Thus, the probability of locating an electron in any particular space is like saying it would be in all the places it probably could be. To assist in understanding his version of the quantum theory, he designed Feynman diagrams, which are simple sketches, similar to vector diagrams, that can be used to trace and calculate the paths of subatomic particles generated in nuclear bombardments that produce new particles and radiation.

At the age of twenty-four, Richard Feynman was the youngest scientist to work on the atomic bomb project during World War II. One of his main contributions took the form of four questions he proposed to the senior scientists on the project for which answers were needed to make the atomic bomb a success. (1) How much fissionable material is needed to achieve a critical mass? (2) What type of material would best make a reflector, or "lens," to focus neutrons on the uranium? (3) How pure must the uranium be? (Less than 1% of uraninite [uranium ore] contains the fissionable U-235.) (4) What would be the expected extent of the damage the explosion would cause (by heat, shock waves, and radiation)? His questions focused the direction of the project and saved time and money in designing the final A-bomb. *See also* Bethe; Fermi; Oppenheimer; Teller; Ulam.

Fischer's Projection Formulas: Chemistry: *Emil Hermann Fischer* (1852–1919), Germany. Emil Fischer was awarded the 1902 Nobel Prize for chemistry.

Projection formulas can be used to describe the spatial relation of atoms in large organic molecules that have the same structural formula.

Emil Fischer demonstrated he could separate and identify sugars, such as glucose, mannose, and fructose, having the same empirical formula. In other words, once he determined the molecular structural formula for one type of organic substance, he could then project this information to synthesize other similar large organic molecules. In addition to his work with carbohydrates (sugars), he contributed to the understanding of the structures of purines, peptides, proteins, and caffeine alkaloids. Fischer used this theory to synthesize polypeptides that contained 18 **amino acids**. He also devised the "lock and key" explanation for high-molecular-weight compounds such as enzymes. Fischer is known as the Father of Biochemistry.

Fitzgerald's Concept of Electromagnetic Contraction: Physics: *George Francis Fitzgerald* (1851–1901), Ireland.

When the light from a body is moving relative to an observer's position, the light contracts slightly in the direction of the motion.

George Fitzgerald and Hendrik Lorentz independently concluded that a fast-moving body will appear to contract according to its velocity as measured by an instrument (or observer). This effect, known as the *Lorentz-Fitzgerald contraction*, described the effect the "**ether**" (in space) had on the electromagnetic (light) forces binding atoms together. This theory proposed to explain the observation made by Albert Michelson and Edward Morley that the speed of light did not depend on the motion of the detector. Therefore this idea could not be used to determine the movement of Earth through space. Einstein used an altered version of the "contraction of light" concept in developing his special theory of relativity, but his theory accepted the concept of space as a vacuum. The concept differs somewhat from the classical Doppler effect and the redshift. *See also* Doppler; Einstein; Fizeau; Maxwell; Schmidt.

Fizeau's Theory of the Nature of Light as a Wave: Physics: *Armand Hippolyte Louis Fizeau* (1819–1896), France.

If the speed of light is known, it can be demonstrated that light travels faster in air than in water or substances denser than air.

Armand Fizeau was the first to measure the speed of light using something other than subjective observations or astronomy. After constructing a device consisting of a rotating disk into which two "teeth" or gaps were cut, he set up a mirror on one hill to return the light sent from another hill about 5 miles distant upon which the "toothed disk" instrument was placed. He sent a light through the gaps in the disk, which acted like a rapid on-off switch that dissected the light into small "bits," similar to a series of light dots. As Fizeau increased the speed of the rotating disk, the reflected light from the mirror on the other hill was blocked off by the solid portion of the disk, but some light would shine through the toothed gap. The faster the disk rotated, the dimmer the light became, until it was blocked entirely by the solid parts of the slotted disk. Conversely, as he slowed the disk, the light would again brighten. By measuring the speed of rotation and the brightness of the light coming through the disk and knowing the distance between the two hills, he calculated the speed of light in air. (See Figure M5 under Michelson.) Both Fizeau and Jean Foucault are

credited with proving that light behaves as waves. With an improved instrument, they measured the speed of light in water and compared it with the speed of light in air to confirm that light travels more slowly in denser mediums. (See Figure F3 under Foucault.) Fizeau also used his instrument to determine that the Doppler effect is the change in the wavelength of light as relative to speed. This is known as the *Doppler-Fizeau shift*. *See also* Doppler; Einstein; Fitzgerald; Foucault; Maxwell; Michelson; Schmidt.

Fleischmann's Theory for Cold Fusion: Chemistry: *Martin Fleischmann* (1927–   ), England.

Nuclear fusion can be achieved at room temperatures by the process of electrolysis using palladium as an electrode in an electrolyte of heavy water.

In 1989 Martin Fleischmann and Stanley Pons (1943–   ) announced they had sustained a controlled fusion reaction at room temperatures in a laboratory that produced 100 percent more energy than was used by the electrolytic process. If this proved to be possible, it was estimated that the discovery would be worth at least $300 trillion and provide an unlimited supply of energy. Other laboratories around the world tried to duplicate this experiment. All failed. Fleischmann claimed other scientists were not using the correct materials or procedures and that he was not going to reveal the exact nature of his experiment. The majority of scientists do not believe cold fusion is possible—at least at this time. Nuclear fusion requires an extremely high temperature and pressure as achieved in the sun and thermonuclear H-bombs in order to "fuse" light hydrogen nuclei into helium nuclei producing a great deal of energy in the process. The debate continues, but only a few scientists believe Fleischmann's cold nuclear fusion can take place as he described the process. Pons and Fleischmann are no longer collaborating on the project. *See also* Bethe; Teller; Ulam.

Fleming's Bactericide Hypothesis: Biology: *Sir Alexander Fleming* (1881–1955); England. Sir Alexander Fleming, Baron Florey, and Ernst Boris Chain shared the 1945 Nobel Prize in physiology or medicine.

If the Penicillium notatum *mold growing in a laboratory dish can destroy staphylococcus bacteria, then it can be tested in order to destroy other selected harmful bacteria.*

Sir Alexander Fleming, a bacteriologist in the Royal Army Medical Corps, was familiar with the use of chemicals in the treatment of wounds. He devised lysozyme, an enzyme found in human tears and saliva, which proved to be a more effective bactericide than the chemicals he was using. Even so, lysozyme was limited in its effectiveness. A few years later while examining a dish containing a culture of staphylococcus bacteria, he noticed several clear areas where the bacteria did not grow due to the presence of a mold identified as *Penicillium notatum*, which seemed to kill the harmful bacteria. Based on his experience with bacteria, Fleming recognized the potential for a new antibiotic but neglected to follow up on this discovery. It had been known for many years that some molds kill "germs," but no one had acted on it until Fleming recognized the importance of this phenomenon. Fleming's mold was isolated, developed, and tested for effectiveness by other scientists in Great Britain. By 1943 Fleming's discovery led to the production of limited amounts of penicillin, which saved the lives of many Allied service men and women during World War II. Since then, a large number of similar "molds" have been identified and produced to provide a much wider selection of antibiotics useful in the treatment of a variety of diseases. *See also* Florey.

Florey's Theory of Mucus Secretions: Biology: *Howard Walter Florey (Baron Florey of Adelaide)* (1898–1968), Australia. Baron Florey shared the 1945 Nobel Prize in physiology or medicine with Sir Alexander Fleming and Ernst Chain.

Cell walls of bacteria can be destroyed by mucous secretions that contain lysozyme.

Following up on Sir Alexander Fleming's work with the enzyme lysozyme, Baron Florey discovered how it affected the cell walls of bacteria to destroy them. This research led to Ernst Chain's and Baron Florey's idea for developing *Penicillium*, an antibiotic, discovered earlier by Fleming. Although Fleming discovered Penicillium and recognized that it killed staphylococcus bacteria, he did not pursue its commercial development. Chain's and Florey's efforts brought to fruition the success of this and other similar effective antibiotics used to treat numerous types of bacterial infections. *See also* Fleming.

Foucault's Theories of Light and Earth's Rotation: Physics: *Jean Bernard Leon Foucault* (1819–1868), France.

Foucault's Wave Theory of Light: *If light is a wave, it will travel faster in air than in water.*

In order to test his theory, Jean Foucault required a more accurate means of determining the speed of light than resulted from Armand Fizeau's rotating "toothed disk" device some years earlier. He devised a system using a rotating mirror, which provided a measurement very close to that later achieved by Albert Michelson. Foucault's mirrors were located on hills 22 miles apart, which made his measurements more accurate. The turning mirror reflected the light back at a slightly different angle to another mirror. This slight angle of reflection could be measured and compared with the rate of rotation of the mirror to give the approximate speed of light. The formula is $v = c \div n$, where v is the speed of light through a particular medium, c is the constant for the speed of light in a vacuum, and n is the index of refraction of the medium through which the light is being measured. He repeated a similar procedure with light projected through water and found that the reflection angle, and thus the speed of light through water, is less than when it travels through air. (See Figure F3.) For instance, the speed of light is approximately 186,000 miles per second in air. To find the velocity of light through water, divide 186,000 by the index of refraction of water which is 1.33, which results in the speed of light through

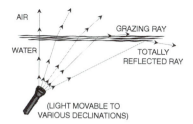

Figure F3: Light rays are "refracted" as they pass through substances of different densities at an angle to the normal (the line perpendicular to the surface). They will change direction when going through the boundary between the two different substances according to their angle of entrance to the boundary. The index of refraction is the ratio of light's speed (in a vacuum) to its speed in a given material.

water at about 140,000 miles per second. This was the proof he needed to arrive at his theory for the wave nature of light, which upset many scientists of his day because of their belief in Newton's theory that light was composed of minute particles (photons). Later scientists accepted the duality of light as having both a wave and particle nature. *See also* Fizeau; Huygens; Michelson.

Foucault's Theory for the Rotation of the Earth: *Since a pendulum swings in an unchanging plane, its apparent progressive movement out of the plane must be caused by the rotation of the earth beneath the pendulum.*

Jean Foucault was familiar with the work of Galileo, who applied the periodic motion phenomenon of a pendulum to measure time. Galileo used his heart pulse to time the frequency of the pendulum "bob," which he determined is inversely proportional to the length of the string suspending the bob. This means that the shorter the string, the greater the frequency of the swinging bob, and, conversely, the longer the string, the lower the frequency. Needing an accurate timing device to measure the speed of light, Foucault considered a pendulum. He knew the pendulum executed a form approximating simple harmonic motion where the force that "drove" it was outside the system (i.e., gravity). He noticed the pendulum appeared to stay in the same plane (compass direction) even when the platform holding the pendulum apparatus was rotated. Recognizing the significance, he determined that since the pendulum always moves in the same plane, it must be the earth beneath that turns. He performed a demonstration during which he suspended an iron ball weighing about 60 pounds from a wire attached to the top of the inside of the Pantheon in Paris. This was the first earth-based demonstration that proved earth actually rotates on its axis, but not at the same rate at all latitudes. For instance, at the equator, an east-west swinging pendulum will show no rotation motion of earth, while at the North Pole earth makes a complete rotation every twenty-four hours under a pendulum swinging in the same plane. The formula for determining the period of rotation at latitudes other than at the North Pole or the equator is $P = 23^h 56^m \div \sin(\text{latitude})$. The pendulum used to measure earth's rotation is called *Foucault's pendulum*. Pendulums are also used to measure acceleration due to gravity and velocities (the *ballistic pendulum*), as well as for accurate clocks. Seventy years after Galileo's study of the pendulum, it was incorporated into an accurate time-keeping instrument by Christiaan Huygens; called the grandfather's clock, it used a weight system to continue the "force" to overcome friction and thus maintain the pendulum's constant swing. There is also a *compound pendulum* where a rigid body swings around a central point. One model, designed by Henry Kater, measured acceleration due to "free fall," which can then be used to calculate the force of gravity. *See also* Galileo; Hooke; Huygens.

Fourier's Theories of Heat Conduction and Harmonic Wave Motion: Physics: *Jean-Baptiste-Joseph, Baron Fourier* (1768–1830), France.

Fourier's Theorem of Heat Conduction: *The rate at which heat is conducted through a body, as related to that body's cross-section area, is proportional to the negative of the temperature gradient existing in that body.*

Baron Fourier worked out the mathematical expression for conduction of heat through different types of solid materials. The theory explains why excellent heat conductors are also good electrical conductors. Conduction relates to the motion of free electrons in solid matter when a temperature difference exists from one end of the matter to the other. The proportionality constant of his equation is referred to as the *thermal conductivity* of the solid. For instance, glass, wood, paper, and asbestos all have thermal conductivities of a much lower value than do metals. They are referred to as *insulators* of both heat and electricity. Conversely, most metals have a high value of thermal conductivity and are excellent conductors of heat and electricity. Conduction is one of three forms of heat transfer described in physics. The other two are convection and radiation. Modern technology and industry use the mathematics of heat transfer and the temperature differential for various materials. Fourier's theory was the beginning of dimensional analysis, which requires any expression of unit quantities to be balanced in equations just as are numbers. Today, computers calculate these properties for the different substances.

Fourier's Harmonic Analysis: *Any periodic motion (wave pattern) can be separated mathematically into the individual sine waves of which the pattern is composed.*

Robert Hooke's law of elasticity states that the change in size of an elastic material is directly proportional to the stress (force applied per unit area) applied to the material. Hooke applied this concept to thermal expansion and wave motions of metal spiral springs. Baron Fourier believed that complicated wave motions were not all that complex and could be solved mathematically. His work, referred to as *wave analysis*, is applied to music. Pythagoras was the first to determine that certain musical notes blend together to produce pleasant sounds. He also ascertained these notes represent a ratio of small whole numbers. Pythagoras used strings of different lengths to define "nice" sounds (e.g., 1:2; 2:3; and 3:4 string-to-length ratios). Musical sounds consist of a number of separate sine waves that when combined display some order of interrelationship. These ordered sets of sine waves, which may or may not reinforce each other, produce musical notes. If the sine waves are randomly selected and combined or interfere with each other, then we call it *noise*. Fourier's wave analysis techniques established a firm physical and mathematical basis for modern music and the development of musical instruments. *See also* Hooke; Pythagoras.

Fowler's Theory of Stellar Nucleosynthesis: Physics: *William (Willy) Alfred Fowler* (1911–1995), United States. William Fowler shared the 1983 Nobel Prize in physics with Subrahmanyan Chandrasekhar.

Thermonuclear fusion reactions in stars produce enough kinetic energy to overcome the electrostatic repulsion of hydrogen nuclei to form helium nuclei. Additional fusion reactions produce even more kinetic energy to overcome the repulsion between other nuclei, thus producing the heavier elements.

A number of scientists have proposed their version of the big bang theory, including Georges Lemaître, George Gamow, and Sir Fred Hoyle. William Fowler expanded on this theory, which propounds that the explosion of a tiny "seed" of energy, which created tremendous radiation and energy, forming in a few seconds mostly hydrogen atoms, which soon resulted in the formation of the entire universe. None of the scientists have claimed to know what caused the "explosion" or the origin of the "seed" of energy/matter. The big bang theory further states the colossal amount of energy provided the force for two hydrogen nuclei ($^+$H + $^+$H) to combine to form a helium nucleus ($^{++}$He) creating **thermonuclear** energy in the stars. This reaction occurred at temperatures of about 10 to 20 million kelvin. The reaction did not produce enough energy to overcome the mutual electrostatic repulsion of the double-charged helium nuclei (alpha particles); thus this temperature was insufficient to form nuclei of the heavier elements. Following the lead of several other physicists, Fowler developed the theory that this temperature in the core of the sun became much greater, up to 100 to 200 million K, and produced enough kinetic energy to overcome the mutual repulsion of the positive alpha particles, which then combined to form the nuclei of heavier elements. First, a fusion reaction occurred, which combined three alpha nuclei (3 $^{++}$He) to form a carbon nucleus with 6 positive protons ($^{++++++}$C). This thermonuclear reaction increased the temperature to over 500 million K, which provided the energy to fuse an alpha nucleus ($^{++}$He) with a carbon nucleus ($^{++++++}$C) to form an oxygen nucleus ($^{++++++++}$O). These stellar thermal nucleosynthesis (fusion) processes continued, reaching over 1 to 3 or 4 billion degrees, thus forming the other heavier elements that make up the earth and universe. Fowler's work aided the understanding of the composition of stars, our solar system, and the nature of the universe, that is, its age and future. *See also* Gamow; Hoyle; Lemaître.

Franck's Theory of Discrete Absorption of Electrons: Physics: *James Franck* (1882–1964), United States. James Franck and Gustav Hertz shared the 1925 Nobel Prize for physics.

Only electrons at specific velocities can be absorbed in precise amounts.

In 1914 James Franck and Gustav Hertz collaborated to demonstrate experimentally that energy is transferred in selected quantized amounts as it reacts with atoms and other particles. They used electrons at different velocities to bombard mercury atoms and discovered the electrons could be absorbed by the mercury atoms only at exactly 4.9 electron volts of energy. If the electrons had less energy, they were lost on collision with the nuclei of the mercury atoms. If the energy was greater than 4.9 eV, they were not absorbed. It was only at the discrete 4.9 eV of energy that electrons were permitted to enter the orbits of the mercury atoms. This was the first experimental evidence for the quantum (Latin for "how much") theory of energy and was later confirmed for the "quantum leap" of electrons for other atoms. A simplified explanation of the quantum leap states that it is a tiny, discrete amount of energy emitted by an electron when it jumps from an inner orbit to an outer orbit (energy level). The closer an electron's orbit is to the nucleus, the greater is its energy. Therefore, as it jumps from an orbit of greater energy to an orbit of lesser energy (further from the nucleus), it must give up "quantized" energy. The energy emitted is a photon (light particle). Conversely, when an atom absorbs a specific level of energy, an electron in an outer orbit can take a quantum jump down to an inner orbit. (See Figure D1 under Dehmelt.) Franck's and Hertz's experimental proof of the quantum theory was an important step in understanding the physics of matter and energy. *See also* Dehmelt; Heisenberg; Hertz; Planck.

Frankland's Theory of Valence: Chemistry: *Sir Edward Frankland* (1825–1899), England.

The capacity of the atoms of elements to combine with the atoms of other elements to form molecular compounds is determined by the number of chemical bonds on the given atoms.

Sir Edward Frankland is considered the father of the concept of *valence* (Latin for "power"). In his work with organic compounds, Frankland discovered that atoms of different elements would chemically bond within fixed ratios with other groups of atoms. From this observation in 1852, he developed an explanation for the maximum valence for each element. The theory of valence explained the relationships of atomic weights with the ratios of atoms combining with each other. (See Figure S2 under Sidgwick.) While valence is the "combining power" of atoms to combine with each other to form molecules, the electrovalence of an ion (atoms that lost or gained electrons) is the numerical value of the electrical charge on the ion. The concepts of valence and electrovalence are important for the understanding and advancement of chemistry. *See also* Abegg; Arrhenius; Berzelius; Bohr; Dalton; Langmuir; Lewis; Sidgwick.

Franklin's Concept of DNA Structure: Physics: *Rosalind Franklin* (1920–1958), England.

The complex organic DNA molecule is a helix structure with phosphate chemical groups situated on the outer boundaries of the helix spirals.

Rosalind Franklin, an expert crystallographer, made x-ray photographs of a form of DNA that clearly indicated the helix nature of its molecular structure. As the story goes, in 1952 James Watson viewed her x-ray photographs and recognized the importance of the obvious helix structure of the DNA substance to the DNA research he and Francis Crick were conducting. Crick obtained copies of Franklin's x-ray photographs from her boss (some say without her permission). These photographs clearly provided the information required for Watson and Crick to develop an acceptable helix structure of the double helix of DNA. (See Figure C2 under Crick.) Rosalind Franklin's photographs of DNA did not indicate the inclusion of base pairs of nucleotides, but they did indicate the phosphate groups on the outside of the strands were responsible for holding the units together. In the publication that reported the structure of the DNA molecule, Franklin was not given credit for her important work or her x-ray photographs. The 1962 Nobel Prize for physiology or medicine was awarded to Watson, Crick, and Wilkins (Franklin's boss), but since Rosalind Franklin was deceased, she was ineligible for the award. *See also* Crick; Watson.

Franklin's Theories of Electricity: Physics: *Benjamin Franklin* (1706–1790), United States.

Franklin's Single Fluid Theory of Electricity: *Electricity is a single fluid with both attracting and repelling forces.*

Benjamin Franklin knew of other scientific experiments that demonstrated that when substances called "electrics," such as amber and glass, were rubbed with wool or silk, static electricity was produced. Conversely, when "nonelectrics," such as metals, were rubbed, no static electricity was produced. Based on this evidence, scientists concluded that since rubbing different substances could produce a repelling or attracting force, electricity must be composed of two different types of fluid. Charles Du Fay (1698–1739) proposed the concept of two kinds of electricity; positive and negative. He called the positive electricity *vitreous* and the negative *resinous*, which seemed to confirm the "two fluid" nature of electricity. In 1747, after experimenting with a **Leyden jar**, Benjamin Franklin advanced a single fluid concept of electricity, but he still considered it a "flowing" substance.

Franklin's Concept of Electric Charges: *Since electricity is a single fluid substance, two types of forces must be present to cause attraction and repulsion.*

Benjamin Franklin, as well as Du Fay, is credited with the terms *positive* and *negative* to explain the attraction and repulsion characteristics of "fluid" electricity. Franklin's single fluid electricity was on the right track, even though he interpreted the terms incorrectly. He reasoned that "positive" would be the direction of the current flow, when in actuality the negative electrons determine the direction of the electric current toward the positive, which lacks electrons. This has caused much confusion ever since.

Franklin's Concept of Lightning: *Lightning is a form of electricity that is more strongly attracted to points, particularly high altitude and metal points.*

For many years, scientists related lightning to static electricity since both produced a jagged spark of light and could cause shock. But there was no proof they were the same phenomenon. Benjamin Franklin's experiments with the Leyden jar illustrated that electricity was more strongly attracted to point sources than to flat surfaces. From this concept, he believed it possible to demonstrate that lightning was an electrical discharge by attracting it to a metal tip on the end of a kite. The result was his famous kite experiment during a thunderstorm in the year 1752, when lightning was attracted to the kite and was conducted to a silk ribbon attached to a metal key. When he brought his knuckle close to the key, a spark jumped to his hand, producing a mild shock. This was a very dangerous experiment, and several scientists were electrocuted when trying to repeat Franklin's demonstration that proved lightning is electricity. However, this experiment forged Franklin's development of the lightning rod, which has prevented lightning damage to many homes and commercial buildings.

Fraunhofer's Theory of White Light: Physics: *Josef von Fraunhofer* (1787–1826), Germany.

White light projected through a prism produces a continuous color spectrum that is crossed by dark lines.

Josef Fraunhofer was an expert lens maker familiar with Newton's studies that proved white light is composed of colored lights when it is projected through a prism. Fraunhofer was also familiar with William Wollaston's (1766–1828) observation of dark lines within a spectrum produced from a white light source. As Fraunhofer permitted light from a very narrow slit to pass through one of his excellent prisms, he observed a series of narrow bands of light of varying wavelengths. (See Figure F4.) Some images of specific wavelengths (colors) were missing and produced the dark lines reported by Wollaston. Fraunhofer did not know what caused them or their significance, but they were then, and still are, referred to as *Fraunhofer lines*. Fraunhofer went on to identify over 700 of these dark lines. Later, Robert Bunsen (inventor of the Bunsen burner) and Gustav Kirchhoff, using a spectroscope, determined that the dark

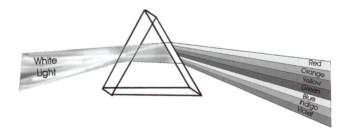

Figure F4: An artist's rendition of the electromagnetic spectrum produced by a beam of visible white light passing through a prism. The white light is composed of separated electromagnetic waves ranging from long infrared to short ultraviolet rays.

lines were the absorption of specific wavelengths of light by a vapor between the source and the prism. This concept of radiation absorption was used to identify numerous new chemical elements by spectroscopy (spectro analysis). The same concept can be used to study the nature of light from the sun and stars to determine their chemical composition. *See also* Bunsen; Kirchhoff; Newton.

Fresnel's Theory for Multiple Prisms: Physics: *Augustin Jean Fresnel* (1788–1827), France.

A single beam of light produces multiple interferences when split into multiple beams by multiple prisms.

Augustin Fresnel considered light to be similar in nature to sound waves. Based on this concept, he worked out the mathematics for light as transverse waves that explained reflection, refraction, and diffraction, which are related to the longitudinal waves for sound. His work with the interference of the beam of light by a prism caused him to consider what would happen if two prisms were used to split the beam of light into two parts. Fresnel conceived of a series of prisms formed as concentric circles on a circular glass lens. This resulted in the Fresnel lens, which, from the front, looks like concentric circles similar to a bull's-eye on a target, but from a cross-section, these circles appear as a series of "sawtooth" tiny circular prisms. The Fresnel lens, first developed to concentrate the light from lighthouses, is now used in a multitude of devices, from overhead projectors, to large-format cameras, to other devices with screens too large to make use of a heavy glass convex lens to concentrate light.

Augustin Fresnel also used transverse waves to explain the phenomenon of polarization of light. The electric field **vector** oscillates in directions perpendicular to the direction of the light. Light is polarized when the electric field oscillates just up and down or right and left. If the electric vector oscillates in all directions, the light is said to be unpolarized. (Today, sunglasses use lenses that transmit the one-directional polarized light waves, while blocking out most of the light with the opposite polarization.) This seemed to settle the controversy of the nature of light as a wave, at least until the particle theory was developed. See *also* Einstein; Fizeau.

Frisch's Theory of a Chain Reaction: Physics: *Otto Robert Frisch* (1904–1979), England.

A sustainable chain reaction can be obtained by using just a few pounds of fissionable isotopes of uranium-235.

In the late 1930s, Otto Frisch was involved in research with other scientists who discovered that uranium would decay into lighter elements when bombarded with neutrons. This work was confirmed and called *nuclear fission*, which was seen as a process capable of producing large amounts of energy. Frisch and Rudolf Peierls determined the rare U-235 was more likely to fission than other isotopes of uranium, such as U-238. Their additional calculations determined it would take only a few pounds of U-235 to reach a critical mass that would produce a sustainable chain reaction resulting in a massive explosion. This made the production of the atomic bomb a practical reality. Otto Frisch moved to the United States in the early 1940s to work on the atomic bomb at Los Alamos. *See also* Bohr; Hahn; Meitner; Peierls; Teller; Ulam.

G

Galen's Theories of Anatomy and Physiology: Biology: *Galen* (c.130–200), Greece.

Galen's Theory of the Circulatory System: *The arteries and veins carry blood, not air, and both the veins and arteries carry blood.*

Until Galen's time, Erasistratus' (c.300–260 B.C.) theory that the essential body elements were "atoms" that were vitalized by air (*pneuma*) circulated throughout the body by the arteries was accepted. Galen, one of the early experimenters who paid attention to his own observations, studied the structure and functions of organs and attempted to disprove this "air" theory. Experimenting with various small mammals, he discovered that blood, not air, flowed through the arteries. But Galen considered the liver to be the main organ of the circulatory system, and his theory stated blood was distributed to the outer parts of the body from the liver by the veins and from the heart by the arteries. Galen also believed blood "seeped" through the intraventricular septum (central wall) of the heart through minute pores and that the heart had three chambers, each with its own function: the anterior or lateral ventricles (sensory information), the middle or third ventricle (cognition and integration), and the posterior chamber or fourth ventricle (memory and motor motion). He did not understand the role of the lungs in the circulatory system and believed the venous system (not the arteries) responsible for the distribution of food from the stomach to all parts of the body. Galen is considered by some historians to be the first to use the pulse of the heart as a diagnostic aid.

Galen's Theory for the Nervous System: *The brain controls the nervous system.*

Through dissection of animals (never humans, except wounded gladiators), Galen demonstrated the distinction between sensory nerves (soft) and motor nerves (hard) and correctly placed the medulla as part of the brain rather than as part of the arteries. He correctly identified the nerves responsible for breathing and speech, as well as demonstrating that different muscles were controlled by specific nerves in the spinal cord.

Galen's Concept of the Kidneys: *The kidneys, not the bladder, produce urine.*

Up to this time it was believed the bladder produced urine. By tying off the ureter, Galen proved the bladder did not produce urine but was just a holding area for it. He also diagnosed several illnesses, including liver disease, by observing the urine of patients.

Galen's Philosophy: *Nature does nothing in vain. God endowed every organ with a special purpose to perform special functions.*

Although eclectic in his acceptance of the doctrines of earlier philosophers, Galen's main beliefs were based on the humoral pathology of Hippocrates (c.460–377B.C.) and Aristotle. For example, he based his theory of circulation on a three-part system of the liver, heart, and brain, each with its own spirits— natural, vital, and animal. His concept of preventive medicine was based on hygiene as well as "critical days," which were days when treatment would be more successful. He believed prevention was better than treatment and thought that diseases could be prevented if the "critical" days were observed. He was an excellent diagnostician for his era and was able to discern the source of many complaints. In addition to prescribing many different types of drugs, he used cold to treat hot diseases, and hot to treat cold diseases, and often used bleeding, purges, and enemas. Galen's philosophical outlook on nature was responsible for his success as a physician and scientist. He believed that the form of an organ was designed by a supreme being to perform a specific function, now known as "form follows function." Galen's medical knowledge and writings were accepted for over 1,500 years. Although his medical knowledge was advanced for this time, later physicians accepted his teachings without question and did little further investigation of the human body. It is believed by many that this respect for Galen's authority (the so-called tyranny of Galen) impeded medical progress for several centuries.

Galileo's Theories: Physics: *Galileo Galilei (1564–1642), Italy.*

Galileo's Theory of Falling Bodies: *Discounting air resistance, two bodies of different sizes and weights will fall at the same rate. Both will increase in speed of descent and land at the same time.*

From the time of Aristotle, it was believed a force could not act on a body from a distance. In other words, for an object to continue to move, something physical needed to continue to push it; otherwise its movement would cease. In addition, Aristotle and others believed a body of greater weight would fall faster than a body of lesser weight, but they had never experimented with bodies heavy enough to overcome air resistance. Very light objects, such as a feather, would descend more slowly than would a rock, which seemed proof enough. It is most likely a myth that Galileo dropped objects of different weights from the leaning tower of Pisa. Actually, it may have been Simon Stevin (1548–1620), not Galileo, who first dropped two rocks of different weights at the same time to demonstrate they would land at the same time. What we do know is that Galileo contrived a method of making accurate measurements and arriving at a reasonable explanation for the phenomenon of free-falling objects. We know that he assembled an inclined plane that allowed two balls of different weights to roll slowly down the incline, which enabled him to measure their rates of descent by using the pulse of his heartbeat. The only other timing devices available at that time were the sundial, time candles, and dripping water clocks. None was accurate enough for Galileo's purposes. He also ensured that the balls were of sufficient weight so they would be only minimally affected by the resistance of air or the surface of the wooden planks of the inclined plane. His measurements confirmed that not only did the balls roll down to the bottom of the plank in equal time, but their rate of descent increased as they passed equally spaced marks on the planks. When he experimented with planks raised higher and lower to form different degrees of inclination, he discovered an interesting factor: No matter at what angle the planks were positioned, the balls covered a single unit of distance on the plank for the first unit of time based on his heartbeat. But for the second unit of time, the balls rolled three times faster than the first unit's distance. He discovered that the ratio of distances covered by the balls increases by odd numbers. This means that for the total time of descent of 4 seconds, the balls covered a distance sixteen times greater than is covered in 1 second. This relationship of the ratio between time and distance is further explained as acceleration acting uniformly on a falling body, where the descending distance covered is directly proportional to the square of the time. From these data, Galileo formulated the law that states $s = \frac{1}{2} at^2$, where s is the distance the ball travels, a is the acceleration, and t is the time lapsing of the ball's descent. Galileo's experimental results illustrated the uniform accelerating force of gravity, which Sir Isaac Newton later developed as part of his concept of inertia and the three laws of motion. The other consequence of this experiment was that Galileo was now able to correct the Aristotelian idea that the push of "angles" was required to maintain planetary motion. Once friction was removed from consideration, the constant pull of the sun's gravity sustains the planets' orbiting the sun.

Galileo's Concept of the Pendulum: *The square of the period (oscillation) of a pendulum varies directly with the length of its suspending string.*

While attending church services in the town of Pisa, a youthful Galileo noticed that as a large chandelier swayed in the breeze (sometimes in longer arcs and sometimes in smaller arcs), the time period of the swing seemingly was the same regardless of the sweep of the chandelier. He used the pulse of his own heartbeat to count the time that lapsed for each swing, which provided him with an idea for an experiment. Upon returning home, he designed a pendulum with a bob on a short string and another bob of a different weight on a longer string. (See Figure G1.) Again timing them with his pulse, he confirmed his theory. He summarized his ideas as follows: (1) The air resistance (friction) prevents the pendulum from returning to its exact starting position. However, if there is

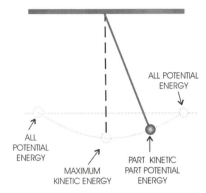

Figure G1: Concept of potential and kinetic energy applied to Galileo's pendulum.

no air resistance, the bob will always return to its original position. Thus, sooner or later, all pendulums come to rest. Pendulums with lighter bobs come to rest sooner than do ones with heavy bobs. (2) The period of swing or sweep of the pendulum is not related to the weight of the bob. (3) The time period for each sweep of a pendulum is not dependent on the length of its sweep (this observation was later proved incorrect). (4) The square of the period for a pendulum is directly proportional to the length of the pendulum.

Once set in motion, a pendulum oscillates with a constant frequency that is inversely proportional to the length of its string. Although Galileo recognized the importance of this phenomenon, he was unable to develop his pendulum into a practical timepiece. (He continued to use a water clock and his pulse). However, it became an important concept in the design of accurate clocks. Some years later, Christian Huygens fabricated a workable pendulum clock similar to a grandfather's clock that used weights to maintain the movement of the pendulum. *See also* Huygens.

Galileo's Concept for the Measurement of Temperature: *There is a direct relationship between the temperatures of air and water and their volumes.*

From the beginning of time, people understood the concept of hot or cold, but until Galileo, there was no objective way to measure the exact temperature for either. Galileo devised a thermoscope, a crude, and not very accurate, instrument for measuring temperature. He used a long, thin, stalklike tube of glass, open at one end and with a closed bulb at the other end. Placing his hands on the bulb until it was warm, he then inverted the open tube into a pan of water. As the bulb cooled, some water was drawn up in the narrow tube toward the bulb. Also as the temperature of the surrounding air changed, so did the level of water in the tube. This furnished Galileo the means to measure the level of water in the tube and to make some calculations. (See Figure G2.) His instrument, however, was quite inaccurate due to the effects of atmospheric pressure on the water in the pan, which was open to the air. Even so, this was the first thermometer, which Galileo later redesigned. He enclosed the water in a sealed

Figure G2: Depiction of Galileo's air thermometer.

tube containing "floats" constructed of small, hollow glass balls adjusted for different water densities. As the temperature changed, so would the density of the water, causing one or more balls to rise or fall, thus indicating the air temperature. (See Figure G3.) Today these fascinating instruments are sometimes referred to as *Thermometro Lentos*.

Galileo's Astronomy Theories: *(1) Dark "spots" on the surface of the sun appear to move around the sun; therefore the sun must rotate, and so must the earth and other planets revolve around the sun. (2) Jupiter has several of its own moons similar to earth's moon. (3) Saturn has bulges on its side as well as its own moons. (4) The Milky Way is composed of a multitude of stars clustered together.*

The telescopes that Galileo constructed enabled him to view objects never before seen by humans, and thus he conceived many theories about the planets and stars. The credit for the development of the first telescope is usually attributed to either the Dutch spectacle maker Hans Lippershey (1570–1619) or Zacharias Janssen (1580–1638), who is also credited with inventing the microscope in 1608. Galileo learned of this "secret" device and then constructed his own telescope. An excellent lens maker, he improved the curvature of his lenses to reduce optical aberration. Galileo built three telescopes, the last of which was improved to approximately 30 power, or about the power of a good pair of modern binoculars.

One of Galileo's first telescopic viewings was of the surface of the sun, where he observed the movement of darker areas or "spots" and concluded the sun must be rotating. Based on knowledge of moving bodies, he surmised the planets and earth are not only spinning on their axes, but are also revolving about the sun in circular paths. This was the first confirmation of the Copernican heliocentric concept of the solar system. (See Figure G4.) Galileo disagreed with, or ignored, Kepler's laws, which state that planets move in ellipses. Since the concept of gravity was unknown in his time, Galileo believed the paths of plan-

Figure G3: A modern version of Galileo's Thermometro Lento.

ets were based on inertial circular movement. This erroneous concept prevented him from completely developing his law of uniform acceleration into the Newtonian-type laws of motion. *See also* Copernicus; Kepler; Newton.

Using the telescope he constructed, Galileo observed two tiny objects that appeared to move around the planet Jupiter, and he tracked and recorded the changes in their position. Later he discovered two other moons of Jupiter, for a total of four larger moons. (A total of sixteen satellites of Jupiter have been subsequently discovered.) His records of the eclipses of Jupiter's satellites aided sailors in determining longitudes at sea.

After viewing Saturn at different times, Galileo noted "bulges" on each side of the planet that periodically became larger, then smaller. His telescope was not powerful enough to resolve these "bulges" into the many rings around the planet that change their apparent shape as the orientation of the planet changes when viewed from Earth. Galileo also identified several of Saturn's moons.

Always fascinated by the multitude of stars that could be seen with his telescope, Galileo observed that when aiming it at the Milky Way, it became obvious this huge area of the sky was composed of many millions of stars. He recorded there were more stars, some very faint, in this area of the sky than in all the other areas combined. Up until this time, the Milky Way was considered to be just a large cloud in the sky.

Although Galileo did not understand gravity or inertia, he had a firm concept of the mechanics of force, and his theories on falling bodies were a forerunner

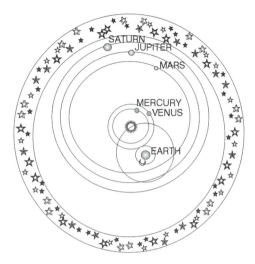

Figure G4: Galileo's concept of a heliocentric solar system.

to Newton's three laws of motion. His "thermoscope" was a precursor to more accurate instruments for temperature and pressure measurements, including the modern mercury thermometers and barometers.

His work with fluid equilibrium, as in a working siphon, led to a new concept of pumps. At one time it was believed that by reducing the air pressure above water, the water would be "sucked up" into the pump (similar to a drinking straw). Galileo understood that normal air pressure outside the pump "pushed" the water up into the pump to the area of reduced air pressure (just as normal air pressure "pushes" the liquid up a straw used for drinking).

He also developed the mathematics that explained the flotation of solids in liquids and studied the magnetism of lodestones that would influence later scientists such as William Gilbert. Galileo was the first to demonstrate that a magnet broken into smaller pieces retained its magnetic properties, since each piece, no matter how small, is still a magnet with its own north and south poles.

Galileo, along with Janssen, is credited with developing the first practical microscope after adapting his concept of a telescope to produce a crude but workable microscope. In addition, he tried to measure the speed of light by flashing a lantern from one hill to an assistant flashing his lantern back from another hill. Although he could not detect the speed of light, nevertheless he was convinced that light travels with great and measurable speed.

At six-month periods, as Earth revolved around the sun, Galileo attempted to measure the parallax of stars to determine their distance from Earth. Although his instruments were not accurate enough to accomplish the task, his concept was correct since he used parallax to measure the distance of the moon to earth.

Most likely Galileo will be remembered for his theory that challenged Ptolemy's Earth-centered universe concept, which was then accepted by the Church of Rome. Because of his theory for and belief in the Copernican sun-centered system, Galileo was secretly denounced to the Inquisition for blasphemous utterances. He was later forced to recant and was sentenced to house arrest until his death at age seventy-eight. Near the end of the twentieth century, the Roman Catholic church removed the charges and exonerated Galileo. *See also* Copernicus; Fahrenheit; Gilbert; Kepler; Newton; Ptolemy.

Gallo's HIV-AIDS Theory: Biology: *Robert Charles Gallo* (1937–), United States.

The HTLV-3 retrovirus suppresses the immune system lymphocytes, thus causing acquired immune deficiency syndrome (AIDS).

Robert Gallo was familiar with the process of how, when under attack, the body's immune system produces interleukin-2, which stimulates special lymphocytes identified as T-cells to fight viral infections in leukemia patients. In the early 1980s he surmised a virus was responsible for a similar suppression of the immune system that led to opportunistic (AIDS) infections. Gallo then hypothesized the retrovirus he identified as HTLV-3 was reacting with the immune system in a similar manner as it did for the blood cancer disease leukemia. At the same time, Luc Montagnier, a researcher in France, made a similar deduction and sent Gallo a sample of his virus, which he called lymphadenopathy associated virus (LAV). Gallo's assistant discovered a particular T-cell type that could be invaded but not killed by these viruses, which then could be used to develop a test for the virus in AIDS patients. Gallo proceeded to secure a patent for his new AIDS test. This resulted in an international argument about the discovery of the AIDS virus, since the American HTLV-3 and French LAV viruses were the same. It was also claimed that the French LAV virus was used in Gallo's laboratory to develop the test. It was settled when Gallo and Montagnier agreed that both their names would appear on the patent document and 80 percent of all royalties would be given to an AIDS research foundation. The HTLV-3 and LAV virus was renamed the HIV virus by an international committee. *See also* Baltimore; Montagnier.

Galton's Theory of Eugenics: Biology: *Sir Francis Galton* (1822–1911), England.

The human race can be improved by selective and controlled breeding, as is done with domesticated plants and animals.

As an anthropologist, Francis Galton's extensive travels enabled him to observe varied cultures and races and subsequently to conduct research in the areas of human heredity. Galton knew that since the beginning of farming, humans selected not only the best grains as seeds but animals with the most desired characteristics to breed, thus improving the quality and quantity of agricultural products, just as is done today with genetic engineering. He was also aware of the theory of organic evolution proposed by his cousin, Charles Darwin. Based on this background, Galton considered controlled breeding a means to improve the human species just as it had for many species of plants and animals. Galton is credited with coining the term *eugenics*, meaning "good genes" or "good breeding." The first to study identical twins, he discovered identical twins have the same fingerprints, while everyone else's are unique. He also developed the first system for classifying and identifying fingerprints, which expanded the field of **forensic** science. Using his research on identical twins, he attempted to resolve the distinction between environmental versus inherited factors that influence intelligence. His research, the premise of which was to determine what is most important in the development of intelligence—nature or nurture—continues today. Galton was the first to use new quantitative methods for eugenics research, including statistical correlation and regression analysis. Although Mendel's work with genetics was not yet "rediscovered," Galton's research indicates an understanding of its basic principles. *See also* Darwin; Mendel; Wallace.

Galvani's Theories of Galvanization and Animal Tissue Electricity: Physics: *Luigi Galvani* (1737–1798), Italy.

Galvani's Theory of Galvanization: *Small electrical currents can be used to "coat" metals that are easily oxidized (iron rust) with other metals that resist oxidation (zinc).*

Galvani was honored for his discovery of the galvanization process that he discovered and called the *metallic arc*, which used an electric current to bind a coating of zinc to iron. Galvani's process is similar to electroplating of metallic and nonmetallic items. Today, galvanized iron or steel also can be produced by dipping the item made of iron into a hot bath of molten zinc or by "spraying" very small zinc particles onto hot iron or steel. The resulting iron products will be more rust resistant than nongalvanized metal. The galvanometer, an instrument used to measure small electric currents, was also named for Galvani by André Ampère, who is credited with the discovery of current electricity. Galvani pursued his work with electric currents into investigating whether animal nerves carry electricity.

Galvani's Theory of Animal Tissue Electricity: *Electricity is present in animal tissue that can be discharged when in contact with two different metals.*

Luigi Galvani's experiments involved touching a dissected frog's leg with a spark from a machine that produced static electricity. His famous experimental error of animal tissue electricity was the result of clamping a dissected frog's spinal cord with brass hooks onto an iron railing during an electrical storm. The frog's muscles twitched, as they did also when the spinal cord was touched by two different types of metals. He incorrectly concluded that the electricity was generated by the frog's tissue, while rejecting the possibility that the electricity that caused the twitching came from another source. Later, Alessandro Volta demonstrated that the electricity was not derived from the tissue but rather from the brass and iron coming into contact with each other under moist conditions. *See also* Ampère; B. Franklin; Ohm; Volta; W. Watson.

Gamow's Theories of the Universe and DNA: Physics: *George Gamow* (1904–1968), United States.

Gamow's Big Bang Theory: *The universe was created more than 14 billion years ago from a single point source in space and time.*

Many civilizations over many generations theorized that the genesis of the universe was based on the egg or seed concept, where everything "grew" from a minute and rudimentary source. This incredibly dense source of energy exploded and rapidly expanded in all directions in just a microfraction of a second, to form all the energy and matter in the current universe. There are several current theories for the creation, nature of, and demise of the universe—the static universe, the ever-expanding universe, and the regeneration concepts. Many scientists proposed theories similar to the big bang for an expanding universe, which they based on Einstein's theory of relativity. In 1948 Sir Fred Hoyle proposed that matter was continually generated by existing matter and is spread throughout the expanding universe. This is considered the model that George Gamow used to revise earlier theories and advance another based on mathematical concepts. Gamow believed there was evidence not only of an expanding universe but also that if the universe is continuing to expand, it must have had a beginning. In other words, if the universe is forever expanding in all directions, then it must have started at a central point from something extremely small that contained all the energy and matter required to form it. As evidence to substantiate his big bang theory, Gamow answered Olbers paradox, which raised the question of why the sky of the universe appears dark rather than full of light, as does the area illuminated by our sun. Gamow explored several possible answers to this paradox, but felt the best is that the universe is ever expanding. Thus, stars cannot shine enough light to fill up all that space because their radiation is not in equilibrium with their surfaces. One of the main factors supporting an expanding universe and opposing a static universe is that a self-contained, nonexpanding universe could not be capable of disposing of the energy produced by all the stars, resulting in a very hot static universe, whereas an ever-expanding universe would reach a balance for star energy. Thus, equilibrium would be established and a stable temperature would exist, or there possibly could be a decrease in overall temperature. According to the second law of thermodynamics (entropy), the final temperature would be absolute zero. George Gamow calculated that the leftover uniform background radiation from the big bang is equal to about 5 degrees kelvin. This "leftover" energy was detected as **primordial** radiation in the 1960s by AT&T scientists and is the best evidence for the big bang theory. Currently, there are new theories that may lead to a reconsideration of an infinite universe: that an ever-expanding universe is an illusion, and that it is finite after all. *See also* Doppler; Hale; Hubble; Lemaître.

Gamow's Theory for the Beginning of Life: *The amino acids, which form proteins, are constructed from the four nucleic acid bases of DNA.*

George Gamow based his theory for life on Francis Crick's and James Watson's proposed structure of the DNA double helix. The nucleic acid–based pairs connecting the two sides of the double helix are the nucleotides of adenine plus thymine (A+T) and guanine plus cytosine (G+C). Gamow realized this sequence of codes for these four nucleotides (A, T, G, and C) could produce only four amino acids, not the twenty or more existing in humans, and thus would be inadequate to produce the multitude of proteins necessary for life. Therefore, he concluded there needed to be at least three sequences of the base pairs present in order to produce codes necessary for the required number of amino acids. Using this code of the nucleotides, at least sixty-four amino acids could be produced ($4 \times 4 \times 4 = 64$). Gamow's mathematical code explained the sequencing required for amino acids to produce proteins.

Gamow's Theory of the Living Cell: *Cells in plants and animals are structured to carry out functions analogous to those processes and procedures related to running a factory.*

George Gamow used the analogy of an industrial factory to explain the functioning of a living cell. The manager's office represents the nucleus of the cell, while the chromosomes are the file cabinets where information, production plans, and diagrams are stored. When a new cell (factory) is to be opened, the secretary and staff produce an exact copy of what is in the file cabinets. As the new factory grows, a new manager's office takes over, and the process is repeated. The "workers" and "machinery" of the factory are abundant and represent the enzymes, protoplasm, and other cell components. The chromosomes and their genes, which are stored in the file cabinets are very limited, but can be used to replicate the factory and start all over. *See also* Crick; Einstein.

Gassendi's Theories: Physics: *Pierre Gassendi* (1592–1655), France.

Gassendi's Atomic Theory: *God created the atoms as immaterial souls that could exist and interact in a void. He then gave them to man.*

Believing that atoms could exist only in a void in which the tiny particles could interact with each other and religious spirits, Gassendi tried to make the atomism of Lucretius, Epicurus, and Democritus agreeable with Christianity, but he opposed Aristotelianism. Gassendi's concept of a void was very much like the modern concept of the vacuum of space. He disagreed with Aristotle's idea that a void did not, and could not, exist. Robert Boyle and other scientists were influenced by Gassendi's "Epicureanism" philosophy. This corpuscular concept states that in order for atoms to exist, they must be surrounded by a vacuum. Thus, if the atoms were removed, only the vacuum would remain. *See also* Aristotle; Atomism Theories; Boyle.

Gassendi's Theories for Falling Bodies, Sound, and Astronomy: Pierre Gassendi, an experimenter and philosopher, propagated his ideas by incorporating his moderate skepticism, which became part of his philosophy.

Gassendi was the first to test Galileo's contention that a ball dropped from the mast of a ship would fall at the base of the mast, not at some distance aft of the mast. Although ancient sailors who dropped rigging tools could attest to this fact, no experiment had been conducted prior to Gassendi's.

It is also reported that Gassendi was one of the first to measure the speed of sound. It is unclear how he made his measurements, but it is assumed he fired a cannon while someone on a far hill at a known distance from the cannon timed the smoke pouring out the barrel until the sound was heard. His figure of 1,473 feet per second was about 50 percent greater than the current figure of 1,088 to 1,126 feet per second in dry air at sea level. The speed of sound depends on the density of the substance through which the sound is traveling. For example, sound travels at the speed of 4,820 feet per second in water, 11,500 feet per second in brass, and 16,500 feet per second in steel.

Gassendi studied **comets** and eclipses and recorded the first observed transit of Mercury. He was the first to describe the northern lights, which he named the aurora borealis. *See also* Aristotle; Descartes; Galileo.

Gauss' Mathematics and Electromagnetism Theorems: Mathematics: *Karl Friedrich Gauss* (1777–1855), Germany.

Gauss' Theory of Least Squares: *A circle can be divided into a heptadecagon by using Euclidean geometry.*

Karl Gauss, a child prodigy in mathematics, was considered a "human calculator" who could solve all kinds of complicated problems in his head. Gauss demonstrated this seventeen-sided polygon (heptadecagon) could be drawn using only a compass, ruler, and pen. All seventeen sides were of equal length when laid on arcs of a circle. Earlier Greek mathematicians could never accomplish this exercise, which was considered an advancement in geometry. Gauss also demonstrated there were a limited number of polygons (many-sided figures) that could be constructed using these tools. An example of a polygon that cannot be so constructed is the heptagon (a seven-sided polygon). *See also* Archimedes.

Gauss' Theory of Errors: *Successive observations and measurements made of the same event by the use of instruments are never identical, but their mean value can be calculated.*

This theory is related to probability. Gauss claimed that the distribution of errors for the **mean** differences of measurements by observations (particularly astronomical observations) is as accurate as the probability (odds) when throwing dice. This statistical technique has been, and still is, used by most scientists who make a series of measurements and calculate the means of these measurements. They can be reasonably certain that the difference between two means is a meaningful representation of their observational measurements. Gauss' work in statistical probability distributions is referred to as *Gaussian* statistical distribution. This concept is used for the statistical treatment of data for most research experiments.

Gauss' Theory of Aggregates: *Properties of individual units of populations can be accurately observed and studied in large groups (aggregates).*

This is another theory used by most scientists and involves the study of large populations of particles, such as atoms, molecules, chromosomes, and genes. An example is the Brownian movement, which is the observed movement of tiny microscopic particles of a solid, such as pollen, that is caused by molecular motion in solution. The concept of aggregates explains the kinetic theory of gases as well as the gene theory for inheritance. Although this theory is based on and accepted as an assumption, it does work. *See also* Ideal Gas Laws.

Gauss' Theory of Magnetic Strength: *The greater the closeness (density) of the lines of force of a magnetic, the stronger the magnetic field.*

Gauss and Wilhelm Weber (1804–1891) collaborated on studying the nature of magnetism. They calculated the number of lines of force and the closeness of those lines, which determine the "flux density" representing the strength of the magnetic field. The International System of Units (SI) for this flux density is called a *gauss* in his honor. In the SI, CGS units (using centimeters, grams, and seconds rather than MKS—meters, kilograms, and seconds), a unit area of 1 square centimeter with a flux density of 1 maxwell per square centimeter equals 1 gauss. The gauss is equal to 1 maxwell per square centimeter, or 10^{-4} weber per square meter, or 10,000 gauss equal 1 weber. Gauss and Weber developed the magnetic-electric telegraph and a new instrument called the magnetometer. Magnetic field strength is rated in gauss units, an important concept for modern technology utilizing all types of magnets.

Gay-Lussac's Law of Combining Volumes: Chemistry: *Joseph-Louis Gay-Lussac* (1778–1850), France.

The volume of gases that react with each other, or are produced in chemical reactions, are always expressed in ratios of small, whole numbers.

In addition to the law of combining volumes, Gay-Lussac discovered that all gases expand equally when the temperature rises. This is a modification of Charles' law. Both of these gas laws, including Boyle's gas laws, are considered the ideal gas laws since they are really approximations. Gases exhibit only the relationships of *P, T,* and *V* (*P* = pressure, *T* = temperature, and *V* = pressure) as expressed in the laws, at ordinary (moderate) temperatures. In 1808 Gay-Lussac published his law, usually called the law of combining gases, when referring to chemical reactions where the number of atoms are constant. This law confirmed the work of Dalton, who missed the importance of the relationship between temperature and volume of gases. In essence, the law states that for any gas, the temperature and pressure are directly related at a constant volume for that gas. The equation is: $P/T = K$, where *P* is the pressure directly related to *T*, the temperature for *K*, a given constant (i.e., volume). Conversely, if the gas is heated, its volume increases as long as the pressure on the gas is constant; and if the pressure increases, so does the temperature of the gas for a given volume of a contained gas. Another way to state it is that the volume of gases expands equally when subjected to the same changes of temperature provided that the pressure remains the same.

Gay-Lussac, a French engineer, physicist, chemist, and accomplished experimenter, made several other contributions to science. He collaborated with several other Frenchmen on a number of projects, including one where he used balloon flights for scientific purposes. In 1804 he and Jean-Baptiste Biot (1774–

1862) ascended 4 miles (about 7 km) in a balloon, the highest attitude attained by humans as of this date. They made the first high-altitude measurements of atmospheric pressure and the earth's magnetism.

Gay-Lussac discovered the poison gas cyanide (HCN), and in 1815 he made cyanogen (C_2N_2), a toxic univalent radical used for the production of insecticides. His experiments with compound radicals were a precursor to the development of organic chemistry. Gay-Lussac and Louis Jacques Thenard (1777–1857) produced small amounts of the reactive metals sodium and potassium. When Gay-Lussac mixed metallic potassium with another element, it exploded, wrecking his laboratory and temporarily blinding him. He also discovered a new **halogen** similar to chlorine. He named it *iode* (iodine), which means "violet." *See also* Avogadro; Boyle; Charles; Ideal Gas Laws.

Geller's Theory of a Nonhomogeneous Universe: Astronomy: *Margaret Joan Geller* (1947–   ), United States.

A map of the redshifts of the light from galaxies indicates a nonuniform distribution of galaxies in specific sections of the observable universe.

By using the Doppler effect, Margaret Geller and John Huchra (1948–   ) observed the distribution of over 15,000 galaxies. They recorded the longer light rays toward the red end of the spectrum, indicating that the galaxies are receding. Light from some of the galaxies started its journey to earth about 650 million years ago, and these galaxies are still receding from us, as well as from each other. According to the big bang theory of an ever-expanding universe proposed by Lemaître, Hoyle, and Gamow, the universe should be rather uniform, or at least galaxies should be randomly distributed throughout all sections of the heavens. When Geller plotted her data for one section of the sky, she discovered very large groupings or clusters of galaxies rather than a random or uniform distribution. Some clusters were many hundreds of millions of light-years across in size. She also noted there were a few galaxies between these clusters, but the clusters contained the majority of all visible galaxies. The implication of this information is unclear, as it relates to future cosmological theory. More recently superclusters composed of clusters of galaxies have been discovered, which seems to support Geller's theory of a nonhomogeneous universe. Geller suggests a revision may be needed for the current big bang model. *See also* Doppler; Gamow; Hoyle; Hubble.

Gell-Mann's Theories for Subatomic Particles: Physics: *Murray Gell-Mann* (1929–   ), United States. Murray Gell-Mann received the 1969 Nobel Prize for physics.

Gell-Mann's Quark Theory: *The heavy particles of atoms (electron, protons, and neutrons) are composed of three fundamental entities, called quarks.*

Quarks, as proposed by theoretical particle physicists, are considered the most fundamental building blocks of matter yet discovered. The string theory proposes a more basic particle, but it has not yet been discovered. Quarks come in groups of threes (originally named red, blue, and green by Gell-Mann), have a fractional electric charge, and have not yet been detected in a free, uncombined state in nature but are being investigated by using a supercollider (RHIC) at Brookhaven National Laboratory. By "smashing" nuclei of gold atoms together at 99.9 percent the speed of light, they hope to create some "free" quarks. Being submicroscopic, quarks have never been observed, but they are considered to be bound up within the interior of the **subatomic particles**. For instance, a positive proton is composed of two up quarks and one down quark held together by gluons. Gluons and their quarks are responsible for the strong interactions that hold nuclei together. There are three types of quark pairs (for a total of six types of quarks): the u quark has a charge of $+2/3$, and the d quark and s quark both have a charge of $-1/3$; thus symmetry is preserved.

Looking for some way to express the triad nature of these theoretical particles, Murray Gell-Mann invented the word *quark* from James Joyce's *Finnegans Wake* (1939, p. 383), "Three quarks for Muster Mark!" (This might be interpreted as three quarts of ale for Mister Mark for a job well done.) Gell-Mann liked the sound of this, and it seemed to fit the concept of his triplet of quarks: u for up, d for down, and s for strange, which was intended to explain the organization of the myriad subatomic particles. Since Gell-Mann first advanced his three quarks theory, many other subatomic particles have been proposed. The **hadrons** are a series of heavier quarks, referred to as the c-quark (for *charmed*), which is many times heavier than, but related to, the moderately heavy s-quark (for *strange*). Today the six quarks are named up, down, charmed, strange, top, and bottom. The u-up and d-down quarks are thought to be just about massless, but they make up almost 100 percent of all the matter (protons and neutrons) in the universe while the others, produced in particle accelerators, are unstable and have a very short existence.

Gell-Mann's Theory of Strangeness: *All fundamental particles are characterized by the property of strangeness.*

The concept of strangeness evolved because of the odd or "strange" manner in which some elementary particles strongly interacted. Strangeness is conserved in the strong and electromagnetic interactions of hadrons and the s-quark, but not so for weak interactions. If for ordinary particles we assign $S = 0$, then we can allow $S \neq 0$ to represent "strange." Therefore, S equals "strangeness." This concept of strangeness led to the development of a new concept for symmetry, used to classify subatomic particles that interact strongly, such as the c- and s-quarks. The new concept of symmetry, also referred to as the *eight-fold-way*, resulted in the discovery of several new particles, including the omega minus. *See also* Feynman.

Gilbert's Theory of Magnetism: Physics: *William Gilbert* (1544–1603), England.

Gilbert's Theory for Electric and Magnetic Forces: *The amber effect (static electricity), which can attract small particles when certain materials are rubbed with certain types of cloth, such as silk, is not the same phenomenon as natural magnetism, which exists in lodestones (magnetic iron ore).*

The phenomenon of rubbing amber with cloth to cause the amber to attract bits of straw and other small particles was known since the days of the Greek philosophers. They related it to some magic or spirit, not to static electricity. William Gilbert experimented with both amber to produce static electricity and lodestones to magnetize iron bars. He was the first to distinguish these two forces of attraction and the first to use the terms *electric attraction* and *magnetic attraction* to make this distinction.

Gilbert's Theory for the Rotation of Earth: *Since a magnetized needle will swing horizontally as it points to the poles of Earth and also dip down toward the vertical, Earth must be a giant spinning lodestone.*

William Gilbert's experiments formed his magnetic philosophy, eliminating much of the superstition and false information about magnetism existing at that time. He constructed a globe from a large lodestone to demonstrate how a compass needle behaves on the lodestone and then related this to Earth. Because of the action of a compass needle, he assumed that the "soul" of Earth was also a spherical lodestone with both a north and south pole. In addition, he demonstrated that the compass needle would dip down at different angles as related to the different latitudes, and the needle would point straight down at the north pole of his lodestone globe. Thus, it would do the same for the North Pole of Earth. This "magnetic dip" phenomenon had already been observed by seamen, but Gilbert was the first to relate Earth's magnetism to latitudes. Gilbert concluded Earth acts like a large, spherical bar magnet, which is spinning on its axis once every twenty-four hours. However, he continued to believe earth was the center of the universe. His theory was the first reasonable explanation for a rotating Earth, but Gilbert did not go as far as Copernicus, who claimed Earth moved through the heavens around the sun. Up to this time, scientists believed Earth was stationary and the canopy of stars was in motion. Historically, the magnetic compass was a reliable instrument that aided in navigation. Gilbert's magnetic philosophy included the belief that earth's magnetic influence affected everything in the solar system that led to the modern concept of gravity. One gilbert (Gb), a unit of electromotive force, named for him, is equal to the magnetomotive force of a closed loop of wire with one turn in which the flowing current is 1 ampere. In the CGS (centimeters, grams, seconds) system, 1 Gb is equal to $10/4\pi$ ampere turns. *See also* Ampère; Coulomb; Faraday; Maxwell; Oersted.

Godel's Incompleteness Theorem: Mathematics: *Kurt Godel* (1906–1978), United States.

The consistency for any formal system, including formal arithmetic and logic systems, cannot be proved by using axioms that are true within that system. A stronger system is required that has an assumed consistency.

The explanation of Kurt Godel's incompleteness theorem is the opposite of a complete theorem, which involves a system where all logical statements made by a formal system can be proved by **axioms** of that very system. Godel claimed no formal system can meet that criterion. The incompleteness theorem may also be stated for mathematics as: *No finite set of axioms is adequate to form the basis for all true statements concerning integers, and there will always be statements about integers that cannot be proved to be either true or false.* The Godel incompleteness theorem also holds for nonmathematically based systems of logic, where a proposition can neither be proved nor disproved based on axioms of that particular system. Godel's incompleteness theorem disturbs many scientists since a noncontradictory system of mathematics may never be constructed, and the means to understand the physical nature of the universe fully may never be found.

Gold's Cosmological Theories: Astronomy: *Thomas Gold* (1920–   ), United States.

Gold's Theory for the Nature of the Universe: *The universe exists in a steady state. It is unchanging in space and time, with no beginning or ending.*

Thomas Gold and other astronomers rejected the big bang concept for the origin of an ever-expanding universe. In 1948 Gold referred to his theory as the *perfect cosmological principle*: a universe with no beginning and no ending, and thus the steady-state universe. In addition, his universe appears the same regardless of where it may be viewed, which was just the opposite of Margaret Geller's concept of a nonhomogeneous universe. In other words, Gold's theoretical universe has a density that is not only constant, but all matter and energy will always be maintained in the same relative proportions. Since this theory conflicts with the laws for the conservation of matter and energy (laws of thermodynamics), the proponents of the steady-state universe needed to produce an idea for the continuous creation of matter and contended that this is just what occurs. Gold postulates a continuous production of one hydrogen atom per cubic kilometer of space that occurs every ten years. Although this amount of "new" hydrogen is undetectable and the vacuum of space is better than any vacuum that can be produced mechanically on earth, it is an adequate "regeneration" of hydrogen to confirm the laws of conservation and maintain a steady-state universe. Currently it is estimated that "empty" space contains 1 proton (hydrogen nucleus) per cubic centimeter. *See also* Geller.

Gold's Theory for Neutron Stars: *The detectable high-frequency, periodic radio signals originating in pulsars are caused by rapidly rotating neutron stars.*

In 1968 Susan Jocelyn (Burnell) Bell (1943–   ) and Antony Hewish (1924–   ) discovered that very short bursts of radio waves were being emitted every second from a new type of star they referred to as *pulsars*, since the signals seemed to "pulse" on a very regular basis. Gold contended these signals originated from rapidly rotating neutron stars, which are extremely dense, spinning bodies that underwent gradational collapse. Neutron stars, through nuclear fusion, have exhausted their nuclear fuel. The result is that gravity becomes so extreme that the stars collapse into themselves, to the point where their protons and electrons combine to form neutrons. At the end of their "lives," these neutron stars collapse into a *black hole*. Since these stars are both very small and very dense, their rotation period can be short enough to produce the detected high-energy radio signals. Gold's theory was substantiated when a new pulsar neutron star with a more rapid rotation was located in the Crab Nebula, even though this pulsar is slowing down about 3.5 seconds every trillion years.

Gold's Theory of Life on Earth and Mars: *Conditions on and below the surfaces of Mars and Earth are very similar and both may support life.*

Thomas Gold made many comparisons between the structure and conditions existing on Mars and Earth—for example:

- Several **meteorites** found in Antarctica appear to have come from Mars. Some contain several rare noble gases (e.g., neon and xenon), as well as nitrogen isotopes, unoxidized carbon, and petroleum-like hydrocarbon molecules.

- The heat on both Mars and Earth increases with depth, and both must have water at some depth below surface levels. This deep liquid water came to the surface on Earth, but not on Mars, where the subsurface temperature keeps the water frozen.

- Although the surface chemicals on Mars and Earth are different, their subsurface chemistry seems similar. Both planets exhibit leftover debris, including meteorites, from their formation billions of years ago.

- Stable hydrocarbon molecules under great pressure are found at great depths on most planets. Earth's carbon-containing liquid and gas petroleum help maintain the carbon cycle by seeping to the surface where carbon dioxide from hydrocarbons is transported to the atmospheric-ocean-rock environments and then recycled by green plants. A similar process occurs for many planets but without the green plants as we know them. All the major outer planets (Jupiter, Saturn, Uranus, and Neptune) have enormous amounts of hydrocarbons, mainly methane and ethane, plus ammonia in their atmospheres, that are recycled as *petroleum rain*. Different forms of hydrocarbons have also been detected on the surfaces of asteroids.

- Evidence suggests that living organisms did not form these deep hydrocarbons. Rather, these primordial hydrocarbons were mixed with biological molecules and are a prerequisite for life. A dilemma exists because primitive microbes must in some way oxidize the hydrocarbon molecules to obtain energy, but the deep interior of planets does not contain such "free" oxygen. A possible means of oxidizing hydrocarbons would be for the deep microorganisms to use sulfur and iron sulfate compounds to oxidize and dine on the hydrocarbons. Organisms that thrive on sulfur have been found at the site of sulfur "vents" discovered at great depths on the Earth's ocean floor.

- Life on the Earth most likely started internally, not from transport of "organic seeds" to the surface from space or by other surface phenomenon. Therefore, life may be found at a primitive stage below the surface of Mars and some other planets. *See also* Gamow; Lemaître.

Gould's Hypothesis of "Punctuated Equilibrium": Biology: *Stephen Jay Gould* (1941–   ), United States.

The evolution of a species is a series of episodic changes within relatively isolated populations.

In 1972 Stephen Jay Gould and Niles Eldredge arrived at their theory of catastrophic evolution by examining fossil records. They disagreed with Darwin's concept of slow but continuous organic evolution involving natural selection, which is a ladder-like, smooth progression of changes from one species to variations of species. Gould and Eldredge detected many gaps in the fossil records, which they claimed disputed this smooth progression. Gould believes such a smooth transition was very rare in nature due to a continuous "pruning" of the branching tree of evolution, which results in the extinction of species. He admits transitional evolution may exist for larger populations or groups of species, but not necessarily for individual species. On the other hand, although the patterns found in fossil records indicate that many species are stable, catastrophic ecological events resulted in the abrupt appearance of new species, which adapted to their new environmental conditions. One problem with the Gould-Eldredge punctuated equilibria theory is that fossils by their nature are never complete, and therefore recognizing species from their fossil remains can only be speculative. One commonly accepted example of the catastrophic theory is the extinction of the dinosaurs approximately 65 million years ago, about the same time that a massive asteroid struck Earth. It is assumed this collision created a world-covering cloud of dust and smoke that blocked out the sun for several years, resulting in the elimination of plant life and great numbers of animal life as well, including dinosaurs.

In the 1970s and 1980s Stephen Jay Gould reexplored the concept of recapitulation that has been out of favor with biologists for many years. It is based on the idea that for a variety of animals, the maturation of embryos in the early fetal stages advances through similar stages of basic structural development. At certain stages of development, the embryos of fish, reptiles, birds, mammals, and humans appear very similar in structure. In other words, very early growth for animals represents a recapitulation of eons of evolutionary development from one species to another. Another way to say this is that ontogeny (individual development) follows phylogeny (evolutionary history). *See also* Eldredge; Haeckel.

H

Haber's Theories: Chemistry: *Fritz Haber* (1868–1934), Germany. Fritz Haber was awarded the 1918 Nobel Prize for chemistry.

The Haber Process (Synthesizing Ammonia): $N_2 + 3H_2 \leftrightarrow 2NH_3$.

During World War I Germany was cut off from its supply of nitrate salts from Chile, an important source of the chemical used for fertilizers and explosives. Fritz Haber developed the process of using "free" nitrogen from the atmosphere and converting it into "fixed" nitrogen. This synthesis, under normal conditions, produced very limited amounts of useful ammonia. To increase the yield to industrial proportions, Haber ran hot steam over hot **coke** at 250°C under high pressure (250 atmospheres; 1 atmosphere equals normal air pressure at sea level) while using a catalyst to increase the rate of the reaction, thus producing more ammonia. This is known as the Haber water gas process, where the nitrogen of the air and the hydrogen from the steam are "fixed" as ammonia. Ammonia, in its various forms, is one of the major industrial chemicals produced worldwide. Without this process for the industrial production of fertilizer, the world would be incapable of producing adequate food. In addition to its use in the manufacture of fertilizers, ammonia is used as a refrigerant and in synthetic fabrics, photography, the steel and petroleum industries, and explosives.

Haber's Idea for Extracting Gold from the Oceans: *Gold can be removed from seawater through the use of high pressures and temperatures and proper catalysts.*

In addition to using very high pressures and temperatures with catalysts to increase the speed of extraction of the gold, Haber used an electrochemical reaction to extract the large amount of gold present in seawater—without much economic success. An estimated 8,000 million tons of gold is dissolved in all the oceans. Although his efforts were unsuccessful, Haber's techniques did lead to the current process of extracting bromine from seawater.

Haeckel's Biological Theories: Biology: *Ernst Heinrich Haeckel* (1834–1919), Germany.

Haeckel's Law of Embryology: *Ontogeny recapitulates phylogeny.*

Ernst Haeckel contended that the embryological stages of all animals were a recapitulation of their evolutionary history. *Ontogeny* is the embryonic development of organisms, while *phylogeny* is the evolutionary history of a species. Haeckel also believed that at one time there were animals on earth that resembled all the developmental stages of animal embryos. In other words, he incorrectly believed there were prehistoric mature animals that resembled the embryos of modern animals. Haeckel hypothesized that the mechanism for inheritance of traits was present in the nuclei of cells, even though he was unaware of chromosomes, genes, and DNA. There is some relationship to the similarities of animal embryos in that the cell nuclei of all animals contain a very large proportion of the same DNA. Mammals, including humans, share approximately 95 to 97 percent of the same DNA, and chimpanzees and humans share over 98 percent of the same DNA. Haeckel's recapitulation theory is no longer accepted by most biologists, although it is partially supported by Stephen Jay Gould. *See also* Gould.

Haeckel's Theory of Evolution: *All animals are derived from inanimate matter.* Haeckel developed a hierarchy evolution design based on the premise that all animal matter was derived from inanimate matter and then progressed upward to humans. He came to this conclusion by the scientific concept of symmetry and proposed his inorganic origin of animal life by relating the symmetry of the simplest animals to the symmetry of inorganic crystals. Although this theory as related to evolution is not now accepted, there is some acceptance of the idea that at some point in history, chemical elements combined to become self-replicating, prebiotic molecules. These reproducing molecules are assumed to have developed into primitive DNA and cells. This is one theory for the origin of life.

Haeckel also contended all higher multicellular animals (Chordata) with three-layered cells (ectoderm, mesoderm, and endoderm) were derived from animals, such as jellyfish and sponges and other marine invertebrates (Gastrea), that are composed of only two layers of cells (ectoderm and endoderm). He was the first to distinguish protozoa (single-celled animals) from metazoa (multicellular animals).

Haeckel's Concept of Social Darwinism: *Human culture and society conform to the laws of evolution.*

Ernst Haeckel, an ardent supporter of Darwin's concept of organic evolution, predated current social Darwinists, such as Edward O. Wilson, by several decades. Haeckel's concept of social Darwinism asserts that human society is as much a product of nature as is human anatomy and physiology, and thus is no different from other organisms. In other words, natural selection applies as much to human behavior, cultures, societies, and religions as it does to the organism.

This concept is not fully accepted by all biologists, partly because it is seen as antireligious and places humans in the same category as other animals and partly because it is used as a rationale for **eugenics**. *See also* Aristotle; E. O. Wilson.

Haeckel's Theory of Ecology: *There is a direct relationship between and among plants and animals and their biological environment.*

In 1868 Ernst Haeckel coined the word *ecology* to describe the study of plants and animals and their relationship to each other and their environment: the soil, water, atmosphere, light, heat, oxygen, carbon dioxide, and all the other elements in their biological surroundings. The term *ecology*, from the Greek word oikos, meaning "household," was not a recognized science until the early part of the twentieth century. Many people confuse environmentalism (an ideological political movement) with ecology (a science). *See also* Darwin; Gould; Wallace; E. O. Wilson.

Hahn's Theories of Nuclear Transmutations: Chemistry: *Otto Hahn* (1879–1968), Germany. Otto Hahn received the 1944 Nobel Prize for chemistry.

Hahn's and Meitner's Nuclear Isomerism: *Nuclei of different radioactive elements exhibit identical properties.*

Otto Hahn discovered a radioactive form of the element thorium (Th), and Lise Meitner discovered protactinium (Pa). The concept of radioactive isomerism, which occurs when some radioactive elements with the same atomic number (protons) and neutrons in their nuclei nonetheless possess different properties and behave in different ways, was the result of their collaborations. In other words, an isomer of an element has exactly the same atomic number and mass but not the same characteristics. This discovery for radioactive elements led to their more important discovery.

Hahn's, Meitner's, and Strassmann's Theory of Nuclear Fission: *Nuclear transmutation occurs in heavy as well as lighter elements.*

The transmutation of the nucleus of one element into the nucleus of another element—one that has a nucleus with a different atomic number and mass—was first discovered by Ernest Rutherford when he observed oxygen being transformed into nitrogen. Marie and Pierre Curie discovered similar reactions with uranium and radium. Hahn and his collaborators theorized there are two important aspects to nuclear transmutation. First, these types of nuclear reactions always emit light beta particles, high-energy electrons, and/or heavy alpha particles (helium nuclei). Second, the transmutation of one element's nuclei to form a different nuclei occurs only between elements that are more than two places apart on the **Periodic Table of the Chemical Elements**. Hahn noticed these rules did not apply to heavy elements when he used heavy atomic nuclei (neutrons and alpha particles) to bombard uranium, which turned into lead. He also bombarded the heavy element thorium with neutrons. He then theorized there must be an intermediate emission of particles and lighter elements involved in stages between the uranium and lead. In essence, Hahn split the uranium atom by bombarding it with neutrons, thus opening the door to research in developing

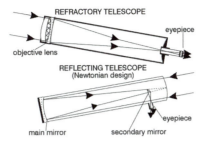

Figure H1: Diagrams of the refractory and reflecting telescopes. The size of the refractory telescopes is limited by the distortions in large glass lenses. Reflecting telescopes collect more light from objects by using a parabolic mirror, which can be made larger than a glass lens.

a fissionable chain reaction and later the atomic bomb. *See also* Curie; Frisch; Meitner; Rutherford.

Hale's Solar Theories: Astronomy: *George Ellery Hale* (1868–1938), United States.

Hale's Sunspot Theory: *The spectra of sunspots can be used to identify the elements in the outer reaches of the sun.*

George Hale spent most of his life seeking funds to build larger and larger telescopes. His efforts resulted in the 40-inch telescope at the Yerkes Observatory in Wisconsin; the 60-inch reflector telescope and the 100-inch Hooker telescope, both installed at the Mount Wilson Observatory in California; and the 200-inch Pyrex mirror telescope installed at the Palomar Observatory in Pasadena, California. (See Figure H1.) Hale also advanced the use of spectrometers for astronomical viewing. One model, the spectroheliograph, which he invented, enabled him to photograph and make measurements of sunspots. He viewed and photographed small bands of wavelengths of the sun and identified the double spectral lines emitted by the element calcium, which led to the study of the chemical makeup of the outer layers of the sun. Hale used his progressively more powerful telescopes to prove that the Martian "canali" (canals) reported by Schiaparelli are nonexistent.

Hale's Hypothesis for Extraterrestrial Magnetic Fields: *Sunspots exhibit strong electromagnetic fields.*

About ten years after using his spectroheliograph to describe the chemical elements in sunspots, George Hale identified the Zeeman effect, which occurs when the magnetic field splits the spectral lines of light from the sun, enabling the viewer to detect specific elements in the sun. This was the first time that a magnetic field was detected as originating from another object beyond earth. *See also* Lorentz; Schiaparelli; Zeeman.

Halley's Theories for Comets and Stars: Astronomy: *Edmond Halley* (1656–1742), England.

Halley's Theory for the Return Visits of Comets: *Gravity that affects the planets also affects the paths of comets.*

Edmond Halley accepted Newton's laws of motion and gravity as well as providing funds for the publication of Newton's *Principia*. Halley theorized that comets must react similarly to planets since comets are also affected by the gravitational attraction of the sun. He collected data for several past comet sightings and noted there appeared to be repetition in the number of years for the sightings. He then calculated that the comets of the years 1456, 1531, 1607, and 1682 were actually the same comet that traveled in a large, elongated ellipse around the sun. Using this seventy-six-year cycle, he predicted the same comet would reappear in the year 1758. Halley died before he could see his comet return, but it was named after him and ever since has been known as Halley's comet. The same comet was first viewed and recorded in the year 240 B.C. The most recent visit was in 1986, and the next is expected in 2061 or 2062.

Halley's Theory of Stellar Motion: *Stars are not fixed, and their movements can be detected over long periods of time.*

Ancient stargazers, as well the Greek astronomers, recorded the positions of the brightest stars. Edmond Halley, many years later, also cataloged the positions of three of the brightest; Sirius, Procyon, and Arcturus. He noted their positions had changed, and thus stars must not be in fixed positions. Halley felt the great astronomer Tycho Brahe had made the best measurements of these same stars' positions, but Brahe was unaware of their slow apparent movement due to the tremendous distances of the stars from earth. Comparing his own records with Brahe's data, Halley established the fact that the real motion of the stars was detectable only after long periods of viewing. Later astronomers used the concept of parallax, where the position of a star is measured at six-month periods as earth revolves around the sun, but these measurements made over a short period of time did not detect star motion. Thus, it was determined the movement of stars was random. It was not until the early 1900s that two different streams of stars moving in opposite directions were observed by Cornelius Kapteyn (1851–1922). This proved the movement of stars was not random, but rather proceeded with some order, as was discovered in galaxies that continue to move away from us.

Edmond Halley, the first to study the stars in the Southern Hemisphere, also identified two "cloudy" areas, one named the Magellanic Cloud, and the other a star group referred to as the Southern Cross. He developed meteorological maps and found the subtropical trade winds blow southeasterly toward the equator in the Southern Hemisphere and northeasterly in the Northern Hemisphere. (This was understood by early sailors, including Columbus, who made use of the trade winds as he crossed the Atlantic Ocean in both directions.) Halley correctly related this phenomenon to a major circulation in the atmosphere due to the difference in the heat produced by the sun in the areas above and below the equator.

Hall's Effect of Electrical Flow: Physics: *Edwin Herbert Hall* (1855–1938), United States.

An electric force is generated when a magnetic field is perpendicular to the direction of an electric current. This force is perpendicular to both the magnetic field and the current.

A more technical explanation is that a transverse electric field will develop in a conductor that is carrying a current while it is in a magnetic field. The Hall effect occurs when a magnetic field is at a right angle to the direction of the current flow. This both modulates and multiplies voltages and can act as a magnetic switch for semiconductors, which is important for electronic applications. These devices, also referred to as *Hall generators*, are used in brushless induction motors, tachometers, compasses, thickness gauges, and magnetic switches that do not require contact points. Devices employing the Hall effect can detect minute hairline cracks in the wings and bodies of airplanes, as well as in very sensitive instruments that can measure weak magnetic fields. *See also* Ampère; Faraday.

Hamilton's Mathematical Theories: Mathematics: *Sir William Rowan Hamilton* (1805–1865), Ireland.

Hamiltonian Functions: *A special function of the coordinates and the moments of a system is equal to the rate of change of the coordinate with time.*

Sir William Hamilton worked out the mathematics expressing the sum of kinetic and potential energies of any dynamic (changing) system. It is related to the Hamiltonian principle, which states that the time integral of the kinetic energy minus the potential energy of a system is always at a minimum for any process. The equation involves both the motion and time for the system. The Hamiltonian function was an important step in developing the theory of quantum mechanics.

Hamilton's Quaternions: *By ignoring the products of multiplication, a system of internal algebra for "hypercomplex" numbers can exist for "quaternions" (Latin for "four").*

In developing his theory, Sir William Hamilton applied algebraic functions to non-Euclidean geometry involving n-dimensional (multiple dimensions) analytic geometry, which involves more than the three dimensions $[x, y, z, \ldots, n]$ of space. The quaternion is a hypercomplex number in algebra. It has always been accepted that $A \times B = B \times A$, but for quaternions, the commutative law of multiplication does not hold. Therefore, $A \times B$ does not necessarily equal $B \times A$. The quaternion theory is important in the field of abstract algebra, but it has little use in physics, where vectors are more applicable. However, it did lead to the development of mathematical matrices, such as matrix mechanics as applied to Werner Karl Heisenberg's quantum theory. Hamilton made other contributions to optics, mathematics, and abstract algebra. *See also* Heisenberg; Lagrange.

Harkins' Nuclear Theories: Chemistry: *William Draper Harkins* (1873–1951), United States.

Harkins' Rule for Isotopes: *The isotopes of elements that have odd-mass numbers are less abundant than are the isotopes of elements with even-mass numbers. Both are represented by whole number ratios for their atomic numbers and weights.*

William Harkins proposed the whole number rule about the same time as Francis William Aston. The rule determined that the mass and the occurrence of isotopes of stable elements are related. When isotopes of stable elements are considered, the numbers of neutrons in the nucleus are related as whole numbers—never fractions of neutrons. Harkins predicted the existence of both the neutron and the element deuterium (heavy hydrogen whose nucleus contains a neutron as well as a proton. See Figure O1 under Oliphant). Using a mass spectrometer, which he invented, Aston determined this whole number concept was not always correct. Sometimes the mass of isotopes was slightly more or less than a whole number. Aston called this the *packing fraction*, which occurs when the hydrogen nuclei join to form helium nuclei (fusion) and a small amount of energy is produced. Therefore, Harkins' whole number rule was modified to conform to new experimental data.

Harkins' Theory for the Hydrogen-Helium-Energy Reaction: $4H_1 \rightarrow He_4 + energy$.

The concept of radioactivity produced by the fission of nuclei of radioactive elements, resulting in the production of enormous amounts of energy, was demonstrated by Marie Curie. It was William Harkins who applied his knowledge of isotopes and neutrons to address Aston's "packing fraction" principle to determine how hydrogen could be converted into helium with a tiny fraction of matter converted to energy. This confirmed the basic laws for the conservation of matter and energy. It was later determined that such a conversion reaction maintains the energy output of the sun and other stars. It was not until several decades later that the concept was applied beyond the laboratory, resulting in nuclear energy. *See also* Aston; Curie; Fermi; Rutherford; Urey.

Harvey's Theory for the Circulation of the Blood: Biology: *William Harvey* (1578–1657), England.

The beating heart propels blood through the arteries, where it is then circulated back to the heart through the veins.

From the time of Galen, medical practitioners taught that blood was produced continually by the liver. Then in the seventeenth century, William Harvey concluded that for this to be true, a tremendous amount of new blood needed to be produced continuously by the liver in order to maintain a constant flow. Thus, he theorized the same blood must circulate through the blood vessels. Harvey dissected cold-blooded animals to observe the pumping action of their hearts. Since circulation is slower in frogs and snakes, he traced the blood passing from the right to the left side of the heart through the lungs, not through "pores" in the septum (tissue that divides the two sides of the heart), as was taught and believed from the days of Galen. Harvey observed that as the heart beats, **systolic** contractions occurred, swelling the arteries with blood as it was pushed throughout the arterial system of the body. At the **diastolic** phase, the heart was again filled with blood that was returned by the veins. Harvey declared the heart to be a pump. This established his theory that the heart caused the blood to circulate throughout the body via the arteries and then back to the heart through the veins. Nonetheless, no reasonable explanation existed of how the blood was transferred from the ends of the arteries to the veins to complete the circuit back to the heart. Harvey correctly theorized the presence of very small blood vessels, too small to be seen, that provided the passage of blood from the ends of arteries to veins. This discovery had to wait until the invention of the microscope and Marcello Malpighi's (1628–1694) observance of the tiny web of blood vessels in the thin skin of a bat's wing, which he named *capillaries* (Latin for "hairlike"). *See also* Galen.

Hawking's Theories of the Cosmos: Astronomy: *Stephen William Hawking* (1942–), England.

Hawking's Theory for the Nature and Origin of the Universe: *Applying the submicroscopic theory of quantum mechanics to the macroscopic universe produces a wave function similar to a wave function for elementary particles, thus shaping the nature of the universe.*

Stephen Hawking theorized that according to the quantum wave function theory, as related to Einstein's theory of relativity, there is no past or future because the time function exhibits characteristics of spacial dimensions (relativity). The universe can, by its own accord, be both zero in size or infinite in size. It can have no beginning or no ending. This led Hawking to a new version of "open inflation" that spontaneously created a "bubble" out of nothingness. In other words, the big bang, which is generally accepted as how our universe began, had to have something with which to start. Hawking's use of quantum mechanics provides this answer by showing how the wave function starts as a *singularity* (a region of space-time where one or more components of curved spaces become infinite.)

These ideas preceded the three different geometries or shapes for the universe. The *first* concept maintains the universe is flat space as related to the rules for Euclidean geometry, in which parallel lines never meet or cross, a triangle always has exactly 180 degrees as the sum of its angles, and the length of a circle is its circumference ($2\pi r$). Although the space in the universe generally is considered flat, space-time is not. For most of our science and measurements on Earth, three-dimensional space involves L for length, L^2 for area (length $\times$ height), and L^3 for volume (length $\times$ height $\times$ width, or depth). Astrophysicists add time *(T)* as the fourth dimension to the three dimensions of space in order to describe matter in relativistic terms. In other words, the structure of matter changes in space over long periods of time–thus, the term *space-time*. Contrary to older theories, the universe is not static, but rather it is a dynamic, changing physical entity. This also means that over great distances and long periods of time, the extended universe will not remain flat. At its outer reaches, the total curvature of space-time is determined by the density of matter as well as the time involved for the continued expansion of the universe. An analogy: When standing on the surface of the earth and looking toward the horizon, for about a mile the surface seems flat, just as the space near us that we can see seems flat. But earth from a spacecraft looks curved–just as do distant sectors of the universe.

The *second* theory for the geometry of space is that space is spherical similar to a globe. This represents a closed universe, which is similar to Einstein's original static universe theory. Curved surfaces are positive in the sense that space curves back on itself similar to the surface of earth. Since a closed universe is finite, parallel lines would always converge (e.g., line of longitude for earth); the sum of the angles for a triangle laid out on the surface of a sphere is always greater than 180 degrees. (The sum of the angles for a triangle can total up to 540 degrees when projected on the surface of a sphere.) And the circumference of a circle inscribed on the globe's surface is always less than $2\pi r$.

The *third* geometry is a hyperbolic "open" universe whose negative surface would look something like a horse saddle. In an open universe, parallel lines would diverge and never meet, the sum of the angles for a triangle would be less than 180 degrees, and the circumference of a circle inscribed on its surface would be greater than $2\pi r$. Before astronomy became a science, the closed geometry theory was usually accepted, but this implied a finite, static, unchanging universe, which is not consistent with the observable evidence. According to the latest research, the flat universe theory is the most unlikely geometry for the actual universe.

Hawking's Black Hole Theory: *Mini black holes, also known as primordial black holes, were formed soon after the beginning of the big bang.*

Stephen Hawking was not the first to advance the black hole theory. John Archibald Wheeler named this phenomenon, which describes the results of what occurs when a very massive star has "burned up" its nuclear fuel. The dying star cools down to almost absolute zero and shrinks below the critical size, meaning its radiation can no longer overcome its internal gravity, resulting in a "singularity" as a bottomless hole represented as a point in space-time. This concept was first known as the *collapsed star phenomenon*. No nearby light or matter can escape the strong gravitational attraction from this bottomless hole— thus, the name *black hole*. Einstein's theory of general relativity asserts that gravity is related to the curvature of space-time, and massive objects, such as giant dead stars, distort space-time. This distortion causes an event horizon as the boundary for a black hole. Once an object moves close to this ever-expanding boundary, extreme gravity pulls it into the hole for which there is no escape, not even for light. Most black holes seem to be located near the center of galaxies. The more massive a black hole is, the larger it is. It can capture and compress many massive stars and may weigh as much as 10^{26} kilograms (10 followed by 36 zeros). On the other hand, there may be smaller black holes. (It might be remembered that this concept is not referring to a hole as in an empty hole in the ground; rather, it is an area of compressed mass where light cannot travel fast enough to escape the overwhelming gravity created by one or more collapsed stars.)

Hawking expanded the concept of black holes by considering thermodynamics and quantum gravity. Ordinary space is not really empty. It is filled with pairs of particles. Some are positively charged particles that annihilate partners that have negative charges. (In space there are pairs of particles and **antiparticles**.) Hawking theorized that for black holes, the partner particles become separated, with the negative particle falling into black holes just as all other matter does and the positive antiparticle escaping and giving off energy that can become a real particle or even approach infinity. These particles and antiparticles can exist because of *vacuum fluctuations*, which are fields similar to light, magnetic, and gravity fields. They are also referred to as *virtual particles*. Hawking also theorized that soon after the big bang, many mini black holes existed. They were no larger than the nuclei of hydrogen atoms (protons) and yet weighed billions of tons and produced tremendous energy. Hawking claims that black holes are not 100 percent black because quantum mechanics caused them to produce particles and forms of radiation at a regular rate. John Wheeler, a theoretical physicist, stated that what falls into a black hole, including ordinary matter such as buildings, cars, or people, is independent of the radiation and thus will come out the same as it went in. Hawking stated that the emission of particles and radiation from a black hole will reduce its mass to the point where it will have zero mass. At this point, the buildings, cars, and people, including information, who fell into the hole cannot come out because there is no longer adequate mass for this to happen.

Hawking's Theories for Inflation and Singularity: *The original expansion of the big bang was at first rapid, slowing down as the universe aged. The densities of black holes are singularities.*

Before Einstein's theory of relativity, the general concept was that the universe stayed the same; it was static. In 1929 Edwin P. Hubble published his measurements that indicated galaxies were receding from earth and from each other at an ever-increasing rate. Later, Roger Penrose, an astronomer, theorized that a collapsing star is a singularity, as are all black holes. *Singularity* means a single point where the density of matter is infinite, just as the curvature of space-time is infinite. It is a region in space where one or more parts of the space-time curvature becomes infinite. It is also known as a single point of a specific function. The concept of singularity is consistent with and conforms to physical laws. On the other hand, space-time, being infinite with no boundaries, is relative; thus there is no beginning or ending to space-time.

Hawking combined quantum theory with gravitation as expressed by Einstein to propose his concept of *quantum gravity*. He then reversed Penrose's singularity theory to explain how the big bang came to be. In other words, there was a singularity—a point in nonlinear time when a single massive point started the expansion of the universe. This expanding universe is explained by the inflation theory, which states that this three-dimensional "spreading" was extremely rapid, causing great heat and rapidly expanding gases and energy that overcame gravity at this point. There are several concepts that explain what occurred after this singularity event created the nascent universe. One states that the early rapid inflation of gases slowed and cooled down, and in time these gases (mostly hydrogen) formed helium and the galaxies, stars, planets, and other objects in space. The left-over radiation and matter, galaxies, and stars are still overcoming gravity and thus expanding but possibly slowing down. The quantum gravity theory states that many billions of years from now, there will be a period of deflation caused by gravity's again taking over. This scenario is called the big crunch, which will end in a new singularity. It somewhat represents the geometry for a closed universe, where gravity finally wins and a collapse occurs. Another theory states that inflation will continue at an ever-increasing rate of expansion, resulting in a lower density of matter in the universe, or new matter will be created continually to maintain the current density of the universe. This scenario represents the geometry for the open universe. The geometry for a flat universe is somewhere between these two extremes, but it best fits the *standard inflationary theory*, which suggests there is much more matter in the universe than we can see or even realize. This so-called dark matter may not be adequate to justify a flat universe theory, but research and speculation continue to find other types of matter or energy in the universe. During the first decade of the twenty-first century, several high-resolution instruments will be placed in orbit to make measurements that will address the many questions and theories related to the origin, nature, and fate of the cosmos. *See also* Einstein; Feynman; Hubble; Penrose; Wheeler.

Haworth's Formula: Chemistry: *Sir Walter Norman Haworth* (1883–1950), England. Sir Norman Haworth shared the 1937 Nobel Prize for chemistry with Paul Karrer.

Carbon and oxygen atoms are linked in specific ratios to form carbohydrate rings of sugar molecules.

Organic chemistry had succeeded in synthesizing a number of different forms of sugar molecules, mostly linear open-chain structures. Sir Norman Haworth was the first to demonstrate his *prospective formula*, which was his hypothesis leading to closed chains of sugar molecules. He proposed two types of closed rings for molecular sugars. One was the sugar molecule that formed a ring composed of five carbon atoms and just one oxygen atom, while the other consisted of furanose, which is the molecular structure for a sugar ring of only four carbon atoms and one oxygen atom. By demonstrating that a chain of carbon-oxygen atoms for several types of sugars could be synthesized into rings,

Haworth synthesized a variety of polysaccharides (a sugar composed of many monosaccharides) that are produced today.

In addition, Sir Norman Haworth and Tadeus Reichstein independently succeeded in analyzing and synthesizing a variety of vitamins. Their first success identified the substance found in orange juice, which was first called "hexuranic acid." Later, it was identified as vitamin C and named "citric acid" by Haworth. Their discovery enabled the production of synthetic vitamins with the same chemical makeup as natural vitamins. Their work led to a decrease in both malnutrition and diseases caused by vitamin deficiencies.

Heisenberg's Uncertainty Principle and Theory of Nucleons: Physics: *Werner Karl Heisenberg* (1901–1976), Germany.

Heisenberg's Uncertainty Principle: *The simultaneous measurement of the position of an electron affects the measurement of its momentum, leading to the uncertainty of either its observed position or movement.*

Werner Heisenberg used matrix mathematics to develop a formulation of quantum mechanics that was equivalent to the Schrödinger formulation but differed in form. (This is analogous to saying the same thing in English and in French.) The Heisenberg formulation perhaps emphasizes the uncertainty principle more strongly than does the Schrödinger formulation. Nature at the very small scale has a wavy nature that can be described only by quantum mechanics. Natural physical laws hold true for everyday objects in the universe, including us. However, in the submicroscopic universe, these same natural laws do not always apply. It is possible for observers to measure accurately the *position* of a minute, submicroscopic, subatomic particle independently, and it is possible to measure a particle's *momentum* (a vector quality for a particle's velocity $\times$ mass) accurately. But both cannot be accurately measured at the same time, thus creating uncertainty. Rather, the small size of their joint uncertainties must be larger than the number referred to as Planck's constant ($\hbar$). Heisenberg insisted the only way to interpret the atom and the subatomic particles that compose them was to observe the radiation (light) they emit. Both Heisenberg and other physicists developed systems and equations to accomplish this feat. Heisenberg's equation for the uncertainty principle is: $\Delta X \cdot \Delta p \geqslant \hbar/4\pi$, where Δx is the uncertainty of the particle's position in any direction, Δp is the uncertainty in determining the momentum related to that position, and $\hbar$ is Planck's constant.

Heisenberg's Nucleon Theory: *Nucleons are isomers of protons and neutrons.*

Werner Heisenberg also developed the theory that nucleons were composed of both neutrons and protons, which make up the nuclei of atoms. He proposed that both the proton and neutron had approximately the same mass but different spins and thus were forms of **isomers**. Based on his theories of matter, Heisenberg attempted to devise a unified field theory for all elementary particles that related their characteristics and energies. Neither he, Albert Einstein, nor several other scientists were able to finalize such a grand unification theory (GUT) or a theory of everything (TOE) that would include all the submicroscopic particles from the quark up to large molecules, ordinary matter, all forms of energy, gravity, and the endless universe. [But when such a theory is produced, it will be expressed as a simple equation.] *See also* Einstein.

Helmholtz's Theories and Concepts: Physics: *Hermann Ludwig Ferdinand von Helmholtz* (1821–1894), Germany.

Helmholtz's Theory for Heat and Work: *Heat and work are equivalent, and one can be converted to the other with no loss of energy.*

Hermann Helmholtz was familiar with James Joule's concept of work and heat, and Lord Kelvin's work with thermodynamics. Lord Kelvin later proposed a concept for the conservation of energy. However, Helmholtz was not the first to relate mechanical work to heat. That honor goes to Julius Mayer (1814–1878). It might be helpful to define some terms as used by scientists. *Heat* is the result of molecules bumping into each other in random motion. The greater their total motion, the greater the amount of heat. The mechanical equivalent (molecular motion) of heat is measured in calories: 1 calorie = 4.185 joules, while 1 joule = 10,000,000 ergs (the joule is a unit of energy or work). Heat results from mechanical *work*. Try rubbing the palms of your hands together very fast. Note the heat generated from the mechanical motion (work) of your hands. Neither the heat produced by this rubbing friction nor the heat generated by muscle energy you used to produce this work is lost. Rather, the heat soon disperses into the air around your hands, slightly increasing the motion of air molecules (warming up the air), and is thus conserved. Heat is a form of energy transferred from a body with a higher temperature to a body with a lower temperature— never in the opposite direction (*entropy*). Work is accomplished when a force on a body transfers energy to that body. Work (W) = applying a force (F) over a distance (d) ($W = F \times d$). Work, in the scientific sense, is not exactly the same as in the common sense, which usually refers to some physical or mental effort—although both involve energy and heat. *Energy* is the capacity to do work. Both heat and work are forms of energy and are interchangeable. There are many forms of energy, heat, light, sound, radioactive, mechanical, work, and so forth, and they are all conserved, but not in the sense that environmentalists use the word *conservation*, usually meaning "saved" and not to be used. The scientific meaning for *conservation of energy* or *mass* means that it can be changed from one form to another without any total loss of the original energy or mass.

As a physician, Helmholtz believed animal heat generated by muscle contractions was related to his theory that heat and work are convertible from one to the other in a *quantitative relationship*, with no loss of total energy. This led to the law of conservation of energy, one of the fundamental laws of physics. Imagine yourself raising a baseball from the top of a table to the height of your head. In doing so, you have converted mechanical work to potential energy in a gravitational field. However, you have done the mechanical work at the expense of the energy stored in your body. One might say you did 1 joule of *work*, which is equal to about 0.25 calorie of *heat*, since work and heat are both forms of energy and interchangeable. You also increased the potential energy of the baseball, which would be released as kinetic energy if dropped, resulting in a conservation of energy (none was lost or gained). Helmholtz also explained his theory on a physical basis by speculating that the total energy of a large group of interacting particles is constant, basing his ideas on the perceptions of Robert Hooke. Using a microscope, Hooke observed the action and interaction of minute particles in a fluid. He related the motion of the particles to heat resulting from what we now know as the molecular action within a body, even before molecular motion was known. Nonetheless, it provided Helmholtz with the idea that work and heat are related. As the law of conservation of energy was refined, it stated that the total amount of energy in the universe is constant. Energy can neither be created nor destroyed, but it can be changed from one form to another. For example, mechanical, magnetic, electrical, chemical, light, sound, kinetic energy, and others are all interconvertible. Also known as the first law of thermodynamics, it is one of the most basic universal physical laws of nature. As did many other scientists, Helmholtz tried unsuccessfully to develop a unified field theory based on his concepts of work/heat/energy, electrodynamics, and thermodynamics. *See also* Carnot; Clausius; Hooke; Joule; Kelvin.

Helmont's Theory of Matter and Growth: Chemistry: *Jan Baptista van Helmont* (1579–1644), France.

Plants grow, increasing their weight, but without reducing the weight of the soil in which they grow.

Helmont was one of the early scientists who carefully observed and measured well-planned experiments, the most famous of which was the observation of the growth of a specific plant over a five-year period. He noted the specific weight of soil in which it was planted, the plant's growth weight, and the weight of the amount of water it used. At the end of five years, Helmont weighed the plant as well as the soil in which it grew and noted the plant weighed about 165 pounds, but the soil lost less than $1/4$ pound in weight. However, he incorrectly attributed all the growth weight to water, which led to a mistaken concept of matter. In another experiment, he burned about 60 pounds of wood, weighing the remaining ashes, which equaled less than 1 pound. He concluded that the lost weight was composed of water and four "gases," which he named *pingue* (methane), *carbonum* (carbon dioxide), and two types of *sylvester* gas (nitrous oxide and carbon monoxide). He was correct that carbon dioxide resulted from combustion as well as fermentation, and he correctly identified the other gases. Helmont also believed there were several different types of air, just as there were different forms of solids and liquids. He related these "airs" to chaos since they had no specific volume as did liquids and solids, and he coined the term *gas*, which soon was accepted along with the common terms *solid* and *liquid*, providing a general concept for the three forms of matter.

Henry's Principles of Electromagnetism: Physics: *Joseph Henry* (1797–1878), United States.

Electromagnetic Induction: *By moving a conductor through a magnetic field, an electric field is induced in the conductor.* (See Figure F1 under Faraday.)

Solenoid-type magnets were developed and improved in the early part of the nineteenth century. These devices consisted of coils of insulated wire, which produced a magnetic field within the coil when electricity passed through the wire. Further improvements were made by William Sturgeon (1783–1850), who placed an iron rod inside a coil and observed that while electricity flowed through the loops of wire forming the coil, the rod became magnetized. However, when the current was turned off, the rod no longer retained its magnetism. Thus, by turning the current on and off, the rod became an on-and-off magnet with the ability to control various devices. The source of electricity at this time was generated by simple voltaic wet cells, which produce direct current electricity of low voltage (about $1^1/_2$ volts). However, this amount of electricity was adequate to produce electromagnetism.

The number of turns of wire loops used to form a coil was limited because if any of the bare wires came in contact with each other, they would short out, breaking the electrical circuit. Henry solved this problem by using insulated wires, which could be wrapped close together and overlapped without creating a short circuit, thus producing much stronger electromagnets. He lifted as much as a ton of iron by using a small battery of voltaic cells to supply the electric current. The idea that magnetic properties from this coil could be induced to an iron bar was independently proposed by both Michael Faraday and Joseph Henry. Joseph Henry, who is said to have developed the theory first, did not receive credit for the principle of magnetic induction since Michael Faraday's findings were published a few months before Henry's. Joseph Henry is immortalized by having the universal SI constant for inductance named after him. The Henry SI unit for induction occurs when an electromotive force of 1 volt is produced when the electric current in the circuit varies uniformly at a rate of 1 ampere per second.

Henry's Concept of an Electric Motor: *A moving wire that cuts across a magnetic field induces an electric current. Therefore, the process should be reversible.*

Faraday's device used a copper wheel to cut through a magnetic field and thus induced a current to flow through the wire. This later became the concept for the dynamo or electric generator. Joseph Henry questioned what would happen if this process was reversed. He devised a machine with a wheel that would turn when an electric current was sent through the electromagnet. The result was the first electric motor, which over many years of development has become one of the major technological developments of all times. It might be said that Faraday, Henry, and other scientists who worked with electricity and electromagnetism created the labor-saving conveniences of the modern world.

Henry's Concept of "Boosting" Electricity over Long Wires: *If a method for maintaining the strength of an electric current over long wires can be found, an electromagnetic communication device could be developed.*

Several people, in addition to Joseph Henry, imagined an electromagnet solenoid that could open and close a "clicker" circuit using a "key" to send signals through wires. The problem was the natural resistance to the flow of electricity in copper wires. Henry solved the problem by developing the electric relay, which used a small coil as an electromagnetic switch. This turns large amounts of electric current on and off without the circuit's actually being continually connected. This relay enabled electricity to be sent over long stretches of wires at intervals, and thus the success of the national system of telegraph communication. The problems related to the early electromagnetic telegraph were money and support for implementation, not the scientific problems. Joseph Henry was so admired in the United States that in 1846 he was named the first secretary of the new Smithsonian Institution. *See also* Ampère; Faraday.

Hertzsprung Theory of Star Luminosity: Astronomy: *Ejnar Hertzsprung* (1873–1967), Denmark.

The luminosity of stars decreases as their color changes from white to yellow to red due to a decrease in the star's temperature.

Previous to Ejnar Hertzsprung's theory, a star's distance from earth was determined by the Doppler effect, the result of sound or light waves lengthening (lower frequencies) as the object producing the sound or light waves moves away from the observer. Vice versa, the waves shorten and become higher in frequency as the object producing the waves approaches an observer. Hertzsprung improved this technique for determining the distance of stars by using photographic spectroscopy to measure the inherent brightness or luminosity of stars. He determined there were two main classes of stars: the supergiant stars that were very bright and the much more common and fainter stars referred to as the main *sequence type stars*. Hertzsprung's theory was confirmed by Henry Russell, who charted the stars according to Hertzsprung's classification, and both developed the H-R diagram. Their diagram arranged stars on the x-axis according to a classification using both temperature and color, and also on the y-axis according to their magnitudes and luminosities. (See Figure H2.) It revealed several clusters or groupings of stars: (1) the most massive supergiants—the largest and brightest; (2) the giants, not quite as large and bright as the supergiants; (3) the main sequence of stars, ranging from very large dim stars to small, bright stars; and (4) the comparatively small white dwarfs. The large bright and cooler stars are found in the upper right of the diagram, while the smaller white dwarfs, which are somewhat dimmer and hotter, are located in the lower left of the x-y coordinates of the diagram. The sun is found among a group of small stars known as red dwarfs, which are located near the middle of the main sequence. It is somewhat more brilliant than are its neighbors. Hertz-

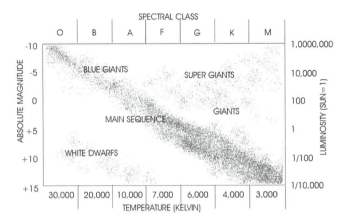

Figure H2: The Hertzsprung-Russell diagram depicting the color and magnitude of stars. It indicates how stars vary according to size, temperature, and brightness. Our sun is an average size bright star located about the middle of the main sequence.

sprung's work and the H-R diagram led to a better method for determining the distance of both stars and galaxies from earth, as well as aiding in the study of the evolution of stars.

Hertz's Theory for Electromagnetic Waves: Physics: *Heinrich Rudolf Hertz* (1857–1894), Germany.

Electromagnetic waves, produced by electric sparks, behave the same as light waves.

Heinrich Hertz expanded the electromagnetic theories of James Clerk Maxwell and Hermann Helmholtz by erecting an apparatus consisting of a metal rod with a small gap cut in the center to detect different wavelengths of the electromagnetic spectrum. When electricity was sent through the rod, a rapidly vibrating spark was produced at the gap, generating high frequencies in the rod. Not only was Hertz able to detect these high frequencies (electromagnetic waves), but he determined these electromagnetic waves would reflect and refract off surfaces just the same as light waves. More important, they traveled at the same speed as light. He also showed their wavelengths were quite a bit longer than gamma radiation or even light rays. They were named *Hertzian waves*. Today we know them as radio waves with much longer wavelengths than light waves. Hertz's work not only confirmed Maxwell's electromagnetic spectrum theory, but also paved the way for radio communications. (See Figure M2 under Maxwell.) Hertz also detected the photoelectric effect where ultraviolet light (wavelengths shorter than light) "knocks" out electrons (photons) from the surface of particular types of metal plates. *See also* Helmholtz, Maxwell.

Hess' Sea-Floor Spreading Hypothesis: Geology: *Harry Hammond Hess* (1906–1969), United States.

The sea floor "split" near the middle of the Atlantic Ocean provides an opening for deeper magma to protrude, thus renewing and spreading the ocean floor under land masses, which eventually separated the continents.

At one time earth's entire land mass was connected. Over millions of years, this mass separated to become distinct continents. How this occurred has perplexed geologists, because it did not seem possible that land masses could pass over layers of solid rock. Hess' hypothesis provides a possible answer. Hess based his hypothesis on evidence that fossils found in the "newer" ocean beds are much younger than those found on continental land masses. As the ocean floor spread out, it approached the land masses and dipped under the continents, while at the same time forming a ridge at the origin of this expansion. This movement of ocean floors resulted in the development of rifts or breaks near the center of both the Atlantic and Pacific ocean beds. This produced weak spots where the magma (molten rock) protrudes through what is known as the Great Global Rift, a deep canyon running along the center of the midocean ridge. This spreading of the sea-floor eventually was forced underneath land masses and is responsible for the North and South American continents moving farther apart and westward, while moving Europe and Asia eastward. The continents did not actually drift or float but were rather fixed to plates that were forced apart and, in some places, forced together. This led to the science of plate tectonics, which means "carpenter" in Greek. Plate tectonics, an important geological theory, is as important to geologists as evolution is to biologists. In essence, it states that, worldwide, there are six large plates and some smaller ones that were moved by the spreading of the ocean floor. The ocean floors and plates move apart about 2 inches each year as new magma rises into the midocean rifts. Earthquakes and volcanoes are in evidence where some of the major plates clash, such as on the west coast of the United States and South America and in the plates bordering Japan and much of the Pacific Ocean. This movement produced, and still produces, mountains and island chains. Hess' hypothesis was later confirmed by studies of the magnetic properties of the new magma material compared to the older magnetic properties of the ocean floor. *See also* Suess; Wegener.

Hess' Theory for the Ionization of Gases: Physics: *Victor Francis Hess* (1883–1964), United States. Victor Hess shared the 1936 Nobel Prize for physics with Carl Anderson.

As altitude increases, so does the level of ionizing radiation that affects gases.

Victor Hess noted that over time, **electroscopes** exposed to air lost their charge. He theorized that the upper atmosphere emitted some kind of radiation. First, he timed the discharge of the leaves in his electroscope at different levels as he ascended the Eiffel Tower in France. Not much difference was noted, so he next tried his experiment in a balloon. At low altitudes, the rate of discharge was much the same. But as he increased altitude from several thousand feet to over 15,000 feet, a much more rapid discharge of the electroscope became evident. His original theory proved correct. There was some form of radiation coming from outside the earth's atmosphere that was absorbed by the more dense gases at low altitudes but was not impeded at higher altitudes. At first he believed the ionizing radiation came from the sun, but at night and during total eclipses of the sun, the rate of discharge was always the same as during the daytime. Hess then theorized that the radiation came from outer space. The scientific community remained unconvinced. At the time, most people continued to believe this form of radiation came from either the sun or some unexplained phenomenon on the earth. His work led to the discovery of cosmic rays by Robert Millikan a few years later. *See also* Millikan.

Higgins' Law of Definite Composition: Chemistry: *William Higgins* (1763–1825), Ireland.

When elements combine to form more than one type of molecule of a compound, they do so in a ratio of small, whole numbers.

William Higgins claimed to be the first to conceive that simple and multiple compounds composed of the same elements are combined in a ratio of small, whole numbers. Despite having no experimental evidence for his proposition, he based his law of definite or constant composition on the concept that for any molecular compound that contains atoms of the same element, these atoms combine in the same fixed and constant proportions (ratio) by weight. Higgins' theory was speculative. Shortly after, John Dalton expanded Higgins atomic theory by adding the statement that atoms are neither created nor destroyed in chemical reactions and that the molecules of compounds are composed of one or more atoms of the elements forming the compound. Further, these elements always combine to form compounds (molecules) in ratios of small, whole numbers of the atoms of the constituent elements. Higgins' law is best understood when considering molecules of oxygen: O_2 (oxygen) and O_3 (ozone). Dalton's contribution to the law is better understood when comparing the five oxides of nitrogen—for example, for NO, NO_2, N_2O, N_2O_3, and N_2O_5 each compound molecule has a different ratio of small, whole numbers by weight combining the two elements, nitrogen and oxygen. Higgins' law of definite composition relates to the ratio of the small, whole numbers of atoms of the *same* element combining to form a molecule, while Dalton's law of multiple proportions relates to the ratio, in small, whole numbers, of the weights of the atoms of *different* elements combining to form a compound. Today, most scientists consider Higgins and Dalton among the founders of modern chemistry. *See also* Atomism; Dalton.

Higgs' Field and Boson Theories: Physics: *Peter Ware Higgs* (1929–   ), England.

At very low temperatures, the symmetry of the electromagnetic force breaks down, producing massive particles (Higgs Bosons) from formerly massless particles.

Peter Higgs endeavored to unify electromagnetic waves and the "weak" force into a single "electroweak theory." He and other physicists observed that at high temperatures, electromagnetic photons and weak force W and Z bosons were indistinguishable from each other. Both seemed massless. Bosons are elementary particles that obey a particular type of statistical mathematics. Some examples are photons, pi mesons, and particles with whole number "spin," as well as nuclei of atoms composed of an even number of particles. Higgs subjected these particles to extremely low temperatures, at which point their symmetry no longer existed, and therefore photons could be distinguished from the W and Z bosons. A mathematical expression of this phenomenon became known as the *Higgs field*, which attributed the weightless bosons with mass. This formerly weightless boson, which now had theoretical mass, became known as the *Higgs boson*. An elementary particle, the Higgs boson, derived from the combined theory of electromagnetic and weak interactions, has not yet been detected. It is more massive and has a much higher level of electron volts, meaning that one of the newer, high-energy superconducting supercollider particle accelerators will be needed for its detection.

Hodgkin's Theory of Organic Molecular Structure: Chemistry: *Dorothy Crowfoot Hodgkin* (1910–1994), England. Dorothy Hodgkin was awarded the 1964 Nobel Prize for chemistry.

The structure of complex organic molecules can be determined by use of x-ray analysis.

The use of x-ray diffraction to study subatomic particles too small to be seen by optical microscopes was developed by Max Von Laue. His concept stated that x-rays were electromagnetic waves similar to light waves but with much shorter wavelengths that could "see" particles invisible to ordinary optical microscopes. He developed a technique for passing x rays through a crystal, forming a diffraction pattern of the crystal's structure. Dorothy Hodgkin perfected the technique to produce diffraction patterns that revealed the structures of rather large organic molecules. Working with her teacher, J. D. Bernal (1901–1971), she made the first x-ray photographs of the large protein molecule pepsin. Hodgkin then produced a three-dimensional photograph of penicillin, which aided in understanding its molecular composition. Later Hodgkin used x-ray diffraction to determine the structure of the vitamin B_{12} molecule, which is constructed of over ninety atoms. This enabled the vitamin to be produced in quantities adequate for the virtual elimination of pernicious anemia. Subsequently, with the aid of computers, she successfully determined the structure of the insulin molecule which has over 800 atoms.

Hoffmann's Theory of Orbital Symmetry: Chemistry: *Roald Hoffmann* (1937–), United States. Roald Hoffmann shared the 1981 Nobel Prize for chemistry with Kenichi Fukui.

Using quantum mechanics, it is possible to predict and explain the symmetry of chemical reactions.

Roald Hoffmann and Robert Burns Woodward collaborated in the development of an orbital molecular theory, now referred to as the *Woodward-Hoffmann rules*. Woodward is known for his work on synthesizing natural substances of very complex structures, while Hoffman was more concerned with how chemical bonds were formed and broken during chemical and cyclic reactions. Hoffmann advanced theories related to the electronic (orbital) structure of both stable and unstable inorganic and organic molecules. His work on the transition states of organic reactions led to the concept of bonding and symmetry used for the analysis of complex reactions. The Woodward-Hoffmann rule outlined the stoichiometric paths taken during the total summing up of the many steps required to complete complex chemical reactions. This rule enabled organic chemists to understand the structure of complex natural substances and synthesize them from simpler chemicals, resulting in chemical synthesization in the laboratory of many substances found only in nature. A few examples are cholesterol, lysergic acid, reserpine, some antibiotics, and, most important, chlorophyll. *See also* Woodward.

Hooke's Laws, Theories, and Ideas: Physics: *Robert Hooke* (1635–1703), England.

Hooke's Law of Elasticity: *The change in size of a material under strain is directly proportional to the amount of stress producing the strain.*

Robert Hooke was the first to apply mathematics to the concept of elasticity and relate this concept to the actions of springs. In essence, his law states that the distance a spring is stretched varies directly with its tension, as long as the spring does not exceed its limits of elasticity—that is, the strain is proportional to the stress. In other words, the more a spring is stretched, the greater becomes its internal tension. Once this was understood, springs could be designed for vehicles to provide smoother ground transportation. Hooke's law of elasticity applies to all kinds of situations and materials, from bouncing balls to the use of an elastic rod or fiber as a torsion balance. Hooke was the first to relate the simple harmonic motion concept for pendulums to the vibrations of micro "hair" springs used in the balance wheels of watches. Hooke also improved the escapement movement in clocks first conceived by ancient Chinese, by devising a method of cutting small but accurate cogs and gears. This special cog, called the *grasshopper escapement*, enabled both pendulum-driven grandfather-style clocks and the pocket watch to be constructed as very accurate timepieces for their time in history. *See also* Huygens.

Hooke's Cell Theory: *All plants are composed of cells surrounded by a defined cell wall.*

Hooke's concept of cells resulted from the construction and use of his practical compound microscope. He observed and drew elaborate diagrams of objects such as feathers, insects, and fossils. His most famous microscopic observation was at the walls of individual cells in a thin slice of cork, nonliving plant tissue. Hooke proposed the concept of cell based on the tiny monks' rooms in monasteries, called cells. Other biologists then examined living plant cells, all of which had well-defined cell walls, while the walls of animal cells were shaped like irregular membranes. Hooke's cell concept led to the development of the modern cell theory, which states that all living plants are composed of cells derived from other cells, as well as the extension of the plant cell theory to animal cells. *See also* Schleiden; Schwann.

Hooke's Theory of Sound and Light Waves: *Sound is transmitted by simple harmonic motion of elastic air particles. Therefore, light must also be transmitted by a similar wave motion.*

Robert Hooke applied his law of elasticity to air particles that are compressed (squeezed tighter together) and rarefied (spread farther apart) as these particles proceed from the source of the sound to a person's ears (the Doppler effect). He concluded that light, both colored and white, was transmitted in a similar wavelike motion through air. (Actually, light waves are not exactly similar to compression-rarefaction–type sound waves. Light has the dual properties of both particles and electromagnetic radiation or waves.) Hooke also extended his law of elasticity to form his own theory of gravity, which he based on mathematics related to the harmonics of planetary motion (a backward approach to the law of gravity as proposed by Sir Isaac Newton). In addition, Hooke claimed that he, not Newton, first conceived the concept that gravity obeyed the universal concept of the inverse square law. A disagreement developed in which Newton disputed Hooke's theories and claims. *See also* Newton; Stark.

Hoyle's Theories of the Universe: Astronomy: *Fred Hoyle* (1915–), England.

Hoyle's Steady-State Theory: *The universe did not originate from the big bang, but rather exists in a steady state.*

Fred Hoyle disagreed with the big bang theory proposed by George Gamow and other cosmologists. The big bang theory states that the universe started as an incredibly dense point or tiny ball that contained all the matter and energy now existing in the universe. Steven Weinberg of the University of Texas and others now claim the first three minutes of the universe's creation was so hot and dense that only subatomic elementary particles and energy existed. This was followed by the production of hydrogen and later helium as stars, galaxies, and other matter rapidly evolved and expanded, and continue to expand today. As the stars evolved, they produced the heavier elements now found on the planets. Hoyle, a respected astronomer, rejected this theory of an expanding universe and proposed a continual creation of atoms and other matter to the extent that for a volume in space the size of a house (about 30,000 to 50,000 cubic feet), only one atom is created each year. He further claims this constant creation of matter explains the formation of new galaxies. Hoyle believes the universe is closed and thus exists as a steady-state universe. Cosmological research, observational evidence, and mathematics over the past forty years or so seem to discredit the concept of a closed steady-state universe, but the debate continues.

Hoyle's Theory for the Origin of the Solar System: *The original sun was a binary star, one of which separated and exploded. Over time, the force of gravity coalesced this exploded matter to form the planets.*

Fred Hoyle made a number of contributions to astronomy and cosmology and provided a great deal of mathematical support for various theories. His theory for the formation of the planets in our solar system states that our sun was once one twin of a two-star (binary) system. One star exploded, but the gravitational attraction of the remaining star (our sun) maintained the pieces of the exploded star in orbits around our sun. In time, these chunks of matter attracted each other and piled up into great masses that became the existing planets, still revolving around our star as they are captured by the gravity of its great mass. This theory is considered a viable account for the formation of the planets as well as comets, asteroids, and meteors.

Hoyle's Theory for the Formation of the Elements: *Hydrogen is "fused" into helium inside stars, which also combined to form heavier elements.*

One of Fred Hoyle's most important theories explains how hydrogen is converted into helium inside the sun by the reaction of atomic (nuclear) fusion, which is the same reaction that occurs when a nuclear (hydrogen) bomb explodes. In addition to this reaction that creates helium plus all the energy output of the sun, Hoyle theorized that a similar process occurred inside the sun to form the heavier elements. One example is the formation of carbon atoms from helium: $^4He + {}^4He + {}^4He \rightarrow {}^{12}C$. A similar reaction formed the lighter elements as well as the heavier elements with which we are familiar on earth. This relates to Hoyle's theory for the formation of the planets of the solar system when the chunks of the exploded twin of our sun agglomerated. These chunks of matter contained all of the known elements at the time the planets were formed.

Hoyle's Disclaimer for the Reptile/Bird Theory: *Reptiles did not evolve into birds.*

Hoyle believed there were interstellar grains similar to bacteria that brought life to earth from outer space. This theory, called *panspermia*, was first proposed by Svante August Arrhenius. The theory is no longer valid since it has been demonstrated that cosmic radiation would kill extraterrestrial life forms of this type. This idea not only influenced Hoyle's concept for the origin of life, but it also affected his ideas related to the evolution of species and the use of fossils to explain evolution. He was unconvinced that fossils represented extinct species. Hoyle became involved in a dispute with a geologist over a type of fossil claimed to represent a species between a dinosaur-type reptile and a bird (Achaeopteryx). Hoyle claimed it was a fake because he insisted the feathers had been glued onto a reptile skeleton to make it seem part bird. The British Museum conducted many tests and found no evidence of glue or of any other deception. Recent studies of DNA, bone structure, and other anatomical comparisons have established an evolutionary relationship between birds and reptiles.

Hubble's Law and Constant: Astronomy: *Edwin Hubble* (1889–1953), United States.

Hubble's Law: *The velocity of a galaxy that is receding from us is proportional to its distance from the earth.*

Even before Edwin Hubble could utilize the 100-inch Mount Wilson telescope, he studied faint "clouds" of gas and dust that appeared as fuzzy images. He considered some of these areas as originating from our Milky Way galaxy. Other images seemed to originate from more distant areas of space, which were called *nebulae.* Once he was able to use a powerful telescope, he identified these more distant dense "clouds" of luminous gases as clusters composed of many millions, perhaps billions, of stars billions of light-years from Earth. He identified two types of these nebulae galaxies—one as spiral, the other as elliptical. He further classified elliptical galaxies as to their shapes approximating a circle. Although not all observed phenomena in deep space fall into these two classifications, Hubble's descriptions are still the basis for galactic classification. He also discovered several cephelids, which are stars that vary in their brightness (period-luminosity). These bright variable stars provide a means for measuring the distance of galaxies relatively near us—about 1 million light-years distant. From these data he proposed Hubble's law, $v = Hd$, where v is the recessional velocity of the galaxy, d is its distance from us, and H is known as Hubble's constant. To develop this law, Hubble measured the distance of about a dozen and half galaxies of several different classifications and related their receding velocities to the degree of redshifts in their light. He then devised the Hubble diagram where the x-axis is the distance and the y-axis is the amount of redshift of the wavelength of the galaxies' light.

Hubble's Constant: *The constant's original value was 150 km/sec/1,000,000 light years.*

Edwin Hubble overestimated the value for his constant by a factor of eight to ten. It has since been corrected as H = 15–30 km per second per Mpc. The symbol H for the constant is sometimes written as H_0, the range of 15–30 kilometers is still not an exact known distance, and *Mpc* is a *megaparsec*, which is equal to 10^6 parsecs. The parsec is a unit used in astronomy to measure very large stellar distances. It is equal to 3.856×10^{13} km, or 3.2615 light-years. Hubble's constant is important for two reasons. First, it provides the factor necessary for relating the redshift from the light of stars to their distances, thus providing a means of calculating the observable size of the universe. And, second, the reciprocal of Hubble's constant provides a means to determine the age of the universe. It is possible to calculate how long it would take galaxies to backtrack (contract) their now-expanding movements to their state at the origin of the universe. The current figure for the age of the universe, as calculated by the reciprocal of Hubble's constant, is between 10 and 20 billion years, with a reasonable estimate of 14 to 16 billion years since the time of the big bang. *See also* Gamow; Hale; Hertzsprung.

Huygens' Theories of Light and Gravity: Physics: *Christiaan Huygens* (1629–1695), Netherlands.

Huygens' Wave Theory of Light: *A primary light wave front acts as a spherical surface that propagates secondary wavelets.*

Christiaan Huygens was the first to conceive that light was propagated as waves and could support his theory by experimental observations. During Huygens' lifetime, most scientists supported the particle theory of light. Huygens, however, demonstrated that when two intersecting beams of light were aimed at each other, they did not bounce off one another, as would be expected if they were composed of minute particles with mass (conservation of momentum). He further theorized that light would travel more slowly when refracted through a denser medium than air. This was in direct opposition to Sir Isaac Newton's concept that light would maintain its speed when fracted toward the normal angle of light's motion. A more detailed statement of Huygens' wave theory is: *At every point on the main spherical wave front, there are secondary wavelets that at some time in their propagation are associated with the primary wave front.* Both the wavelets and wave front advance from one point in space to the next with the same speed and frequency. Since Huygens' time, the wave theory has been refined, and the wave-particle duality is now generally accepted. Electromagnetic radiation (light) waves are now also considered as particles, called *photons*, as explained by the quantum theory for the photoelectric effect and diffraction.

Huygens' Concept of Gravity: *Gravity has a mechanical nature.*

Huygens disagreed with Newton's law of gravity, which was based on Newton's laws of motion as well as the concept of force at a distance. Huygens considered Newton's theory of gravity as lacking the means to explain mechanical principles based on Cartesian concepts. Huygens used the term *motion* to mean "momentum" and considered the center of gravity to extend outward in a straight line similar to centrifugal force. He based his theory on mechanics, which states there is no loss or gain of "motion" following a collision between bodies (conservation of "motion" or momentum). He further developed a mathematical explanation for perfect elastic bodies. Huygens' emphasis on the mechanistic nature of gravity was in direct conflict with his wave theory of light, based on secondary wave fronts he considered as massless waves that would not obey mechanistic rules. Over time, Newton's concept of gravity was accepted but revised to conform to the new theories of relativity and quantum mechanics by Einstein and others. Today's theory of gravity is still based on the relationship of bodies' masses and distances separating them. There is a modern concept for the wave nature of gravity based on yet-to-be discovered wave particles called *gravitons. See also* Einstein; Newton.

Huygens' Concept for Using the Oscillations of a Pendulum for Timekeeping: *Based on the mathematics of curved surfaces and centrifugal force, the motion of an oscillating simple pendulum should maintain periodic time.*

Needing a device for accurate timekeeping for his work in astronomy and relying on the concepts developed by Galileo, Huygens worked out the mathematics of the pendulum, enabling him to develop the first accurate practical clock. (See Figure G1 under Galileo.) Much like today's grandfather clocks with a weighted bob suspended at the end of a long rod, it made use of slowly dropping weights and a crude escapement cog to maintain the swing of the pendulum at regular intervals. Huygens also developed a method for grinding lenses that improved the resolution of telescopes. Along with his pendulum clock and telescopes, Huygens made several discoveries, one of which was to identify a separation in the rings of the planet Saturn and another the discovery of one of Saturn's moons. *See also* Galileo; Hooke; Newton.

I

Ideal Gas Law: Chemistry: *Robert Boyle* (England); *Jacques-Alexandre-Cesar Charles* and *Joseph-Louis Gay-Lussac* (France). This entry is a combination of several laws, theories, and hypotheses developed by several scientists.

The ideal gas law may be expressed as $PV = nRT$, or $PV/T = nR$. ($P = $ pressure, $V = $ volume, $T = $ temperature, n is the number of moles of gas in the system, and R is the gas constant equal to 8.314 joules.)

Guillaume Amontons (1663–1705) was the first to measure and state the relationship between the temperature of a gas and its volume. He accomplished this by using an improved air thermometer similar to one devised by Galileo. (See Figure G2 under Galileo.) Amontons demonstrated that the volume of a gas increased at a regular rate when heated, and, conversely, its volume decreased when it cooled. He also concluded that this relationship applied to all gases. His work was undiscovered, and later, Robert Boyle was credited with recognizing the relationship between the volume and pressure of a gas at a constant temperature.

The ideal gas law is also referred to as the law of perfect gases or the generalized gas law. It is a combination of two laws and a constant to form a *universal gas equation*. The ideal gas law is expressed in the equation $PV = RT$, where P is the pressure exerted by 1 mole of gas in a volume V; R is also sometimes expressed as nK, where n is Avogadro's number and K is Botzmann's constant. It is the gas constant ($p_0 v_0/273.2°C$) at an absolute temperature T in the kelvin scale, or $R = 8.314$ joules per gram-mole-kelvin. The ideal gas law can also be determined in a physical sense from the kinetic theory of gases. There are four assumptions for the use of the kinetic theory to derive the law:

1. The gas within a given volume contains a very large number of molecules, which follows Newton's laws of motion and maintains randomness.

2. The actual volume of the total mass of the gas molecules is very small compared to the total volume the gas molecules occupy (the molecules are very small).

3. The only force acting on the molecules is the force resulting from short-term elastic collision (kinetic energy and conservation of momentum).

4. The gas molecules do not attract or repel each other.

The ideal gas law is a generalized law that relates the temperature, pressure, and volume of a gas under ideal or perfect conditions, which do not exist in nature. It is a combination of Boyle's law (pressure × volume) and Charles' and Gay-Lussac's laws (volume/temperature). The equation describes the behavior of only "real gases," with some accuracy, at relatively high temperatures, and low pressures. For some gases, it is a good approximation at standard temperature and pressure. At extremely high or low temperatures or pressures, the ideal gas law no longer applies to any gas. Understanding the gas laws was the beginning of modern chemistry. These laws have proven invaluable for almost all areas of science. *See also* Avogadro; Boyle; Charles; Gay-Lussac.

Ingenhousz's Theory of Photosynthesis: Biology: *Jan Ingenhousz* (1730–1799), Netherlands.

Plants not only produce oxygen in daylight, but also absorb carbon dioxide.

In 1771 Joseph Priestley, an English chemist, demonstrated that plants in a contained environment of carbon dioxide could make the air breathable in that environment. Jan Ingenhousz not only confirmed Priestly's hypothesis that green plants produce oxygen, but went one step beyond. Ingenhousz's theory stated that green plants absorb carbon dioxide and emit oxygen when exposed to light by a process, now known as *photosynthesis* (meaning combining or synthesizing by light), and that plants also reverse this process in the dark. To date, no one has been able to replicate this process in the laboratory. More recent experiments indicate that an atmosphere rich in carbon dioxide increases the rate of plant growth in nature as well as in greenhouses. Thus, one solution for reducing the amount of carbon dioxide in the atmosphere would be to plant more trees and other green plants. Ingenhousz's theory set the foundation for the concept of the relationship between living things. He demonstrated that animal life is dependent on plant life (oxygen and food), while plants depend on animals for carbon dioxide and the products of decomposition of dead animals and the wastes deposited by animals, as well as dead plants, thus establishing an ecological balance between the animal and plant worlds. *See also* Lavoisier; Priestley.

Ingold's Theory for the Structure of Organic Molecules: Chemistry: *Sir Christopher Kelk Ingold* (1893–1970), England.

If an organic molecule can exist in two different states other than its normal structure, then it can only exist in a hybrid form.

Sir Christopher Ingold called this idea *mesomerism*, which he used to describe a process similar to the resonance or oscillation of molecules in certain types of organic structures as described by Linus Pauling. He was interested in the molecular structures of particular organic compounds that can have two or more different molecular structures. The "mesomerism" molecules of these compounds have the same basic structure but with different arrangements of their valence electrons, which led to Pauling's concept of organic molecules existing in an intermediate (hybrid) form. Isomerization (mesomerism) that forms organic compounds does not fit the octet rule for the **Periodic Table of the Chemical Elements**. Ingold's theory led to a better understanding of special types of organic molecules, which became important in the development of new drugs, particularly antibiotics. *See also* Pauling.

Ingram's Sickle Cell Theory: Biology: *Vernon Martin Ingram* (1924–), United States.

Hemoglobin S (HbS) varies from normal hemoglobin A (HbA) in only one amino acid.

Sickle cell disease was first described in 1910 by James B. Herrick (1861–1954), a Chicago physician. A patient who came from the West Indies suffered from a form of anemia in which some of the patient's blood's cells were sickle shaped—thus the name *sickle cell anemia*. Some years later it was discovered that these special cells were related to a low level of oxygen in the blood. Linus Pauling and his graduate students separated type S hemoglobin from normal type A hemoglobin in 1947. The question still unanswered was how hemoglobin S (HbS), which caused the debilitating disease, was related to the proteins and amino acids of the blood.

In 1956 Vernon Ingram split hemoglobin into smaller units and separated them further by using electrophoresis. This electrochemical process uses a weak electric charge, causing large molecules to separate from each other into different "paths" or tracks according to their individual characteristics, which provides a means of identifying the components of the original substance. This procedure helped Ingram to determine that the sickle cell hemoglobin was caused by changes in only one of over 500 amino acids in the human body. The HbS appeared when the glutamic amino acid was replaced by the valine amino acid. Ingram then determined this was a mutation of the blood cells. The sickle cell disorder, the first known genetic disease, consists of at least two varieties of sickle cells. One form is characterized by a severe slackening of the blood flow, resulting in a reduction of oxygen in the blood vessels, which causes more restriction of the flow of oxygenated blood to the body's organs. This reduction of oxygen is the cause of genetic sickle cell anemia. Another kind is called a trait, which is not as devastating as the anemia form. A test for the presence of sickle cell disorders exists, but it cannot distinguish the two types. *See also* Pauling.

Isaacs' Theory of Proteins Attacking Viruses: Biology: *Alick Isaacs* (1921–1967), England.

When under attack by a virus, animal cells are stimulated to produce a protein (interferon) that interrupts the growth of the virus.

Alick Isaacs' study of various genetic varieties of the influenza virus led to his investigation of how the human body responds to different variations of a particular virus. He discovered a low-molecular-weight protein that had some effect on the way a virus multiplied and mutated. He named this newly identified protein *interferon*. The interferon protein is produced naturally by animal cells and enters the bloodstream automatically when the body is invaded by viruses. Although the body produces only small amounts of interferon, it enables healthy cells to manufacture an **enzyme** that counters the viral infection. As a natural component of an animal's body it is referred to as a *biopharmaceutical*. Each species of animal, including humans, produces its own type of interferon, which cannot be interchanged between species. The human body produces three types of interferon: leukocyte (alpha), fibroplast (beta), and immune interferon (gamma). However, all interferon is produced by only one type of cell. When first discovered, interferon was very expensive to produce since the human body produces only minute amounts. However, it is available now in large quantities due to the genetic engineering of the protein molecule. It is used to treat a variety of diseases, including liver disorders, such as hepatitis C, hairy cell leukemia, Kaposi's sarcoma, genital warts, and diseases of the gastrointestinal tract.

J

Jansky's Theory of Stellar Radio Interference: Physics (Engineering): *Karl Jansky* (1905–1950), United States.

Unexplained radio interference on earth originates in space.

Karl Jansky was assigned the task of solving the problem of radio noise interference in shortwave transatlantic radio-telephone transmission. He devised a rotating linear directional antenna that he mounted on an automobile wheel, enabling him to turn it through 360 degrees. Using this antenna, he identified a number of sources of "noise" or static originating in the atmosphere, some of which came from industrial sources, as well as thunderstorms. However, he continued to believe some of these unwanted signals originated from the sun. After more study he realized this was not possible since the time of the peak noise was shifting as earth revolved around the sun during a twelve-month period. He finally rejected this solar theory when he was unable to detect a signal from the sun during a partial solar eclipse. Jansky's theory, which he published in 1932, stated there were two sources in space from which these radio signals were originating. One source of the signals is the Milky Way galaxy. An even stronger signal comes from the direction of the Sagittarius configuration of stars and galaxies. His work and theory led to the development of the important fields of radio and x-ray astronomy. Currently, man-made satellites use his concept to explore the radiation of deep space. X-ray astronomy measures the leftover radiation from the big bang at the time the universe was created. These observations are expected to explain the age and nature of our universe more accurately. The unit for the strength of transmitted radio waves was named *Jansky* in his honor.

Jeans' Tidal Hypothesis for the Origin of the Planets: Astronomy: *James Jeans* (1877–1946), England.

A passing star pulled off a lump of sun matter, which later solidified into the planets as they revolved around the more massive sun.

Before Jeans' time, the accepted concept for the origin of planets was Pierre Simon Laplace's **nebula** hypothesis, which stated the sun and its planets were formed by a contracting cloud of dust and gas. A different tidal hypothesis was proposed by James Jeans, which asserted a large star passed close to our sun, causing a cigar-shaped protrusion of gas and matter to be pulled off the sun. This oblong formation of gas and other sun chemicals coalesced to form planets as they were sent into orbits around the sun. Some years later, other astronomers rejected Jeans' theory. They resurrected and revised Laplace's condensation hypothesis, which is still accepted by some astrophysicists. Later, Jeans demonstrated that all the outer planets had very cold atmospheres, but residual internal heat keeps them relatively warm. *See also* Hale; Laplace.

Jenner's Inoculation Hypothesis: Biology: *Edward Jenner* (1749–1823), England.

Injecting humans with cowpox fluid will immunize them from smallpox.

An old English wives' tale asserting milkmaids who had contracted cowpox would not be susceptible to the more deadly smallpox led Edward Jenner to hypothesize that deliberately infecting people with cowpox germs would prevent them from contracting smallpox. To do this, he extracted some fluid from a cowpox blister on Sarah Nelmes, a milkmaid. Using a procedure Jenner later named *vaccination* (from two Latin words, *vacca* for "cow" and *vaccinia* for "cowpox"), he injected this fluid into an eight-year old boy, James Phipps. Jenner is also credited with introducing the term *virus*. Six weeks after the cowpox injection, Jenner injected the boy with fluid from a smallpox blister. The boy did not contract smallpox. After some additional trials, Jenner published his results. The public reaction to his experiment was, and still is, mixed, and at one time his vaccinations were banned in England. Then, soon after a serious outbreak of smallpox, all English children were required to be inoculated. Even today, there is criticism of his procedures because of our current laws restricting the use of humans as experimental subjects without proper consent, which did not exist in Jenner's day. Because of the success of Jenner's controversial experiment, the death rate from smallpox dropped from about 40 per 10,000 people to about 1 in every 10,000 during the nineteeth century. The World Health Organization claims that smallpox has been eradicated, but it can still flare up in rural areas of India, Asia, and Africa, where many children are not inoculated. Smallpox is also being produced and stored for use as a biological weapon by some countries. Several industrialized nations are vulnerable because many citizens are no longer vaccinated against smallpox. *See also* Lister.

Jerne's Theory of Clonal Selection of Antibodies: Biology: *Niels Kaj Jerne* (1911–1994), Denmark. Niels Jerne shared the 1984 Nobel Prize for physiology or medicine with Cesar Milstein and Georges Kohler.

Diverse antibodies are present in humans at birth and, when attacked by a virus, can produce additional antibodies.

Niels Jerne was aware of the concept that the body's lymphocytes (white blood cells) produce a wide range of various types of antibodies that attack specific bacterial and viral infections. Jerne based his theory on the belief that each cell that produces a specific **antibody** is present in the body from birth. Bacteria or viruses infect the body by releasing their particular set of chemicals. The infected person's antibodies cause the lymphocytes related to particular bacteria or viruses to divide, producing clone cells that greatly increase the number of antibodies available to fight that specific infection. This theory led to the question of how all this genetic information was included in these original, at-birth cells. Jerne developed the concept of *somatic mutation*. Somatic body cells are the many types of cells that make up the tissues in the body, with the exception of the germ cells (ova and sperm). This cell mutation concept was the forerunner to Susumu Tonegawa's (1939–   ) more complex antibody inter-active control mechanism referred to as the *jumping genes theory*. Niels Jerne proposed a theory on the functioning of the immune system, but he neglected to consider the multitude of chemical compounds involved in modulating the immune system. His work and theory are responsible in large part for the current study of the immune system. *See also* Koch.

Johanson's Theory for the Evolution of Humans: Anthropology: *Donald Carl Johanson* (1943–   ), United States.

*In the pre–*Homo sapiens *species*, Australopithecus afarensis, *the males were larger than the females, indicating they did the hunting while females gathered and cooked food.*

In Ethiopia in 1972, Donald Johanson discovered several bones of a fossilized skeleton, which he identified as a small, three and one-half foot female who was as much a bowlegged, upright-walking, chimpanzee-like creature as she was human. He named this small-brained fossilized female Lucy, after the Beatles' song "Lucy in the Sky with Diamonds," which was playing on the phonograph at the camp where he made the discovery. Johanson named his new species *Australopithecus aferensis*. Previous to Johanson's discovery, Raymond Dart's (1893–1988) assistant found a skull in Africa that was identified as a new species of fossil primate that predated various species of *Homo sapiens*. Dart named this new species *Australopithecus africanus* (meaning southern African ape). Although both discoveries were made in Africa, they were discovered in different regions. Johanson's discovery and theories eventually were challenged by other anthropologists. Johanson stated that bipedalism (walking upright on two feet) preceded the development of the large brain capacity in humans. This contradicted existing theory. He also claimed females of this prehistoric species stayed "home," pregnant, caring for children, and cooking the bounty brought home by the larger and stronger male hunters. This theory was recently challenged by Adrienne Zihlman, an anthropologist and self-proclaimed feminist, who claims that this is a typical male anthropologist's sexist interpretation. Anthropologist Richard Leakey (1944–   ), son of famous anthropologist Louis Leakey (1903–1972), claimed the various human species could be traced back even further in history than could the fossils of *A. afarensis* (Lucy). He also claimed there were probably two or more branches to the ancestral tree of modern humans, not one as Johanson claimed. Others declared that since Johanson gathered the bones from different sites, the partially completed skeleton was not of a single person or even a female. Additional pre–*H. sapiens* fossils found in Africa seem to confirm Johanson's theory. Although Johanson's discovery and theories significantly contributed to science, some doubt remains as to the history of the ancestors of early *H. sapiens*. In addition, it has been claimed that *Australopithecus* is not a separate distinct species in the continued evolution of humans, but may be just another unsuccessful, extinct pre–*H. sapiens* species similar to the Neanderthal man.

Josephson's Theory of Semiconductors: Physics: *Brian David Josephson* (1940–   ), Wales. Brian Josephson shared the 1973 Nobel Prize for physics with Leo Esaki and Ivar Giaevar.

A DC voltage applied across a thin insulator between two superconductors produces a small alternating current whose frequency varies inversely to the voltage.

The BCS theory (named after John Bardeen, Leon Cooper and John Schrieffer) demonstrated the concept of superconductivity at super low temperatures. The BCS theory states that under conditions of near absolute zero, electrons travel in pairs rather than individually, as the result of vibrations of the atoms. Josephson demonstrated this phenomenon by placing an insulator between two electron-conducting plates of metal known as a *Josephson junction*. The effects of these electrons flowing across this partially insulated junction produced a semiconducting flow of current known as the *Josephson effect*. A current can continue to flow across this junction for a short period of time even when the voltage is temporarily removed. In addition, a small current can produce an alternating current on the other side of the junction whose frequency varies inversely with the applied voltage. By using paired electrons Josephson maintained a tunneling effect that allowed the alternating super current to flow across the thin, insulating barrier of the **semiconductor**. This changing of the current's frequency could be used as a means of controlling electronic devices somewhat like a switch. This revolutionized the electronics industry and modern life. Today, the separator between the plates of semiconductors can be applied to a thickness of only one or a few atoms of material. Semiconductors are used in sensitive instruments to make accurate magnetic and electrical field measurements. Some applications of Josephson junctions are the detection of microwave frequencies, magnetometers, thermometers to measure near-absolute zero temperatures, and detection and amplification of electromagnetic signals. Of more importance, semiconductors are used to make high-speed (almost the speed of light) switching devices, which make modern computers possible.

Joule's Law and Theories: Physics: *James Prescott Joule* (1818–1889), Scotland.

Joule's Law: *The relationship for heat produced by an electric current in a* *conductor is related to the resistance of the conductor times the square of the amount of current applied: $H = RI^2$.*

By experimentation, James Joule established the law that states that when a current of voltaic electricity is sent through a metal or other type of conductor, the heat given off over a specific time period is proportional to the resistance of the conductor multiplied by the square of the electric current. The equation for this law is: $H = RI^2$, where H is the rate of the heat given off in watts in joule units, R is the resistance in the conductor in ohms and I is the amps (amount) of the current. The application of this law is important in all industries using electricity as a source of energy. The resistance to an electric current flowing through a conductor is analogous to the friction of air, the movement of engine parts, and tires on the road for a moving automobile. The electrical, as well as mechanical, energy is not just "lost," rather it is converted to heat, just as is friction. Joule was interested in improving the mechanical advantage of electric motors, but since they were very primitive during his lifetime, he devoted more of his work to improving the efficiency of steam engines. He accurately predicted that electric motors eventually would replace most other types of mechanical devices.

Law for the Mechanical Equivalent of Heat: *A fixed amount of mechanical work (expenditure of energy) ends up in a fixed quantity of heat.*

Although earlier, Julius von Mayer developed a less accurate figure for the mechanical equivalent of heat, Joule was the first to include heat as a form of energy. Joule conducted exacting experiments to measure the amount of heat generated not just by electricity but also by mechanical work. He demonstrated that 41 million ergs of work produced 1 calorie of heat, which is now known as the *mechanical equivalent of heat*. Since 10 million ergs are equal to 1 joule, named after James Joule, 4.18 joules are then equal to 1 calorie. Joule's work enabled others to perfect the law for the conservation of energy, which states that energy, like mass, cannot be created or destroyed but can be changed from one form to another.

Joule-Thompson Effect: *When a gas expands, its internal energy decreases.*

James Joule collaborated with Lord William Thomson Kelvin to devise the Joule-Kelvin effect, which is related to the kinetic theory of gases. They measured the change in energy involved when the pressure of a compressed gas is released and then expands. As a gas expands, the motion of molecules is reduced. In other words, its internal energy is decreased as is its temperature. It can be reheated if it can "consume" energy from its surrounding environment, thus providing a cooling effect to the area around it. *See also* Ideal Gas Law.

K

Kamerlingh-Onnes Theory of Matter at Low Temperatures: Physics: *Heike Kamerlingh-Onnes* (1853–1926), Netherlands. Heike Kamerlingh-Onnes was awarded the 1913 Nobel Prize for physics.

Some metals lose their electrical resistance at super-low temperatures.

Heike Kamerlingh-Onnes experimented with the properties of matter at low temperatures and improved on the apparatus and procedures employed by Sir James Dewar (1842–1923) in the late 1800s. This enabled him to use liquid hydrogen and the Joule-Thomson evaporation effect to cool helium to a temperature of 18 kelvin (which is 18 degrees above absolute zero). After cooling the gas still more by allowing it to expand through a nozzle, he determined liquid helium has a boiling point of 4.25 kelvin. When the liquid helium is in an insulated container and the vapors are rapidly pumped away, the liquid helium is cooled still further to just 0.8 kelvin. This was as close to absolute zero as had so far been reached. He was the first to study the nature of materials at this extremely low temperature and ascertain that molecular activity (kinetic energy) almost ceases at this temperature. This was the beginning of the science of **cryogenics**, which led to the observation of superconductivity, where metals lose their resistance to electricity, thus enabling electric currents to pass through wires without generating heat by internal resistance. Superconductivity is now used for supercooled magnets in particle accelerators, but it is not yet practical for most purposes because of the difficulty of achieving and maintaining sufficiently low temperatures. Currently work is progressing on developing warm superconductivity, which will provide the same low resistance to electricity but at temperatures much higher than absolute zero. It is hoped this research may someday prove useful in developing more cost-effective methods of transmitting electricity and be used to produce supermagnets to levitate high-speed trains. *See also* Joule; Kapitza.

Kapitza's Theory of Superfluid Flow: Physics: *Pjotr Leonidovich Kapitza* (1894–1984), Russia. Pjotr Leonidovich Kapitza shared the 1978 Nobel Prize for physics with Arno Allan Penzias and Robert Woodrow Wilson.

Thin film-vapor systems exhibit superflow properties at very low temperatures, with the resistance to flow increasing as the film's thickness increases.

Pjotr Leonidovich Kapitza developed an improved method for liquefying air, which enabled him to study the properties of liquid helium. He determined that liquid helium, known as He-II, behaves as a "superfluid" at near absolute zero and exhibits very unusual flow characteristics. At this temperature liquid helium appears to be in a perfect atomic *quantum* state as a superthin film that manifests some novel forms of internal convection, including its ability to flow up the sides of its container, even when the container is closed. His methods for liquefying gases facilitated the commercial production of liquid oxygen, nitrogen, hydrogen, and helium. Large-scale production of these gases enabled the development of very high magnetic fields used in many areas of research and technology, such as particle accelerators and nuclear magnetic resonance (NMR) instruments. *See also* Kamerlingh-Onnes, Kusch.

Kekule's Theory of Carbon Compounds: Chemistry: *Fredrich Kekule von Stradonitz* (1829–1896), Germany.

Carbon is a tetravalent atom (a valence of 4) capable of forming ring-type organic molecules.

Friedrich Kekule von Stradonitz was the first to propose a structural formula that indicated atoms bonded with each other to form molecules. His study led to the concept of the carbon atom's structure consisting of four (tetravalent) bonds with a central nucleus by which it could form numerous types of molecules. (See Figure V3 under Van't Hoff.) He also related this unique atom to the basic structure of all organic (living) carbon compounds. However, one form of carbon puzzled the chemists of his day. Michael Faraday discovered that the molecule for the aromatic compound benzene contained six carbon atoms with a total of twenty-four bonding electrons, but benzene also had six hydrogen atoms; however, each hydrogen atom had only one bonding electron. When diagramed as a linear or even a branching molecule, this combination was impossible since each carbon atom had four bonds (valence) and each hydrogen atom had just one bond (valence)—$6 \times 4 = 24$ for carbon, and $6 \times 1 = 6$ for hydrogen. Therefore, there were too few electrons to satisfy the octet rule for a linear structure such as C_6H_6. Reportedly, Kekule solved this problem one night as he dreamed. He saw different configurations of atoms forming various arrangements. One arrangement resembled a snake eating its own tail. (See Figure K1.) He woke up, electrified and, working the rest of the night, he came up with the structure of the benzene ring. The ring consisted of each carbon atom sharing two of its four bonding electrons with another carbon atom, one valence electron with a partner on its other side on the ring, and one valence electron with a hydrogen atom outside the ring, resulting in the classical benzene, hex-

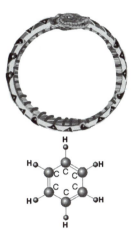

Figure K1: Artist's depiction of Friedrich August von Kekule's dream of a snake eating its own tail that aided him in solving the problem of the structure for the organic compound benzene, composed of a ring of carbon atoms.

agonal ring. This answered many questions and was a revolution for organic chemistry. It was then possible to substitute another atom, a *radical*, or a molecule for one of the hydrogen atoms of the six in the ring. This resulted in what is known as a single substitution to form a derivative of the benzene ring (e.g., C_6H_5X). Specifically, if the free radical (molecular fragment with a single unshared electron and no charge) NH_2 is substituted for one of the hydrogen atoms, $C_6H_5 \cdot NH_2$, the compound aniline would be the result. In addition, it is possible to combine many of the hexagonal benzene rings to form more complex compounds. This answered the questions related to the multiplicity of organic compounds. To some extent, this seems odd since Kekule considered the existence of atoms a metaphysical problem, and he claimed chemistry was concerned only with arriving at hypotheses that explained chemical structures and reactions. *See also* Couper.

Kelvin's Concepts of Energy: Physics: *William Thomson Kelvin* (known as Lord Kelvin) (1824–1907), Scotland.

Kelvin's Theory of Thermodynamics: *When a gas cools, loss of volume is less crucial than loss of energy. Kinetic energy (molecular motion or heat) approaches zero at temperatures approaching absolute zero (zero degrees kelvin = −273.16°C).*

William Thomson Kelvin was knighted by Queen Victoria and given the title of Lord Kelvin of Largs for his work on electromagnetic fields. Kelvin was familiar with the works of Carnot, Joule, and Clausius, early explorers in the field of thermodynamics, where mechanical heat is related to energy. The second law of thermodynamics deals with the concept of entropy, which states that heat always flows from a warmer object to a cooler environment—and never in the opposite direction. For example, a hot cup of coffee always cools and will never get warmer than its surroundings unless an external source of heat is applied to the cold coffee. This means that energy in a closed system is striving to reach a state where there is no transfer of energy—equilibrium. Eventually everything will be the same temperature. Another way of saying this is that everything becomes more disorganized (entropy) and "runs downhill" to a common level, unless more energy is pumped into the isolated system. But then it would no longer be a closed system. For instance, if the earth did not receive most of its energy from the sun, everything would run down, and it would soon be a very cold place. Kelvin provided a mathematical formulation of the second law of thermodynamics. Since entropy (disorganization) always increases, Lord Kelvin believed the universe would sometime in the future have maximum entropy and thus uniform temperature. He called this the heat death of the universe. The law of conservation of energy states that energy can be neither created nor destroyed but can be transferred from one form to another (e.g., mechanical energy to heat energy, as in rubbing your hands together vigorously, generating friction heat), or chemical energy transformed to light and heat (a burning candle). Heat itself is the manifestation of the kinetic energy of molecular motion, while the temperature of a substance is proportional to the average motion of the molecules when they are in thermal equilibrium, that is, the temperature is a measure of the average internal energy. Kelvin proposed a new scale for measuring the absolute temperature of matter, which at its zero point would be the lowest temperature possible. He started with what is known as the triple point of pure water, which is about 0.01°C, where an equilibrium between the water, ice, and water vapor is established. At this temperature, water can exist in its three states at the same time—solid, liquid, and gas. This point was also used to set up the metric temperature scale now referred to as the Celsius scale, where 0 is the freezing point of water (or melting point of ice) and 100 is the boiling point of water, where it attains its gaseous state. Selecting 100-degree units for this scale was arbitrary. Any units could be used, such as the units for the Fahrenheit temperature scale. Kelvin used the same metric (100) scale and, by extrapolation, arrived at −273°C degrees as the absolute zero point. This point was originally called A, for absolute. Therefore, water freezes at 273 K and boils at 373 K. Absolute zero was later refined to equal −273.16 degrees C, or 0 K. This absolute temperature scale, named after Lord Kelvin, provides for the measurement of very cold and very hot temperatures. Since no thermometer has been invented that can measure absolute zero, Lord Kelvin reached this point by theoretical consideration. Some people believe that all molecular motion ceases at this point. This is not quite correct since molecules of solids continue to "vibrate" but not move at random or exhibit any kinetic energy as molecules would in matter at higher temperatures. In other words, the energy has the lowest possible value at absolute zero and the entropy is zero.

Kelvin's Theory of Electromagnetic Fields: *Electromagnetic fields travel through space as do light waves; the electric field vector and the magnetic field vector vibrate in a direction transverse to the direction of the wave propagation.*

Lord Kelvin's theory for electromagnetic fields stated the fields associated with alternating current (**ac**) electricity are waves that travel through space similar to light waves. His theory not only proved that both types of waves are transverse waves, but also travel at the same speed. Kelvin's theory for electromagnetic fields was put to good use and provided the information necessary to lay the first successful transatlantic telegraph cable. Two previous attempts had failed. In addition, his theory led to the electromagnetic theory of light developed by James Clerk Maxwell. *See also* Carnot; Joule; Maxwell.

Kepler's Three Laws of Planetary Motion: Astronomy: *Johannes Kepler* (1571–1630) Germany.

Law I: *All the planets revolve around the sun in elliptical paths, and the sun occupies one of the two focal points for the ellipse (the other focal point is imaginary).*

Kepler's mathematical analysis of Tycho Brahe's data resulted in the concept of planetary orbits as being ovoid (egg shaped). However, after checking his data, he corrected an error in calculation and realized that all planetary orbits, including the orbit of Mars, are elliptical.

Law II: *An imaginary straight line joining the sun and a planet sweeps over equal areas in equal intervals of time.*

Kepler's second law follows directly from the first law. Also referred to as the law of areas, it is probably the most important and easiest to understand. Kepler measured the distance of a short path of a planet progressing along a segment of its elliptical perimeter. In addition, he measured the time elapsed for the planet to cover this short segment of its orbit. Using geometry, he determined the area for the pie-shaped wedge of space formed by the two sides of a triangle originating at the sun (the meeting point of these two lines), which extends to the perimeter. The area of this pie-shaped wedge was related to the distance covered by the planet along its elliptical perimeter in a given period of time. He then made similar measurements as the planet progressed to different segments (chords) of its orbit. (See Figure K2.) When the planet was at its closest to the sun, it traveled much faster in its orbit to cover an area equal to that area covered when it was farthest from the sun. This law of areas was extremely important for Isaac Newton's formulation of his concept of gravity and his laws of motion. Kepler's second law also explains the theory for conservation of angular momentum for bodies in nonlinear motion. For instance, when an ice skater spins rapidly with arms extended and then pulls his or her arms in close to the body, the momentum gained when the arms were extended is now conserved by increasing the rate of spin. The same is true for an earth-orbiting spacecraft. When it drops to a lower orbit, its momentum is conserved by increasing its speed relative to the speed it had obtained in a higher orbit.

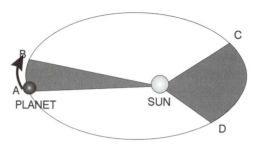

Figure K2: Kepler's second law states that planets revolving around the sun in an elliptical orbit do so in such a manner that equal areas are covered in equal times. (Figure not to scale.)

Law III: *The square of planets' orbital periods is proportional to the cubes of the semimajor axes of their orbits.*

Another way to say this is that the square of the time it takes a planet to complete one orbit (orbital period) is proportional to the planet's average distance from the sun cubed. This may be expressed as: $P^2 = (AU)^3$, where P is the time it takes a planet to complete one revolution around the sun in years, and AU is the astronomical unit, which is equal to the average distance between earth and the sun, or 93 million miles. Kepler developed his third law while attempting to devise a mathematical basis for musical-type harmony as related to his first two laws.

Kepler tried to apply mathematics to Plato's concept of five regular solids and from this derive mathematical harmony for Plato's model of the universe. He continued to apply his mathematics to achieve harmony with Copernicus' concept of a sun-centered solar system. After leaving his home in Germany, Kepler secured a position in Prague with Tycho Brahe, a firm proponent of Ptolemy's earth-centered universe who never accepted the Copernican heliocentric theory. Tycho assigned his new assistant, Kepler, the task of observing and measuring the orbit of Mars. Kepler at first thought it would be a simple task, but it took him over eight years to complete it. Tycho died soon after Kepler joined his staff, leaving reams of data from his own extensive observations, which Kepler put to good use. Many historians give Tycho, not Kepler, credit for discovering that Mars' orbit is elliptical because this discovery was partly due to Tycho's extensive observation records that aided Kepler's accomplishment. This work led to Johannes Kepler's three laws of planetary motion, which remain valid.

Kepler's laws have applications far beyond the orbits of planets. They apply to all kinds of bodies found in the universe that are influenced by gravity: moons orbiting their mother planets, other solar systems, binary stars, and artificial satellites orbiting earth, for example. Kepler was also interested in other sciences besides astronomy. From his study of optics and vision, he developed a theory stating that light from a luminous body is projected in all directions, but when the human eye viewed this light, only the rays that enter the pupil of the eye were refracted, ending up as points on the retina. This is much closer to today's concept of electromagnetic radiation as related to vision than the ancient Greek idea of the eye sending out a signal to the object, which was then reflected back to the eye. After Galileo developed his telescopes, Kepler explained how the lenses of the instruments worked. He did the same for the new eyeglass spectacles. He was very supportive and complimentary of Galileo's work, even when others ridiculed Galileo's observations. It seems Galileo either did not understand or appreciate Kepler's mathematical contributions to astronomy because he ignored his publications. *See also* Brahe; Galileo; Newton.

Kimura's Neo-Darwinian Theory for Mutations: Biology: *Motoo Kimura* (1924–1994), Japan.

Genetic mutations at the molecular level can increase within a population without being affected by Darwinian natural selection.

Motoo Kimura developed data indicating there are certain types of mutations that multiply in a given population without resulting in these mutations being selected out, as proposed by Darwin's concept of natural selection. By using chemical means, he identified several molecules for mutant genes that proved to be not harmful to the individual. In fact, some of these seemed to adapt better than nonmutated genes. He concluded evolutionary change may be caused by a normal drift of selected genes that may have mutated. There is recent evidence of exceptions to his theory, which is based on findings that several mutations at the molecular level are selective (not random) and do cause evolutionary changes. One exception is the mutated gene affecting human hemoglobin in the blood. Another is the genetic decoding study conducted for the population of Iceland. Over the centuries, Icelanders were isolated and their numbers often reduced by natural disasters. Recent researchers found that an ancient survivor carried a mutated gene that was missing five units of DNA. The researchers concluded that the 275,000 modern Icelanders are somewhat inbred, as evidenced by similar physical characteristics (blue eyes, blond hair). Fortunately, Icelanders have maintained excellent genealogical and health records, which assisted in this genetic research. The decendants of this one person, who now compose a large portion of the present population, carry this mutation, which causes a high risk of breast cancer for both men and women. This seems to be an example of a mutated gene at the molecular level that has drifted and was selected according to evolutionary theory, thus disproving Kimura's theory.

Kirchhoff's Laws and Theories: Physics: *Gustav Robert Kirchhoff* (1824–1887) Germany.

Kirchhoff's Electrical Current and Voltage Laws: *(1) The sum of all the currents flowing in the direction of a point is equal to the sum of all the currents flowing away from that point, $\Sigma I = 0$. (2) At any given time the algebraic sum* *of a voltage increase through a closed network loop will be equal to the algebraic sum of any voltage drop,* $\Sigma IR = \Sigma V$.

These two laws are important in the analysis of electrical circuits and for solving problems related to complex electrical networks. They are extensions to Ohm's law. *See also* Ampère; Ohm.

Kirchhoff's Law of Radiation: *A hot body in equilibrium radiates energy at a rate equal to the rate that it absorbs energy. Both the absorbed and radiated energy have the same given wavelength (black body radiation).*

Gustav Kirchhoff's law of radiation states that a perfect black body will absorb all light and other forms of radiation that are not reflected. A black body may be thought of as the perfect radiator where the maximum energy obtained per unit of time is due to the temperature of the "radiator." A radiator that is a perfect absorber is also a perfect reflector. Such a device is never found in nature. When heated, black bodies emit all the different wavelengths of light. This raised the question of how the different wavelengths of light were actually given off and how the individual wavelengths changed with temperature. Important theories related to energy, in particular, the visible light spectra followed, which led to the unification theory of electricity, magnetism, and light, referred to as the *electromagnetic spectrum*. The concept of black body radiation also provided Max Planck the idea needed to develop his "quanta" theory. *See also* Maxwell; Planck.

Kirchhoff's Theory for the Use of the Light Spectrum in Chemical Analysis: *Each chemical element, when heated, exhibits a different line or set of lines in the spectrum of visible light.*

Sir Isaac Newton was the first to use a glass prism to split white light from the sun into a spectrum of colored lights. William Hyde Wollaston (1766–1828), using an improved prism, divided these colors into seven distinct divisions. Von Fraunhofer developed a "diffraction grating" made of fine wires to substitute for a glass prism, which enabled him to detect almost 600 individual lines in the sun's spectrum. These devices made it possible for Kirchhoff to identify specific lines in the spectra of elements, thus identifying them by their particular colored light patterns. Kirchhoff and Robert Bunsen collaborated in the development of the spectroscope, which used the diffraction grating developed by von Fraunhofer. The spectroscope consisted of a tube with a thin vertical slit on the front end, followed by a light collecting lens and either a prism or diffraction grating. An eyepiece lens for viewing the specimen was located at the other end of the spectroscope. A small sample of the element to be analyzed was heated on a glass bead by Bunsen's new type of gas burner, which produced a flame hotter than an alcohol burner. When viewed through the slit/prism, or later through an improved diffraction grating, the light separated into specific spectra lines, unique for each element. These lines were then measured on a scale. Later, photographic plates were used that produced spectrophotographs for chemical analysis. The viewed image of lines could be a bright spectrum, which showed the element's spectrum as bright lines. Conversely, if the light from the heated element was passed through a gas similar in wavelength to the element's wavelength, a dark line spectrum was visible. Spectroanalysis now can be used for analyzing the electromagnetic spectrum beyond the range of visible light, in both the infrared and ultraviolet ranges, as well as x rays from outer space. The technique has been invaluable in the analysis of the chemical makeup of all types of objects in the universe (e.g., the chemical composition of the sun and other stars, as well as the atmospheric composition of the planets). Today, spectroanalysis of the electromagnetic waves given off from elements is an important analytical tool used by all fields of science. *See also* Bunsen; Fraunhofer; Maxwell.

Kirkwood's Asteroid Gap Theory: Astronomy: *Daniel Kirkwood* (1814–1895), United States.

The asteroids located between the orbits of Mars and Jupiter are separated by gaps that correspond to their orbital periods and are integer fractions of Jupiter's orbital period.

From the days of Galileo, chunks of matter smaller than planets yet large enough to be seen with a telescope were found in the region between Mars and Jupiter. Over 100,000 objects, ranging from less than 1 mile in diameter to over 500 miles in diameter, have been identified in this region, called the asteroid belt. Daniel Kirkwood detected a tugging on these asteroids by Jupiter's great gravitational force. This force created a resonance phenomenon (reverberation). Asteroids that were drifting randomly were then pulled toward this region between Mars and Jupiter, thus forming their own orbits around the sun. (See Figure K3.) This resulted in gaps among the thousands of asteroids revolving around the sun, which Kirkwood attributed to the influence of Jupiter's gravity. The areas of depleted asteroids are called the *Kirkwood gaps*, which are the consequence of their orbital separations related to their resonance motions with Jupiter. A similar theory explains the separation of the Cassini division for the rings of Saturn. This division or separation between the rings of Saturn is caused by its moon, Mimas. The inner rings for both the asteroid belt and Saturn's rings follow Kepler's laws of planetary motion. Since the inner sections of the ring must travel faster than the outer sections, density waves or ripples are formed that are called the *spiral density waves*. *See also* Cassini; Kepler; Kuiper; Oort.

Koch's Germ-Disease Postulate: Biology: *Heinrich Hermann Robert Koch* (1843–1910), Germany. Robert Koch was awarded the 1905 Nobel Prize for physiology or medicine.

A set of specific conditions must be met before it can be established that a specific germ has caused a specific disease.

Robert Koch arrived at his postulate after many years of trying to isolate and identify specific bacteria that caused a great variety of diseases for which there was no known cause or effective cure. These included anthrax, tuberculosis,

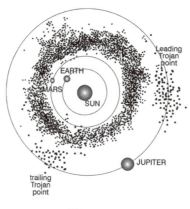

Figure K3: Diagram of the asteroid belt located between the orbits of Mars and Jupiter. Asteroids are leftover matter from the formation of the solar system and are too small to be considered planets. Jupiter's great gravity most likely prevented them from coalescing into planets. This region is the source of most of the meteors and asteroids that intersect earth's orbit.

sleeping sickness, bubonic plague, malaria, cholera, as well as other "germ"-caused diseases. He developed techniques for growing and isolating bacteria in **agar** cultures, staining them for identification and using inoculations to test the suspected germs. His methods and postulate are responsible for bacteriology's becoming a respected science.

Robert Koch's postulate states:

1. *The bacteria (germ) must be found in all cases of the disease that have been examined.*

2. *The bacteria causing the disease must be prepared, cultured, isolated, and maintained in a pure culture that has not been contaminated by other organisms.*

3. *Bacteria grown for several generations removed from the original specimen must still be able to produce the same infection (disease).*

Koch's most famous application of his postulate was his culturing, identifying, and maintaining the rod-shaped bacillus responsible for tuberculosis, a widespread disease that killed hundreds of thousands of people in Europe and Asia. Isolating this particular bacterium was difficult since it was so much smaller than any others with which he had previously worked. Koch spent most of his life traveling the world in an attempt to identify the causes of and develop eventual cures for a number of diseases. *See also* Pasteur.

Kohlrausch's Law for the Independent Migration of Ions: Physics: *Friedrich Wilhelm Georg Kohlrausch* (1840–1910), Germany.

The electrical conductivity of ions in a solution increases as the dilution of the solute increases in the solvent.

Definitions for some terms follow:

• *Electrical conduction* is the passage of electrons or ionized atoms though a medium that in itself is not affected (wires, solutions, etc.).

• *Ions* are atoms or molecules that contain positive or negative charges.

• A *solution* is a mixture composed of two parts: the *solvent* (usually a fluid that effects the dissolving) and the *solute* (usually a soluble solid, such as sugar or salt—electrolyte).

• *electrolytes* are chemicals that, when molten or dissolved in a solvent, conduct an electric current.

Kohlrausch measured the electrical resistance of electrolytes, which are the dissolved substances that conduct electricity by transferring ions in solution. This means that when some substances, such as salt or acids, dissolve in water, an electric current can pass from the negative to the positive electrodes placed in the solution. The reason the current "flows" through the solution is that the electrolyte has formed ions (charged atoms) that carry the current. When direct current (**dc**) is used, there is polarization at the electrodes that interferes with the resistance of the current, thus making measurements difficult. To overcome this obstacle, Kohlrausch used alternating current (**ac**) instead of dc. This enabled him to measure accurately the conductivity of different electrolytes in solution. To some it may seem strange that the weaker the solution, the greater the flow of ions (current), but this is what Kohlrausch's law states. His theory is correct only over a limited range of dilutions because if the dilution increases to the extent that no ions are available in the solution, no electric current will be conducted between the electrodes. (See Figure A2 under Arrhenius.) In other words, absolutely pure water will not conduct electricity.

Krebs Cycle: Chemistry: *Sir Hans Adolf Krebs* (1990–1981), England. Sir Hans Adolf Krebs shared the 1953 Nobel Prize for physiology or medicine with Fritz Albert Lipmann.

All ingested foods undergo a chemical breakdown while going through a cyclic sequence of reactions where sugars are broken down into lactic acid, which is metabolized into citric and other acids and further oxidized into carbon dioxide and water, releasing a great deal of energy.

The Krebs cycle, also known as the citric acid cycle or the tricarboxylic acid cycle, is a sequence of enzyme reactions involving the process of oxidation in the metabolism of food, resulting in release of energy, carbon dioxide, and water. (See Figure K4.) This complicated process is responsible for the source of energy in animals that take in oxygen in the processes of respiration and the metabolism of digested food. A number of scientists studied the metabolism of carbohydrates and the various forms of organic carbon molecules, but they were

Figure K4: Artist's depiction of the Krebs citric acid cycle indicating the complete oxidation of glucose, which releases energy as the cycle is repeated during the metabolizing of food.

unsure as to how these chemical reactions all came together in living organisms. Food is first broken down into smaller carbon groups that combine with a four-carbon compound called citric acid. This citric acid molecule loses its carbon atoms, which frees up four electrons that produce energy in the form of ATP (adenosine triphosphate). Next, another energy-type molecule, GTP (guanosine triphosphate), is formed in the cycle. Still later in the cycle, the original molecules are regenerated and act as catalysts to start the whole cycle over again. Hans Krebs located two organic acids that contained six-carbon atoms, as well as other acids already known to contain four-carbon atoms. From this information he devised the cycle, with the carbon dioxide molecules exiting at the other end of the cycle. The process ends with the release of energy to the organism, followed, by an uptake of more carbohydrates (sugars), enabling the whole cycle to start all over again. This important process is both the source of energy for our living cells, as well as the process that is responsible for **biosynthesis**.

Kroto's "Bucky Balls": Chemistry: *Harold William Kroto* (1939–   ), England.
Graphite subjected to laser beams produces sixty carbon atoms bunched together in the form of a twenty-sided polyhedron.

Harold Kroto collaborated with Richard Smalley (1943–   ) who used laser beams to vaporize metals that, when cooled, formed compact masses or groups of one-of-a-kind metallic atoms. Kroto suggested graphite (a form of carbon) could be used to create new structures for groups of carbon atoms. Before finishing their research, they exhausted their research funds. A few years later, several other scientists repeated the experiment and produced an amount of material adequate for analyzing and verifying its structure. The result was a ball-like molecule that resembled a field soccer ball or a geodesic dome, similar to the ones designed by architect Buckminster Fuller. (See Figure C4 under Curl.) Thus, the C_{60} molecule was given the name *buckminsterfullerene* or, more commonly, *bucky balls*. The unique 60 atoms, 20-sided polyhedron group of carbon molecules is being explored for properties that may improve the structure and functions of other materials. Since the discovery of C_{60}, other polyhedrons with more than 60 atoms have been produced experimentally. *See also* Curl; Van't Hoff.

Kuiper's Theory for the Origin of the Planets: Astronomy: *Gerard Peter Kuiper* (1905–1973), United States.
The planets evolved from condensing gas clouds distinct from the gas that formed the sun.

Gerard Kuiper discovered a new satellite for Uranus and another for Neptune. He also determined that carbon dioxide gas is present in the thin atmosphere of Mars and that Titan, the largest of Saturn's satellites, has an atmosphere composed of methane. Kuiper theorized some comets originated beyond the Oort cloud, at a distance of more than 100,000 AU. He also postulated that the space beyond Pluto contains many comets with solar orbital periods of hundreds of years. In 1992 the first objects were found at about 120 to 125 AU in the Kuiper belt, thus verifying the theory he had advanced forty years earlier. All of these ideas relate to the formation of comets, asteroids, and planets from gas clouds. Like the planets, the objects in the Kuiper belt are primitive remnants left over from the formation of our solar system. Kuiper also theorized that some of our short-term comets originate in the Kuiper belt, while other long-term comets originate in the Oort cloud.

Kuiper proposed two other theories that are not generally accepted. One was his measurement of the planet Pluto. Based on his estimates of the perturbations of Pluto on Uranus, he estimated Pluto's size to be about half the Earth's diameter. (It later proved to be only about one-fourth Earth's diameter.) Second, his estimate of the mass of Pluto was also much too large. These errors were later corrected. Kuiper's theory that the planets formed from their own gaseous clouds is no longer accepted by other astronomers. Before his death, Kuiper was involved in several of the early NASA space missions. *See also* Oort.

Kusch's Theory for the Magnetic Moment of the Electron: Physics: *Polykarp Kusch* (1911–1993), United States. Polykarp Kusch shared the 1955 Nobel Prize for physics with Willis Lamb.

The electron, like all other charged particles, will possess a magnetic moment (field) due to the motion of its electric charge.

Polykarp Kusch used the interacting principle of electromagnetism to demonstrate that the electron, which has an electric charge, is affected by a magnetic field. His measurements, which were extremely accurate, were based on the structure of different energy levels of several elements. He discovered that all particles containing an electrical charge will exhibit a turning effect in a strong magnetic field. Later, it was determined that the neutron, which has zero electrical charge, will also be affected by a strong magnetic field due to its internal structure, which involves a polarity distribution of positive and negative charges. (A neutron is basically composed of a negative electron and a positive proton, or it might be thought of as three quarks held together by gluons.) (*See also* Gell-Mann.) This polarity is why nuclear magnetic resonance and magnetic resonance imaging work. The magnetic field around the patient causes nuclei in the different tissues to resonate (oscillate), with the applied field and are thereby detected by the MRI instrument, producing an image of the tissue. Kusch's work also led to the concept of quarks and the field of quantum electrodynamics. *See also* Gell-Mann.

L

Lagrange's Mathematical Theorems: Mathematics: *Comte Joseph-Louis Lagrange* (1736–1813), France.

Lagrange's Theory of Algebraic Equations: *Cubic and quartic equations can be solved algebraically without using geometry.*

Lagrange was able to solve both cubic and quartic (fourth power) equations without the aid of geometry, but not fifth-degree (quintic) equations. Fifth-degree equations were studied for the next few decades before they were proved insoluble by algebraic means. Lagrange's work led to the theory of permutations and the concept that algebraic solutions for equations were related to group permutations (group theory). Lagrange's theory of equations provided the information that Niels Abel (1802–1829) and others used to develop group theory. *See also* Euler; Fermat.

Lagrange's Mechanical Theory of Solids and Fluids: *Problems related to mechanics can be solved by nongeometric means.*

Before Lagrange, Newtonian mechanics were used to explain the way things worked, as well as to solve problems dealing with moving bodies and forces. By applying mathematical analysis to classical mechanics, Joseph-Louis Lagrange developed an analytical method for solving mechanical problems that used equations having a different form from Newton's law ($F = ma$), by which an acceleration is proportional to the applied force. Lagrange's equations, which can be shown to be equivalent to Newton's law and can be derived from Hamilton's formulation, are, like Hamilton's formulation, very convenient for studying celestial mechanics. In fact, Lagrange himself applied his equations to the mechanical problems of the moon's librations (oscillating rotational movement), as well as those dealing with celestial mechanics. For one example, he solved the three-body problem when he demonstrated by mechanical analysis that asteroids tend to oscillate around a central point—now referred to as the *Lagrangian point. See also* Einstein; Newton.

Lagrange's Concept for the Metric System: *A base ten system will standardize all measurements and further communications among nations.*

Historically, all nations devised and used their own system for measuring the size, weight, temperature, distance, and so forth of objects. As the countries of Europe developed and commerce among them became more common, it was obvious that the jumble of different measuring systems was not only annoying but limited prosperity. At about the time of the French Revolution, a commission was established to solve this problem. Lagrange, Lavoisier, and others were determined to find a natural, constant unit on which to base the system. They selected the distance from the North Pole to the equator as a line running through Paris. This distance was divided into equal lengths of 1/10,000,000, which they called a *meter* ("measure" in Greek). A platinum metal bar of this length was preserved in France as the standard unit of length. Today, a meter is defined as the length of the path light travels in one 299,792,458th of a second and is based on the speed of electromagnetic waves (light) in a vacuum. Units for other measurements besides length were devised, using the base of ten to multiply or divide the selected unit. For instance, a unit of mass is defined as the mass accelerated 1 meter per second by a 1-kilogram force. After several years of resistance, other countries recognized the utility of the metric system, which has since been adopted by all countries, except the United States and Burma. Even so, international trade and commerce have forced the United States to use the metric system along with the archaic English system of measures. Despite several attempts to convert the United States to the metric system, the general public has refused to accept it.

Lamarck's Theories of Evolution: Biology: *Jean Baptiste Pierre Antoine de Monet, Chevalier de Lamarck* (1744–1829), France.

Theory 1: *New or changed organs of an animal are the result of changes in its environmental factors.*

Lamarck proposed that the first requirements for modifying the form or structure of an organism were changes in environmental circumstances. This was the basis for his view that there was a natural tendency for greater complexity and that a change in the environment was responsible for the changes in functions and forms of the organs of animals. In other words, the occurrence of new organs in an animal's body is the result of some new need that made itself "felt" by the animal.

Theory 2: *Those parts of an animal not used will either not develop or will degenerate over time, and those parts of an animal that are used will continue to develop. The "need" responses by animals over many generations are acquired changes in functions and structures that will be inherited in future generations.*

Lamarck believed these changes resulted from environmental factors, which led to changes in the animals' behavior as well as structure, and, in time, this acquired behavior also became habitual. One of his examples was the behavior of antelopes fleeing from predators. As they ran faster, their leg muscles developed, thus passing this escaping behavior and fleetness to offspring. In other words, as the environment changes, so does an animal's behavior, as well as its organs' functions, and structure. The behavior becomes an active agent for the species' evolutionary development, and when this behavior becomes habitual behavior, it determines the extent and nature of the animal's structure. This is usually referred to as the *inheritance of acquired characteristics*, which includes an interpretation of the pre-Darwinian concept of natural selection. But this natural selection, according to Lamarck, resulted in either habitual use or disuse of a particular body part, which carries over from generation to generation. One of the classical examples of the inheritance of acquired characteristics is the theory of how the giraffe acquired its long neck. The giraffe had to stretch higher and higher to obtain tree leaves for food after the lower leaves were consumed. Thus, over time, the giraffes "acquired" longer and longer necks, a characteristic passed on to the next generations. Today, however, it is usually considered that possibly due to genetic mutations, some giraffes with slightly longer necks were able to secure more food, be healthier, live longer, and thus reproduce more giraffes with the altered genes.

Lamarck's ideas were not well received by other scientists, including Darwin. Even so, at one time Darwin accepted some aspects of the concept of acquired characteristics caused by environmental changes and incorporated this into his theory of natural selection. *See also* Buffon; Cuvier; Darwin; Lysenko.

Lamb's Theory for the Quantum States of the Hydrogen Atom: Physics: *Willis Eugene Lamb, Jr.* (1913–   ), United States. Willis Lamb shared the 1955 Nobel Prize for physics with Polykarp Kusch.

Each of the known states of hydrogen is actually two states having the same energy in the absence of a magnetic field. However, the two states exhibit slightly different energies in the presence of a weak magnetic field.

Willis Lamb's theory on the quantum states of the hydrogen spectrum required a slight revision of Paul Dirac's electron theory, which stated, according to quantum mechanics, that the hydrogen spectrum should exhibit two different but equal states of energy. Lamb's research demonstrated that the spectrum of the hydrogen atom was split into two parts, but there was a small shift of the energy level of the hydrogen spectrum from that which Dirac predicted. This discrepancy for the predicted quantum electrodynamics of the hydrogen atom is now known as the Lamb shift. Willis Lamb first demonstrated it by splitting the spectrum for hydrogen into two distinct parts, each with slightly different energy states. His research revealed how electrons act within the influence of electromagnetic fields and is considered important for the electronics and computer industries in the development of new products. *See also* Dirac; Kusch; Lorentz; Zeeman.

Landsteiner's Theories of Blood Groups: Biology, *Karl Landsteiner* (1868–1943), United States. Karl Landsteiner was awarded the 1930 Nobel Prize for medicine or physiology.

Individuals within a species exhibit different proteins in their blood serum (plasma), just as different species also exhibit different blood groups.

Since 1628 when William Harvey explained the circulatory system in animals, scientists and physicians were aware that blood of one animal species was incompatible with the blood of another animal species. When incompatible blood types are mixed during a transfusion, the blood clots, blocking blood vessels, which leads to death. At the beginning of the twentieth century, Karl Landsteiner demonstrated that only blood serum from certain types of patients could be mixed with blood from others whose blood had some similar characteristics. He found that the plasma (liquid portion of the blood) from some donors would form clots in transfusions for person A but not for person B. Thus, person A could provide blood that was safe for another person with A type blood. He also found that some other types of blood were incompatible with person B, that some types of blood would clot for both A- and B-type people, and blood from still other people would not clot either A- or B-type people. This resulted in the classification of blood into the four groups: A, B, O, and AB. Only people with O blood can donate to most people from the other groups, but only in an emergency; a definite match for the other types is required. The understanding of blood grouping has increased and has been used to determine parenthood long before DNA testing. Due to Landsteiner's efforts, blood transfusions are safe.

Langevin's Concept for Use of Ultrasound: Physics: *Paul Langevin* (1872–1946), France.

High-frequency electrical currents can cause piezoelectric crystals to produce short ultrasound wavelengths mechanically.

Paul Langevin built on the work completed by Marie and Pierre Curie as related to piezoelectric crystals. Pierre Curie realized that when a mechanical force was applied to a piezoelectric crystal, an electric current is generated between the two sides of the crystal. He used this process, in reverse, to measure the amount of radiation (strength) of the radioactive element on which he and Marie Curie were working. Langevin theorized that if a variable (alterable) electric current could be sent across one side of a crystal to the other side, the crystal would vibrate rapidly, thus producing sound waves shorter than those that can be heard by the human ear (e.g., ultrasound). He also was aware that sound waves of high frequency travel better under water than do light waves. Therefore, objects under water should be detectable at greater ranges using sound waves rather than light waves. His theory was later applied to the development of a system called echolocation, which during World War II became known by the acronym sonar (SOund Navigation And Ranging). Sonar used ultrahigh sound waves generated by the piezoelectric crystals to detect enemy submarines. Since then, sonar is an invaluable tool in the field of oceanography. Not only can it detect objects, such as schools of fish and sunken ships, but it can measure the contour of the ocean's floor.

Langmuir's Theories of Chemical Bonding and Adsorption Surface Chemistry: Chemistry (Physical): *Irving Langmuir* (1881–1957), United States. Irving Langmuir was awarded the 1932 Nobel Prize for chemistry.

Electrons Surround the Nuclei of Atoms in Successive Layers: *The electrons surrounding a nucleus progress in number from two, located in the closest layer (orbit), followed in additional layers containing eight, eight, eighteen, eighteen and thirty-two electrons successively.*

It was understood for some years that since the atom is neutral, it must have as many negative electrons as positive protons. It was also determined that the electrons in the outer layer (orbit or shell) are held with the weakest force to the positive nucleus. In other words, their "energy level" is less than the electrons located in the inner orbits. Therefore, these outer electrons must be responsible for different atoms combining in specific ratios to form molecules, or for similar atoms to combine to form simple diatomic molecules such as O_2 or Cl_2. (See Figure S2 under Sidgwick.) Models for the structure of atoms were proposed by many chemists. Niels Bohr's quantum concept of a "solar system" atom and Gilbert Lewis' idea that electrons are shared as "bonds" to form molecules were based on Langmuir's "layered" structure for an atom's electrons. (*See also* Bohr, Lewis.)

Langmuir's Adsorption Theory: *The adsorption of a single layer of atoms on a surface during a catalytic chemical reaction is controlled by the gas pressure if the system is maintained at constant temperature.*

While working as a research scientist at General Electric in the United States, Irving Langmuir developed light bulbs containing inert gases (such as argon) that did not oxidize the filaments. He also lengthened the life of the bulbs by using tungsten filaments, which further reduced oxidation. As a result, he theorized that electrons from the metal filament interact with monolayers (single layers) of atoms or molecules adsorbed to the surface. His theory is related to the **adsorption** of single layers of an element (usually a gas such as hydrogen) on the surface of another element. (Note that adsorption is not the same as absorption.) For example, the chemical reactions that take place inside an automobile's catalytic converter occur when the hydrogen compounds formed by the burning of the hydrocarbons in gasoline are adsorbed on the platinum metallic beads inside the converter. This means the exhaust gases resulting from combustion are spread on the surface of the platinum, where they are converted to less toxic gases. These hydrocarbon atoms and molecules are not absorbed as a sponge absorbs water but rather obey the laws of surface chemistry (adsorption).

Laplace's Theories and Nebular Hypothesis: Physics: *Marquis Pierre Simon de Laplace* (1749–1827), France.

Laplace's Theory of Determinism: *What affects the past causes the future.*

In 1687 Sir Isaac Newton published his laws of motion, which were deterministic and mechanical in that they explained the movements of objects on earth as well as celestial bodies. Aware of perturbations and irregularities in the motions of planets and other heavenly bodies, Newton also believed the universe would end if these irregularities were not somehow corrected. The Marquis de Laplace believed these irregularities did not indicate some massive destructive force since they were not cumulative. In other words, forces generated by these perturbations did not combine as one big force, ending in disaster. Rather, they were of a periodic nature and occurred at regular time intervals. He believed the future is determined by past events. This theory became known as *Laplace's demon.* In essence, it states that what has affected things in the past will also cause the future. Laplace thought that if all data were known and analyzed, this information could then be stated in a single formula and there would be no uncertainty. Thus, the future would be caused by the past. This is an old concept that could not be supported once new and more difficult equations were explored. The concept of the past causing the future can be expressed by linear equations or straight line conceptualizing; new nonlinear equations and methods of reasoning provided more branches or alternative causes resulting in a multitude of possible effects (e.g., chaos theory).

Laplace's Nebular Hypothesis: *The solar system was formed by the condensing of a rotating mass of gas.*

The concept of swirling bodies in the universe originated with the ancient Greeks, but it was Laplace who tried to establish his nebular hypothesis by using Newtonian principles and mathematics. His concept stated that a ball of gas formed the sun, and from this the planets were thrown off, which in turn threw off their moons. The nebular hypothesis has been updated as a rotating cloud of gas that cooled and contracted as it threw off rings of matter, which further contracted to form planets and moons. The great mass of leftover condensing gas formed the sun and other bodies in the solar system.

Laplace's Theory of Probability: *Mathematics can be used to analyze the probability (chance) that a specific set of events will occur within the context of a given set of events.*

Probability is the likelihood of a particular cause resulting in the occurrence of a particular event (effect). Today, most scientists apply probability theory to the study of many fields of science, such as thermodynamics. The probability scale ranges from 0.0 which indicates an event is highly uncertain or unlikely to occur (or will not occur) to the probability of 1.0 for high certainty or it is likely that an event will or did occur. The terms *possible* and *impossible*, are not measurable and thus have no meaning when considering probability as related to an event.

Larmor's Theories of Matter: Physics: *Sir Joseph Larmor* (1857–1942), Ireland.

Larmor's Theory of Electron Precession: *An electron orbiting within the atom will wobble when subjected to a magnetic field.*

Sir Joseph Larmor's concept of matter was a synthesis referred to as the electron theory of matter. This was a rather radical description for the structure of atoms, the nature of matter, and the electrodynamics of moving bodies that relate to kinetic energy. The *Larmor precession* describes the behavior of an orbiting electron when moving through a magnetic field. The axis of the electron actually changes its angle (precession) while moving in the field and thus appears to wobble. Larmor then calculated the rate of energy that radiated from an accelerating electron, an important concept for future work in particle physics.

Larmor's Concept of the Aether: *Space is filled with an aether partially composed of charged particles.*

Sir Joseph Larmor was one of the last physicists who attempted to justify the existence of an **aether** (or ether) in space, believing that it was necessary as a medium to provide a means for waves (e.g., light, radio, and other electromagnetic waves) to travel from one point in space to another. He also believed aether must contain some electrically charged particles in order for matter and light to traverse space. These concepts are no longer considered valid, but they did contribute to other ideas.

Laurent's Theories for Chemical "Equivalents" and "Types": Chemistry: *Auguste Laurent* (1807–1853), France.

Laurent's Chemical Equivalents: *A definite distinction between atoms and molecules exists based on their equivalent weights.*

Auguste Laurent classified molecules composed of two atoms, such as oxygen, hydrogen, and chlorine, as *homogeneous compounds*, which could become *heterogeneous compounds* when they "decomposed." Laurent's work established the relationships of the elements' atomic weights to their characteristics. This concept of atomic weight as being related to the element's chemical properties provided a key for Mendeleev's arrangement of the elements in his **Periodic Table of Chemical Elements**.

Laurent's Type Theory: *Organic molecules with similar structures can be assigned to a classification of "types."*

From the days of alchemy, scientists placed different chemicals and minerals into separate classifications. These groupings were usually based on the "types" of color, physical consistency, or reactions with each other, and not on their basic elementary structures. Laurent's work with chemicals provided the distinctions needed to develop types of compounds, which were arranged according to his concepts of their structures. For example, he considered water to be one type of compound and alcohol another. Although both are composed of hydrogen and oxygen atoms, each is different in structure. His theory of types was helpful in classifying organic compounds, but was not adequate for describing their different structures. Nevertheless, it was a step in the right direction toward understanding the structure and nature of organic compounds.

Lavoisier's Theories of Combustion, Respiration, and Conservation of Mass: Chemistry: *Antoine Laurent Lavoisier* (1743–1794), France.

Lavoisier's Theory of Combustion: *The gas emitted when cinnabar (mercury oxide ore) is heated is the same as the gas in air that combines with substances during combustion (burning).*

Antoine Lavoisier was a meticulous scientist who at first believed in the phlogiston theory (*see* Stahl). However, his experiments soon indicated there was another explanation. When he burned sulfur and phosphorus in open air, they gained weight, while other substances lost weight. He concluded that something in the air, not the "phlogiston" in the substance being burned, was involved in combustion. In 1774 Joseph Priestley determined that cinnabar, when heated, emits a gas that made a candle burn more brightly and a mouse placed in this air more lively. Taking this further, Lavoisier believed the gas produced by cinnabar was the same as one of the gases in air that combined with substances as they burned. He later named this gas *oxygine*—Greek for "acid producer." Lavoisier then proceeded with several experiments, one of which was to repeat Priestley's experiment of burning a candle in an upturned jar placed in a pail of water. As the candle burned out, the water level rose in the jar. He also burned a candle in the gas produced by heating cinnabar. The results were the same; thus, his conclusion was that the gas in air was the same as the gas emitted by burning the cinnabar. After the candle burned out, the gas remaining in the jar was inert and comprised a large portion of the volume inside the jar. Since mice could not live in this leftover gas, Lavoisier concluded it would not support life as did his "oxygine." He called this inert gas *azote*, meaning "no life" in Greek. This gas is now called nitrogen.

Lavoisier's Theory of Respiration: *Animals convert pure air to fixed air.*

Antoine Lavoisier was the first to test experimentally Joseph Priestley's concept that normal air lost its phlogiston during combustion and respiration. First, he placed a bird in an enclosed bell jar. When it died, he tried to burn a candle in the air left in the jar. It would not burn, and according to the science of the day, the air was then pure, or as they said, it was "dephlogisticated" air. In other words, the bird had used up the phlogiston. Conducting other experiments with burning candles, Lavoisier hypothesized that as more air was available to the burning candle, more of it would be converted to "fixed air." In 1756 Joseph Black first prepared and named fixed air, or carbon dioxide. Carbon dioxide gas dissolves in water, as Priestley later discovered when he "invented" carbonated soda water. But the proportion of the gas consumed by the candle to the proportion of the gas left in the jar indicated there was some other gas that made up the remaining air in the jar. At the time it was not known that oxygen composes only about one-fifth of a given volume of air, while nitrogen gas makes up most of the remaining four-fifths. Nevertheless, Lavoisier was the first to measure the amount of oxygen consumed and carbon dioxide emitted through animal respiration. He was also the first to measure the heat produced by respiration and determined it could be compared to the amount of oxygen required to burn charcoal. He changed the concept of phlogiston to the concept of caloric, referred to as weightless fire that changed solid and liquid substances into gases. Science was no longer saddled with the misconception of the phlogiston theory.

Lavoisier's Law of Conservation of Mass: *The mass of the products of a chemical reaction are equal to the mass of the individual reactants.*

When some metals are "roasted" at a high temperature in the presence of air, their surface turns into a powder called an **oxide**. In Lavoisier's time this coating of metallic oxide was referred to as *calx*. The old phlogiston theory explained the loss of weight in the air to the loss of phlogiston. Since metals gained weight during smelting, it was believed the calx combined with the charcoal and that the charcoal contained phlogiston. We now know this is not a true concept. In order to make careful measurements of the burned substances, Lavoisier invented a delicate balancing device that could measure a tiny fraction of a gram. Using his balance, he conceived his law of conservation of mass which is still valid today. Lavoisier is known as the Father of Modern Chemistry due to his use of step-by-step experimental procedures, making careful measurements, and keeping accurate records. *See also* Black; Priestley; Scheele; Stahl.

Lawrence's Theory for the Acceleration of Charged Particles: Physics: *Ernest Orlando Lawrence* (1901–1958), United States. Ernest Lawrence was awarded the 1939 Nobel Prize for physics.

When charged particles are accelerated in a vertical magnetic field, they move in accelerated spiral paths.

Since the discovery of natural radioactivity, it was known that alpha particles (helium nuclei with a positive charge) were ejected from the nuclei of radioactive substances and could induce other nuclear reactions. A method to increase the acceleration of charged particles by using electromagnetic forces was needed to penetrate and "smash" atomic nuclei to separate their component particles. To achieve the greatly increased speeds for charged particle "projectiles," John Douglas Cockcroft and Ernest Thomas Sinton Walton developed a low-energy linear (straight line) accelerator in 1929. Their "atom smasher" used a voltage multiplier to build up a high-voltage capacity capable of accelerating alpha particles beyond the speed from which they are emitted naturally from radioactive elements. These early linear accelerators still did not provide the energies required to "smash" the nuclei of atoms to the extent where they produced smaller particles.

Ernest Lawrence proposed a unique design to solve the problem of early linear accelerators' not having adequate energies to interact with heavy nuclei to produce smaller nuclear particles. He constructed two D-shaped metal halves of a hollow circular device with a small gap between the two semicircular Ds. He named his first model the cyclotron, which was only 4 inches in diameter. By shooting charged particles into the semicircular "Dees" and then applying high-frequency electric fields to the particles, they reached tremendous speeds as they continued to circle in increasing spirals inside the Dees, getting an electromagnetic "push" each time they passed the gap (something like repeated pushes making a swing go faster and higher). As the spiraling particles approached the inside rim of the Dees, they achieved maximum acceleration. At that point they were directed to targeted atoms to determine what bits of subatomic particles and energies would result. Lawrence's early device was the precursor of the current circular accelerators that are several miles in diameter and develop very strong electromagnetic fields aided by cryogenic supercondutivity. (See Figure V2 under Van der Meer.) For the current giant atom smasher, the particles (alpha, beta, and other subatomic particles) are accelerated by a powerful linear accelerator, which then feeds the particles into the giant cyclotron, thus combining a linear device to a circular one. These giant particle accelerators generate very high electron voltages (eV), which are a measure of the energy of the particle. Physicists continue to use these powerful devices to produce many different types of subnuclear particles, which may help explain the basic nature of matter, energy, and life. Lawrence's original cyclotron produced particles with only about 10,000 to 15,000 electronvolts (eV). Advanced particle accelerators are being developed that could reach 300 billion electronvolts (GeV) per nucleon (proton or neutron). These will be much larger and many times more powerful than the cyclotrons used in the latter part of the twentieth century. The greater the power of these new accelerators to increase the speeds at which particles slam into each other, the more information will be obtained to answer questions as to the nature of the universe.

Le Bel's Theory of Isomers: Chemistry: *Joseph Achille Le Bel* (1847–1930), France.

The asymmetric quadrivalent carbon atom can form molecules composed of the same atoms but with different structures.

Earlier chemists worked on various theories to explain the structure of atoms and how they bonded (joined) with other atoms to form molecules. It was determined that carbon had a tetrahedron structure. Joseph Le Bel devised his concept of the asymmetric carbon atom at about the same time as did another chemist, Jacobus Van't Hoff. Both of their concepts were based on the tetrahedron carbon atom with its four valence electrons arranged something like a three-legged tripod, with the fourth bond pointing up. (See Figure V3 under Van't Hoff.) This structure enabled the carbon atom to combine with other carbon atoms or atoms of other elements, in pairs or individually, when forming carbon compounds. This would produce isomers of compound molecules that had the same chemical formula but different physical characteristics, such as boiling points, color, and reactivity. This concept ultimately resulted in the development of organic (carbon) chemistry and explained the myriad existing organic molecules. While there are some inorganic compounds that contain carbon (e.g., CO_2 and cyanide [CN]), all organic compounds contain carbon (e.g., all living matter and products of living matter). *See also* Kekule; Van't Hoff.

Le Chatelier's Principle: Chemistry: *Henri-Louis Le Chatelier* (1850–1936), France.

Any change made in a system of equilibrium results in a shift of the equilibrium in the direction that minimizes the change.

Henri-Louis le Chatelier's principle in essence describes what happens within a system that is in equilibrium (symmetry, parity, stability, or balance) when the factors of temperature, pressure, or volume change. If there is an increase in the pressure, the system decreases its volume to bring itself back into equilibrium. This principle includes the law of mass action and the theory of chemical thermodynamics. Le Chatelier's concept provides scientists with both a mathematical interpretation of the system's dynamics and practical physical means to control what occurs within a system where the changes in pressure and temperature cause the system to readjust its equilibrium. Le Chatelier's principle is invaluable for understanding how to control the mass production of industrial chemicals (e.g., ammonia and hydrocarbon products, such as gasoline). *See also* Boyle; Haber.

Lederberg's Hypothesis for Genetic Engineering: Biology (Genetics): *Joshua Lederberg* (1925–   ), United States. Joshua Lederberg shared the 1958 Nobel Prize for physiology or medicine with Edward Lawrie Tatum and George Wells Beadle.

If viruses can inject themselves into the genes of bacteria cells to cause infections, then it should be possible to inject genes into animal cells.

Joshua Lederberg's first experiments demonstrated that bacteria contain genes in their nuclei and at times reproduce by sexual mating, as well as by conjugation. Previously it was believed bacteria reproduced only by "fission," where the mother cells split into two new daughter cells without any interchange of genetic material. This is known as asexual reproduction. Lederberg demonstrated that when crossing different strains of bacteria, a mutant strain would develop randomly, which caused a mixing of genetic material between the two strains. Since the crossed bacteria could develop their own colony of bacteria, sexual mating must be occurring. Occasionally he found that some enzymes were destroyed by what are called *bacteriophages*, which are viruses that enter and infect bacteria, thus causing genetic changes. (See Figure D2 under Delbruck.) Lederberg and Max Delbruck both proved that new strains of viruses result when two different strains are combined in a form of sexual reproduction just as occurs for bacteria. Their work led to the new science of genetic engineering, where genes can be recombined by inserting them into bacteria and other cells. *See also* Delbruck.

Lederman's Two Neutrino Hypothesis: Physics: *Leon Max Lederman* (1922–   ), United States. Leon Lederman shared the 1988 Nobel Prize for physics with Melvin Schwartz and Jack Steinberger.

The two different types of neutrinos are generated by different physical decay processes.

When beta particles (electrons) were ejected during radioactivity, the end particles exhibited less energy than expected. To explain this seeming negation of the law of conservation of energy, the neutrino (Italian for "little one") was postulated to account for the missing energy. Even though neutrinos may be considered "nonparticles," they do exist, as do "antineutrons." Both are important to maintain the symmetry and mathematics related to particle physics. Leon Lederman knew there are two different decay processes controlled by the weak interaction between subatomic particles that produce neutrinos. One decay process occurs when pions decay into muons plus μ neutrinos ($V\mu$). This results in one type of neutrino that Lederman hypothesized is different from the other type. The second decay process is a form of beta decay, where a neutron is converted into a proton by ejecting an electron plus an electron neutrino (Ve). (See Figure F2 under Fermi.) In other words, Lederman was trying to find out if the muon-related neutrino was the same particle as the electron-related neutrino. His experiment with his two colleagues, Melvin Schwartz (1932–   ) and John Steinberger, resulted in the identification of the existence of the muon neutrino ($V\mu$), and the ability to distinguish it as a different subatomic particle from the electron plus neutrino combination emitted when a neutral neutron is converted to a positive proton. *See also* Fermi; Pauli; Steinberger.

Leeuwenhoek's Theory of Microscopic Life: Biology: *Anton van Leeuwenhoek* (1632–1723), Holland.

Multitudes of living "animalcules" exist in water and other fluids.

Anton van Leeuwenhoek is sometimes, and incorrectly, credited with inventing the microscope. Leeuwenhoek is best known for developing improved microscopes in the seventeenth and early eighteenth centuries. During his lifetime, he constructed and sold over 500 models of his microscopes. He based his theory of "little animalcules" on a lifetime of observing the microscopic world around him. He was the first to observe and describe protozoa in water, bacteria in his own feces, red blood cells, nematodes in soil, rotifers, ciliates such as *vorticella*, bacteria with different shapes, *spirogyra* (alga), and human sperm. Some of his descriptions were very accurate and led to further investigations. He examined the plaque and sputum from his mouth and the mouths of others and then described the strong actions of these multitudes of "animalcules" found in spittle as "fish swimming in water." This was the first viewing and written description of bacteria. As early as 1684, Leeuwenhoek calculated that red blood cells were 25,000 times smaller than specks of sand. He also made extensive observations of microscopic fossils, crystals, minerals, as well as tissues from a variety of animals and plants. His descriptions proved valuable for future biologists.

Leibniz's Theory for "The Calculus": Mathematics: *Gottfried Wilhelm Leibniz* (1646–1716), Germany.
Finite areas and volumes for curves can be calculated by use of differential and integral mathematical calculations.

The dilemma of how to determine the area on or within a curved surface had been explored by dozens of mathematicians, philosophers, and scientists since ancient Greece. Leibniz realized a workable notation method (the use of symbols to represent quantities) was required to solve problems related to the areas and volumes of curves. His solution to the notation problem was: $\int y\,dy = y^2/2$, which is still used today. Leibniz published results of his calculus (the use of symbolic notations to perform mathematical calculations for the investigations and resolutions of multiple variables) in 1684, which turned out be a significant event in mathematics. A major dispute as to the discoverer of calculus resulted when Sir Isaac Newton, who had developed his calculus much earlier in 1665, delayed publication until 1687. Therefore, Leibniz is credited with the discovery and development of calculus. *See also* Newton.

Lemaître's Theory for the Origin of the Universe: Astronomy: *Abbé Georges Edouard Lemaître* (1894–1966), Belgium.
Einstein's theory of relativity requires an expanding, not static, universe.

Georges Lemaître was one of the first astronomers to relate relativity to cosmology. He based his nonstatic universe on the supposition that if matter is expanding everywhere within the universe, then there must have been a moment in the past when this expansion began. Although he disputed Einstein's belief in a static universe, Lamaître based his own thesis on Einstein's theory of special relativity of space-time. Lamaître assumed that if we could revert far enough in time, we would see the entire universe as a very compact, compressed point of matter and energy. He also considered radioactivity as the force that caused the original explosion, an idea no longer considered valid. Unfortunately, Lemaître did not completely calculate the mathematics for his theory of an expanding universe. From the late 1920s to the early 1940s, his expansion theory was unpopular with other astronomers, who still considered a static universe the preferred model. The most important aspect of Lemaître's theory was not just the expansion concept (which was well known), but the idea that something started the whole process—that is, there was a physical origin to the universe. In the late 1940s, his theory of an expanding universe was revived and revised by George Gamow, who named it the "big bang." Today it is considered one of the most likely explanations for the origin of the universe. *See also* Einstein; Gamow.

Lenard's Theory for Electron Emission: Physics: *Philipp Eduard Anton Lenard* (1862–1947), Germany.

During the photoelectric effect, the speed of electrons emitted is a function of the wavelength of the light (electromagnetic energy) involved.

Philipp Lenard based his research on the photoelectric effect first detected by Heinrich Rudolph Hertz. Lenard observed that when ultraviolet light "struck" the surfaces of certain kinds of metals, electrons were "kicked" out and could be detected. He designed experiments to determine the cause and found that the speed at which electrons were ejected from certain types of metal during exposure to the light was a function of the wavelength of light used. Further, he found the shorter the wavelength of light used, the greater the speed of the emitted electrons. At the same time, the intensity of the light had no effect on the electrons' speed, but the brighter the light, the greater quantity (number) of electrons emitted. Some years later, Einstein explained the photoelectric phenomenon by relating it to Planck's quantum theory. *See also* Einstein; Hertz; Planck.

Lewis' Theory of Covalent Bonds: Chemistry: *Gilbert Newton Lewis* (1875–1946), United States.

When atoms combine to form molecules, they share a pair of electrons, thus forming covalent bonds.

Ionic bonds, also called polar bonds, were first introduced in the late 1800s and were thought to be one-to-one sharing of electrons from one atom to another. **Ionic bonding** occurs when two oppositely charged atoms (ions) are attracted to each other, thus forming a more stable molecule (See Figure S2 under Sidgwick). However, this concept was not valid for the formation of all molecular compounds. Lewis first conceived the structure of atoms as cubes with the possibility of one electron located at each of the eight vertices (corners) of the cube. More complex atoms were structured with smaller cubes located inside larger cubes. Lewis soon rejected this cube structure, but he still believed that at least eight electrons were required for each neutral atom. This knowledge led to the Lewis-Langmuir octet theory (*see also* Langmuir), which provided information about the atomic structure of rare gases in which all eight vertices of the cube are occupied. Therefore, since the atoms of these gases could neither gain, lose, nor share electrons, they are inert because their eight vertices are all occupied by electrons. This also explained why atoms that did not have a complete octet of outer electrons were available to combine with other atoms, by either ionic bonding or sharing covalent bonds. All atoms have a natural tendency to attain the same octet formation, and thus become more stable (inert). For example, sodium has only one vertex occupied by an electron, while chlorine has seven vertices occupied. Both sodium and chlorine act to establish the stable octet structure; thus sodium gives up its one electron, and chlorine accepts it, satisfying both. ($^{+}Na + ^{-}Cl = NaCl$ is an example of ionic bonding, while $^{--}O + ^{--}O = O_2$ depicts a pair of atoms sharing electrons, i.e., covalent bonding). The most common type of covalent bonding is single bonding where just two electrons are shared—one from each partner. There may be double bonds as well, involving four shared electrons, or even triple bonds, with six shared electrons. Lewis and Langmuir's octet theory resulted in a better understanding of the laws of thermodynamics and the periodic arrangement of elements. *See also* Langmuir; Sidgwick; Thomson.

Liebig's Theory of Isomers and Organic Compound Radicals: Chemistry: *Justus von Liebig* (1803–1873), Germany.

Inorganic or organic compounds (molecules) with the same formula can have different structures and thus exhibit different characteristics.

Justus Liebig was working with Joseph Louis Gay-Lussac when he discovered silver fulminate. At the same time, his friends Friedrich Wohler and Jons Jacob Berzelius had prepared silver cyanate. To the surprise of all, both of these compounds had the same formula, but they behaved very differently. Berzelius named this phenomenon *isomerism*. Today, this concept is used to develop different chemicals that have the same basic formulas but exhibit many different and useful properties. Isomerism is one of the reasons so many different types of organic (carbon-based) synthetic drugs, man-made fibers (nylon), plastics, and numerous other chemical products can be manufactured. Exploring this phenomenon further, Leibig and Wohler used different forms of the benzoyl radical to formulate their theory of compound radicals. This is an example of a family of similar chemicals that can have additional atoms added to the main radical. Chemists can use the basic benzoyl radical to form different compounds by adding, for example, chlorine (Cl), bromine (Br), hydrogen (H), or other atoms to the basic structure. *See also* Kekule, Lewis.

Lindemann's Theory of Pi: Mathematic: *Carl Louis Ferdinand von Lindemann* (1875–1939), Germany.

It is impossible to "square the circle" to arrive at pi (π) by using a straight edge and compass and thus accurately determine the area of a given circle.

Carl Lindemann was aware of Archimedes' method of using geometry (multiple polygons) to determine the value of pi by "squaring the circle" to determine the ratio of the circumference of a circle to its diameter. Historically, mathematicians wished to use this ratio (pi) to determine the area of a circle ($A = \pi r^2$). Archimedes arrived at a ratio of 3.142, which is very close to the current accepted value of 3.14159. . . . Lindemann used algebraic methods to prove that an accurate ratio for pi could not be determined by geometric methods using a straight edge and compass because pi is a transcendental number. This means that pi is not a root for any polynomial equation with rational coefficients, such as $2/3x^3 - 5/7x^2 - 21x + 17 = 0$. *See also* Archimedes.

Linnaeus' Theories for the Classification of Plants and Animals: Biology: *Carolus Linnaeus* (1707–1778), Sweden.

Plants and animals of different species can be classified according to similarities within a species as well as differences between species.

Until Linnaeus' time, plants and animals were classified from the top down, beginning with large classes and working down to smaller groups. Carolus Linnaeus devised the system of taxonomy, still in use today, that is based on the concept of species. A major factor in determining what animal or plant belongs in a particular species is whether reproduction is limited within that species. We know today that the vast majority of DNA is the same for all mammals. However, just a small difference of DNA between species prevents cross-fertilization. For instance, chimpanzees and humans share over 98 percent of the same type of DNA. Linnaeus further classified according to similarities within a species as well as differences between species. His concept required a new terminology, for which he used Latin names. Starting with humans, he used *Homo* as the genera for "man" and *sapiens* for the species of "wise man." From here Linnaeus combined similar *genera* into *family*, from family up to larger groups called *orders*, then divided further into *phyla*, and phyla into the two *kingdoms* of plants and animals. The term *taxon* is meant to encompass all the special traits shared at any of the seven major categories of his taxonomy system. Since then, subdivisions have been added, and with the knowledge of evolution and cellular and molecular distinctions, even finer similarities and differences are used. More recently, biologists have divided the two kingdoms of plants and animals into five major groups. This new classification system consists of three major plant kingdoms and two major animal kingdoms. These distinctions are based on differences in molecular DNA. *See also* Aristotle; Cuvier; Darwin; Lamarck.

Lister's Hypothesis of Antisepsis: Biology: *Baron Joseph Lister* (1827–1912), Britain.

Carbolic acid, used during surgical procedures, can prevent and control subsequent infections.

Joseph Lister based his hypothesis on and credited the work of Francesco Redi and Louis Pasteur. Redi determined that rotting meat developed maggots only when exposed to flies, and Pasteur proved that such putrefaction was caused by microorganisms. Using this knowledge of airborne microorganisms, Lister sprayed the air of operating rooms with carbolic acid (a derivative of benzene called phenol). To his dismay, deaths from infections following surgery were still over 50 percent. He then soaked a cloth in carbolic acid and used it to bind an open wound. The wound healed without any infection. Following this lead, Lister then soaked all his surgical instruments in carbolic acid, had surgeons rinse their hands in dilute acid, and maintained a clean operating room. These procedures reduced surgical mortality in his operating rooms to about 5 percent by 1877. At first, Lister's hypothesis for controlling infection was not well received, but by the late 1800s antiseptic procedures were standard in all hospitals and doctors' offices. Lister made another contribution when he replaced silk thread, which was used for sutures and could not be easily sterilized, with catgut disinfected by carbolic. *See also* Pasteur; Redi.

Lockyer's Solar Atmosphere Theories: Astronomy: *Sir Joseph Norman Lockyer* (1836–1920), Britain.

A unique spectral line from the sun's light is produced by a new and unknown element.

Sir Joseph Lockyer spent much of his life attempting to determine the composition of the sun's atmosphere and its effects on earth. He and Pierre-Jules-Cesar Janssen (1824–1907) devised a method of observing the sun during daylight hours. Up to this time, the only way to view the sun was during a solar eclipse or through glass that was smoked by holding it over a burning candle. In 1868 Janssen, using a solar spectroscope, was the first to view a peculiar spectral line in sunlight that did not match other known spectral lines. Although he was the first to see this spectrum, it was later identified and named by Lockyer, who hypothesized that since it existed only in the sun, it should be called helium, from the Greek word *helios* meaning "sun." Helium was considered a hypothetical element until it was detected on Earth twenty-five years later by Sir William Ramsay. Lockyer made other contributions. Using the Doppler effect (*see also* Doppler), he determined the "wind speed" of solar flares. In addition, he determined the temperature of the surface of the sun and that sunspots have a lower temperature than the sun's surface. Lockyer also was convinced the solar atmosphere affected the earth's weather because its orbit is just at the edge of the sun's outer corona. Additionally, he believed the size and number of sunspots affect the amount of rainfall on Earth. At the time, he was unable to examine these phenomena, which today are partially accepted, but not as Lockyer hypothesized. *See also* Doppler; Ramsey.

Lorentz's Physical Theories of Matter: Physics: *Hendrik Antoon Lorentz* (1853–1928), Netherlands. Hendrik Lorentz shared the 1902 Nobel Prize for physics with Pieter Zeeman.

Lorentz's Electron Theory: *Atoms and molecules are very small, hard bodies that carry either a negative or positive charge.*

James Clerk Maxwell determined that light waves were the result of the vibrations of charged particles (atoms); as these particles oscillated, electromagnetic waves were produced. Hendrik Lorentz's electron theory expanded this theory and was based on the assumption that (1) there is a wave-carrying medium in space known as **aether** and (2) matter (solid, liquids, and gases) was a separate entity from the wave/aether; therefore, (3) only electrons could interact between them. He found that atoms with a positive charge "oscillate" in one direction within a magnetic field, and those with a negative charge "oscillate" in the opposite direction. His mathematical theory was developed before there was any proof that electrons existed, but it indicated that light waves were the result of oscillating electrically charged atoms.

Lorentz Force: *There is a force applied to moving electrically charged particles when they are in the presence of an electromagnetic field.*

Hendrik Lorentz identified charged particles produced in a cathode ray tube as negative electrons. His theory also explained the Zeeman effect (*see also*

Zeeman), which asserted that the spectral lines for sodium atoms split into several closely spaced lines when exposed to an electromagnetic field. This phenomenon was later explained by quantum theory.

Lorentz's Theory for the Contraction of Moving Bodies: *Light from moving bodies traveling through the aether caused these bodies to appear to contract in size in the direction of their motion.*

At about the same time Lorentz proposed his theory for the contraction of moving bodies, another physicist, George Francis Fitzgerald, independently arrived at the same concept. Therefore, the mathematics for this phenomenon is known as the *Lorentz-Fitzgerald contraction.* In essence, the theory states that bodies moving through an electromagnetic field contract somewhat in the directions of their motion in proportion to their velocity. This explains why light appears to move at the same speed in all directions at the same time from its source. Einstein used this concept in developing his theory of special relativity. *See also* Einstein; Zeeman.

Lorentz Invariant: *Natural laws must be invariant to a change in the coordinates (space and time) of any system.*

The Lorentz invariant is sometimes referred to as the Lorentz transformations theory. The theory is based on the mathematics Hendrik Lorentz developed to explain how moving bodies seem to contract. The consequence of these theories is that both space (three dimensions) and time must be equally considered when developing any type of equation that explains the relative motion of matter. The theories of contraction and transformations describe the coordinates that need to be considered for the contraction in the length and the increase in mass of moving bodies at relativistic speeds. They provided the foundation for Einstein's theory of special relativity. Einstein relied on the mathematics of Hendrik Lorentz and also recognized the contributions made by other scientists, that aided him in developing his theories of relativity. *See also* Einstein; Fitzgerald; Maxwell; Michelson; Zeeman.

Lorenz's Theory for Complex/Chaotic Systems: Mathematics: *Edward Norton Lorenz* (1917–   ), United States.

The sensitivity of a dynamic system depends on small initial conditions.

Edward Lorenz, a meteorologist, applied mathematics to weather forecasting and climate changes. Using a computer, he analyzed the initial conditions of a weather system with temperature as the only single variable. His original data were carried out to six decimal places but rounded off to three decimal points, a practice used by most scientists and which Lorenz assumed was such a small difference that it would not affect the outcome. This assumption proved to be false since he obtained a different result each time he ran the computer data. At this point, he realized that small initial differences can have a cumulative effect over long periods of time and thus affect events differently. This phenomenon, which describes the sensitivity of a system as dependent on small differences in initial conditions, became known as the "butterfly effect." The butterfly effect

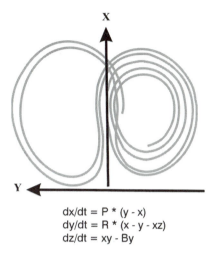

$$dx/dt = P * (y - x)$$
$$dy/dt = R * (x - y - xz)$$
$$dz/dt = xy - By$$

Figure L1: Lorenz's strange attractor depicted as a curve on a single plane, which is actually a curve in three-dimensional space. The line forming the curve is a single unbroken line that never follows the same path and, in three dimensions, never intersects itself.

hinges on a number of weather factors, including the temperature and humidity of the air and how the air is flowing. It aptly describes the chaotic systems in which small perturbations result in very different outcomes. The name *butterfly effect* came from a fable about a butterfly whose flapping of its wings created an airflow in China that added to the cumulative airflows around the world, thus causing hurricanes in Florida and snowstorms in Wisconsin. The principles of complex systems, chaos theory, and nonlinear mathematics are used to interpret dynamic systems as diverse as economic cycles, the stock market, population changes, the dynamics of three-dimensional flow of fluid in pipelines, the dynamics of prehistory archaelogy, and weather predictions. Edward Lorenz originated the *Lorenz attractor*, a mathematical expression using differential equations to describe how a system settles down, based on the three variables of space orientation (x, y, and z). Diagrams of this concept are looping curves on a two-dimensional surface, but the curves are in three dimensions where the single line never crosses itself at any given moment of time. (See Figure L1.) A point on the line represents a variable of the system expressed as a point in three-dimensional space. *See also* Penrose; Wolfram.

Lowell's Theory of Life on Mars: Astronomy: *Percival Lowell* (1855–1916), United States.

Canals and oases seen on Mars indicate that it was once inhabited.

Percival Lowell became interested in the 1877 report by Giovanni Schiaparelli in which he stated he observed *canali* (Italian for "channels") on the surface of

Mars. Lowell constructed a 24-inch reflector telescope atop a 7200-foot mountain in Arizona in order to make use of the area's clear sky. He reported he also saw Schiaparelli's Martian canals and claimed they were built by intelligent beings. He theorized these canals were dug to transport water from melting ice at its poles to the dry central regions of the planet. It is now known that these lines and patches were aberrations in Lowell's lens/mirror system, the "shimmering" of Earth's atmosphere, or the results of a vivid imagination. *See also* Schiaparelli.

Lyell's Theory of Uniformitarianism: Geology: *Sir Charles Lyell* (1797–1875), Britain.

Currently observed geological changes and processes are adequate to explain geological history.

This basic concept was first expressed by James Hutton (1726–1797) and John Playfair (1748–1819), but both neglected to explain fully or examine their concept in detail. Sir Charles Lyell explicitly stated that the same scientific laws and geologic processes apply in the past, present, and future and therefore are responsible for geologic changes. This led to his famous saying, "The present is the key to the past." He explicitly rejected the theory of Abraham Werner (1750–1817), who believed that a huge deluge of water (the "Flood") was responsible for earth's current topography. Lyell believed the action of the wind, rain, the sea, earthquakes, and volcanoes, rather than some great catastrophe, explained geological history. He rejected the concept of catastrophism, which was first believed to conform to biblical history. Lyell based much of his "uniformitarianism" concept on his study and classification of the strata of ancient marine beds. He observed that the layers of sediment closest to the surface contained shells as well as the remains of animal species still living in modern times. Conversely, the deeper, older strata contained more fossils of extinct species. He divided the rocks containing fossils into three groups, or epochs and named them after ancient geological periods—Eocene, Miocene, and Pliocene—terminology still in use today. Charles Darwin, who developed the theory of organic evolution, relied on parts of both Lamarck's and Lyell's earlier works. *See also* Darwin; Eldredge-Gould.

Lysenko's Theory of the Inheritance of Acquired Characteristics: Biology: *Trofim Denisovich Lysenko* (1898–1976), Russia.

Characteristics acquired by parents during their lifetime can be inherited by their offspring.

Trofim Lysenko followed in the footsteps of I. V. Michurin (1855–1935), a Russian who advocated the acceptance of Lamarckism. Michurianism and later Lysenkoism were the biological and genetic party line (ideology) of the Soviet Union under Stalin. Lysenko, a minor agriculturalist, promoted a new theory based on an old farmers' concept, called vernalization, as a means to improve the germination of grain. He claimed that by treating grain with cold water, the flowering of the grain would improve, and it would sprout sooner in the spring. Thus, it would take less time to raise a crop and would increase the production of grain to feed the masses. It did not work. Neither this concept nor any of his other ideas used standard controlled experiments, peer review, or other accepted process of scientific research. His mistake, as well as those of some others in Russia, was in rejecting the science related to Mendelian genetics and insisting that this "cold treatment" would not need to be repeated each year because this "acquired" characteristic could be passed on from one generation of grain to the seeds of the next. Lysenko's ideas seemed to fit the Marxist philosophy, and he soon discredited the president of the Lenin All-Union Academy of Agricultural Sciences, Nikolai Vavilov (1887–1943), who was exiled to Siberia. Lysenko then became head of this all-powerful institution and had complete support from Joseph Stalin and the Communist party. Biologists who disagreed with his new "science" and supported the science of modern genetics and Darwin's theory of natural selection were designated as reactionary, decadent enemies of the Soviet people. *See also* Darwin; Lamarck.

M

Mach's Number: Physics: *Ernst Mach* (1838–1916), Austria.

There is a ratio that expresses the velocity of an object in a fluid to the velocity of sound in that fluid.

Ernst Mach believed if information about nature cannot be sensed, it was useless. In addition, he thought discoveries can be made by intuition and accident as well as by using mathematics and scientific methods. These ideas and his concept of motion, which states that the inertia of a body arises from interactions with all of the mass within the universe, influenced the field of quantum mechanics and Einstein's formulation of his theory of relativity. His experimental work with vision and hearing led him to use high-speed photography to detect the shock wave produced in air by a high-speed bullet. This so-called barrier is also created when an airplane approaches the identical speed of sound traveling in cold air, which is about 750 (more or less) miles per hour. This "sound barrier" was first believed to be similar to a wall that must be overcome. However, there is not now, nor has there ever been, a wall to overcome. For example, it is well known that artillery shells, bullets, and thunder all travel faster than the speed of sound and produce shock waves. Air molecules are compressed and produce a shock wave. "Sonic booms" are heard when two shock waves are so close together that they are heard as a single "boom" by an observer on the ground. The wave front for an airplane is a V-shaped area of compressed air analogous to the V-shaped bow wave produced by a speeding boat in water. The exact speed of the object traveling through a fluid (e.g., air) required to break this "barrier" will depend on the temperature of the air as well as the air's density and moisture content. The denser the medium through which sound travels, the faster it travels; at room temperature, sound travels 1126 feet per second in air, 4820 feet per second in water, and 16,800 feet per second in iron. The greater the density of the medium, the faster the sound proceeds through that medium. Whenever the speed of an object exceeds the speed of the sound in a particular medium through which the object is traveling, a shock wave is produced. At 0°C the speed of sound traveling though dry air is 331.4 meters per second at sea level. The Mach number varies for airplanes flying at different altitudes. At higher altitudes, the air is colder, thinner, and dryer than at sea level; thus the sound barrier is reached at different speeds at different altitudes.

The Mach number Ernst Mach devised is the ratio of the velocity of an object, such as an airplane, to the speed of sound in air through which it travels. Mach numbers below 1 are referred to as subsonic flows of fluid; numbers greater than 1 are supersonic flows of fluid. An airplane flying at a speed lower than a Mach number of 1 will be traveling in subsonic flight. Once the airplane exceeds Mach 1, it has reached supersonic velocity, and the so-called sound barrier will be broken. As an example, if an airplane travels 1500 miles per hour and the speed of sound in the air through which the airplane is flying is 750 miles per hour, the ratio is 1500/750 = Mach 2. The airplane overtakes the wave fronts in the front as well as in the rear of the airplane, producing overlapping wave-fronts.

Mansfield's Theory of Magnetic Resonance: Physics: *Peter Mansfield* (1933–   ), England.

The nuclei of atoms have a magnetic moment that can be detected and be used to form an image of living tissue.

Peter Mansfield based his work on that of Felix Bloch (1905–1983) and Edward Purcell (1912–1997), who discovered that the nuclei of some atoms have a **magnetic moment**. The protons and neutrons of nuclei have spins that are not paired. This creates an overall spin on the particles, generating a magnetic dipole along the spin axis, which is a fundamental constant of physics referred to as the *nuclear magnetic moment* (μ). This can be compared to a wobbling spinning top whose axis circumscribes a precessional path. This wobble is created by an unbalanced spin of the nucleus' axis, which results in a resonance. Absorption of electromagnetic radiation by atomic nuclei in response to strong magnetic fields causes the nuclei to radiate detectable signals. The process became known as nuclear magnetic resonance (NMR) since it involved the nuclei of the atoms of the substance exposed to the electromagnetic radiation. This process was first used as a spectroscopic method for analyzing the atomic and nuclear structure and properties of matter. Later, the radiation produced by the resonating nuclei was detected and recorded by computers to form a two-dimensional spectroscopic image of living tissue. Godfrey Hounsfield (1919–   ) used NMR to develop a computer-aided tomography (CAT) scan that could form an image of human tissue and was less intrusive than x rays. Mansfield improved the process by altering the manner in which magnets affected the spin of the nuclei so that a three-dimensional image could be produced. This became known as magnetic resonance imaging (MRI), which was designed to produce three-dimensional images of cross-sections of any part of the human body. Still there is confusion between NMR/MRI and nuclear energy. They are *not* the same. The NMR and MRI procedures use powerful magnetic influences to "excite" the atomic nuclei of various types of tissue in bodies, causing them to resonate (oscillate) differently and thus emit distinct frequencies (signals). These signals are recorded by computerized instruments, thus enabling physicians to distinguish between healthy and diseased tissue. The word *nuclear* refers to the oscillating of the nuclei of atoms of these various elements that make up living cells. It does not mean "radioactive," as with nuclear radiation produced by either nuclear fission or fusion. *See also* Rabi; Ramsey.

Matthias' Theory of Superconductivity: Physics: *Bernd Teo Matthias* (1918–1980), United States.

Superconductivity of a material depends on the number of outer electrons on the atoms of that material.

Bernd Matthias' theory related to attempts to cause the electrons located on the outer shell (orbit) of atoms, or even free electrons, to flow as an electric current without resistance at temperatures much higher than absolute zero. Superconductivity occurs when the electrical resistance of a solid disappears as it is cooled to the "transition temperature," which for most metals and alloys takes place below 20 K or about −253°C. In 1911 Heike Kamerlingh-Onnes, when liquefying helium, also discovered that mercury, the only metal existing in a liquid state at room temperature, became superconductive at about 4 K (−269°C). At normal room temperatures, electrons in metal conductors collide as they flow through wire, causing a resistance to the flow of electrons (current). Thus, the wire heats up (as in the filament of an incandescent light bulb). What caused supercooled metals to lose their resistance to electricity was not known until the early 1950s, when Matthias began experimenting with various metals and alloys. By observing the behavior of several samples, he determined the number of electrons in the outer orbit of atoms was one factor and the crystalline structure of the material was another. Matthias made a compound of niobium and germanium (Nb_3Ge), which became superconductive at the unexpected high temperature of 23 K. Since then, physicists have attempted to determine the transition temperatures of numerous alloys and compounds, thereby causing superconductivity at much higher temperatures, since cooling to near absolute zero Kelvin is extremely difficult and expensive. Because liquid helium is used to cool metals to near absolute zero, its use is limited to small applications (e.g., cooling magnetic resonance imaging supermagnets). Experiments were conducted using the less expensive element nitrogen, which becomes liquid at −196°C instead of helium. Several ideas have been proposed to develop high-temperature superconductivity. One is to work with films of newly developed materials that can pass on free surface electrons with little or no resistance to the flow of current. A possible application of the film technique is a high-temperature superconductor for computer switches and components that can run exceedingly fast without producing much heat. Reports of the discovery of so-called high-temperature superconductivity are very promising, but not yet confirmed. Low-cost electrical transmission lines, without the conversion of electricity to heat by resistance within the wires, is one goal for the future. Another is to use supercooled magnets to levitate magnetically driven trains. *See also* Kamerlingh-Onnes.

Maunder's Theory for Sunspots' Effects on Weather: Astronomy: *Edward Walter Maunder* (1851–1928), England.

When there is a minimum of observed sunspots, there is a corresponding long, cold period on earth.

While examining ancient reports of dark spots on the sun by astronomers, Edward Maunder realized there were no reports of similar activity on the surface of the sun for the period from the mid-1600s to the early 1700s. It was also determined, from other reports, that this was a period during which earth experienced lower temperatures than usual. Using this material, Maunder developed a statistical analysis demonstrating that when there is a dearth of sunspots, a prolonged cold spell on earth occurs. Maunder's "minimum" is still used, along with more sophisticated techniques, to aid in determining long-term climate changes on earth.

Maupertuis' Principle of Least Action: Physics: *Pierre-Louis Moreau de Maupertuis* (1698–1759), France.

Nature chooses the most economical path for moving bodies.

Pierre-Louis Maupertuis' principle of least action was a forerunner of all later theories and laws dealing with conservation, such as the conservation of energy and the concept of entropy. Maupertuis conceived this principle to explain the path that light rays travel, but it seemed applicable to all types of moving bodies. It basically states that nature will take the easiest, shortest route to move things from one point to another. Maupertuis' principle was widely applied in the fields of mechanics and optics. A similar principle was proposed by Leonhard Euler in his form of calculus dealing with mathematical variations. Pierre de Fermat too used Maupertuis' principle to describe how light is refracted according to Snell's law, which states that light takes the least time possible when traveling from a medium of one density to a medium of a different density. *See also* Euler; Fermat.

Maxwell's Theories: Physics: *James Clerk Maxwell* (1831–1879), England.

Maxwell's Kinetic Theory of Gases: *All gases are composed of large numbers of particles (atoms or molecules), all of which are in constant random motion.*

James Clerk Maxwell determined that the stability of Saturn's rings could be explained only if the rings consisted of a multitude of very small solid particles, a theory that has been accepted ever since. From this reasoning, he formulated his kinetic theory of gases, which states that heat is the result of molecular movement. His theory is also accepted today with modifications that incorporate relativity and quantum mechanics. Maxwell realized he could not predict the movement of a single tiny molecule. But on a statistical average basis, the laws of thermodynamics, first proposed by Nicolas Carnot, could be explained as molecules at high temperatures (rapid movement) having a high probability of movement in the direction of other molecules with less movement (lower temperatures). Maxwell and Ludwig Boltzmann arrived at a theory that relates the "flowing" motion of gas molecules to heat being in equilibrium. They also formulated a mathematical expression that indicated what fraction molecules had to a specific velocity. This expression is known as the Maxwell-Boltzmann distribution law. *See also* Boltzmann; Carnot.

Maxwell's Demon Paradox: *The demon is a tiny hypothetical entity that can overcome the second law of thermodynamics, thus making possible a perpetual motion machine (i.e., its energy will never escape into useless heat).*

James Clerk Maxwell's work with the kinetic theory of gases led him to speculate on what became known as his "demon paradox." He created his mythical demon from the Maxwell-Boltzmann statistical distribution law that describes the properties of large numbers of particles that, under certain statistical conditions, were inconsistent with the second law of thermodynamics. That is, heat does not flow from a colder body to a hotter body without work (energy) being expended to make it do so (e.g., an air-conditioner). Maxwell proposed an intellectual hypothesis where an **adiabatic** wall separates each of two sealed, equal-sized enclosed compartments (see Figure M1). The trapdoor separating the left and right sides of container number 1 is open and the molecules can freely go from side to side. Thus equilibrium is established, and the second law of thermodynamics is upheld. The trapdoor for container 2 is controlled by the "demon," which allows only fast (hot) molecules to enter the right side of the container and only slow (cold) molecules to go through the trapdoor to the left side. (These are closed systems, no gas can enter or leave the boxes.) The "presence" of the tiny "demon" that sits by the trapdoor in container 2 opens the door for molecules of specific high speeds to go in only one direction and the slow-speed molecules to only go in the other. The object of the experiment is for the demon to collect in container 2 all the faster/hotter (on the average) molecules inside the room on the right side and the slower/colder ones (on the average) in the room on the left side. The paradox is there are more fast-moving, hot gas molecules in one room than the other without the input of any outside energy. Thus, the concept of conservation of energy has been satisfied since no kinetic energy was lost, but the concepts of entropy and thermal equilibrium have been violated. However, in the demon's experiment, we now have imbalance of kinetic (heat) energy because the demon allowed only the high-energy (hot) molecules to go into one side of the number 2 container. This appears to be a violation of the second law of thermodynamics.

The paradox is resolved by recognizing that the demon is doing work by opening the doors and that he requires information about the speed of the molecules in order to open the door at the right time. The input of information is

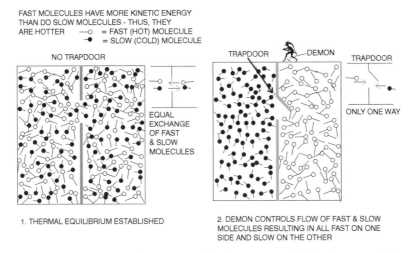

FAST MOLECULES HAVE MORE KINETIC ENERGY
THAN DO SLOW MOLECULES - THUS, THEY
ARE HOTTER   ─○ = FAST (HOT) MOLECULE
              ─● = SLOW (COLD) MOLECULE

NO TRAPDOOR

TRAPDOOR ─ DEMON   TRAPDOOR

EQUAL
EXCHANGE
OF FAST
& SLOW
MOLECULES

ONLY ONE WAY

1. THERMAL EQUILIBRIUM ESTABLISHED

2. DEMON CONTROLS FLOW OF FAST & SLOW
MOLECULES RESULTING IN ALL FAST ON ONE
SIDE AND SLOW ON THE OTHER

Figure M1: (1) The trapdoor is open to molecular movement by kinetic energy for both fast (hot) and slow (cold) molecules. Therefore, there is no change in temperature for the right or left side. Energy is conserved and equilibrium maintained. (2) Maxwell's demon using the trapdoor allows only fast (hot) molecules to enter the right-side compartment and only slow (cold) molecules to enter the left-side compartment. Energy is conserved, but there is no equilibrium between the two sides, and the second law of thermodynamics is violated. No work (energy) was exerted to achieve this situation, resulting in perpetual motion, and heat in the right side can now be used to do work.

responsible for the decrease in entropy of the system. Thus, the second law of thermodynamics is not really violated. Of course, perpetual motion is unobtainable, but this was not understood until the laws of thermodynamics became known. Maxwell's treatment of entropy on a statistical basis was an important step in realizing that to obtain knowledge of the physical world, one must interact with it.

Maxwell's Theory for Electromagnetism: *Both magnetism and electricity produce energy waves, which radiate in fields with differing wavelengths.*

Familiar with Michael Faraday's theories of electricity and magnetic lines of force and expanding on the mathematics developed by Faraday, Maxwell combined several equations that resulted in the establishment of direct relationships in the fields produced by magnetism and electricity and how together they affect nature. Once the equations for magnetic and electric fields were combined, he calculated the speed of their waves. Maxwell concluded that electromagnetic radiation has the same speed as light—about 186,000 miles per second. At first, Maxwell accepted the ancient concept of an aether in space. (It was thought that light and other electromagnetic waves could not travel in a vacuum; therefore the concept of "ethereal matter" in space was invented but never verified.) Maxwell believed electromagnetic radiation waves were actually carried by this aether and that magnetism caused disruptions to it. Later, in 1887 Albert Mi-

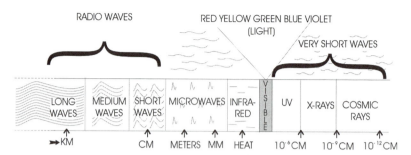

Figure M2: James Clerk Maxwell hypothesized the existence of electromagnetic radiation with longer and shorter wavelengths than visible light, which is located near the middle of the scale. This idea developed into the electromagnetic radiation spectrum for frequencies from very long radio waves to extremely short x-rays and cosmic radiation.

chelson demonstrated that an aether in space was unnecessary for the propagation of light. Maxwell's equations were still valid, even after the aether concept was abandoned. Maxwell concluded there were both shorter wavelengths and longer wavelengths of electromagnetic radiation next to visible light wavelengths on the electromagnetic spectrum (later named *ultraviolet* and *infrared*). He further concluded that visible light was only a small portion of a "spectrum" of possible electromagnetic wavelengths. Maxwell then speculated and predicted that electromagnetic radiation is composed of many different (both longer and shorter) wavelengths of different frequencies. This concept developed into the electromagnetic spectrum, which ranges from the very short wavelengths of cosmic and gamma radiation to the very long wavelengths of radio and electrical currents. (See Figure M2.) The theory of electromagnetic radiation is one of the most profound and important discoveries of our physical world. It has aided our understanding of our world and resulted in many technological developments, including radio, television, x rays, lighting, computers, and electronic equipment. Maxwell combined his four differential equations ("Maxwell's equations") describing the propagation of electromagnetic waves (radiation) to form the wave equation. This was the first time the constant for the velocity of light waves (c) was used; it later became an important constant in Einstein's theories of relativity and his famous equation, $E = mc^2$. See *also* Einstein; Faraday; Michelson.

McClintock's Theory of Cytogenetics: Biology: *Barbara McClintock* (1902–1992), United States. Barbara McClintock received the 1983 Nobel Prize for physiology or medicine.

Genes can move around within cells and modify chromosomes, thus restructuring the genetic qualities of a species.

Barbara McClintock's moving **genes** were referred to as "jumping genes" because they could transpose (change position or the order of genes) within chromosomes. Much of her work was done by using stains to aid in identifying the ten chromosomes found in maize (corn). These chromosomes were large enough to be viewed by a microscope, enabling her to identify and distinguish the different chromosomes from each other. By planting seeds from corn growing one year to the next, she tracked mutant genes over several generations. She found that in addition to single genes that were responsible for color (pigmentation) in the corn, she found groups of genes linked together that caused other mutations. She referred to these as "controlling elements," which dictated the rate for the on-off switching action of other genes. McClintock discovered that these controlling genes could move within a single chromosome, or they could "jump" to other chromosomes and control their genes. McClintock's work was ignored to some extent because it was very advanced for the 1940s, a period during which Mendel's principles of heredity were generally not accepted. Today, McClintock's concept of groups of genes working together and controlling other genes has advanced our understanding of evolution. *See also* Mendel.

McMillan's Concept of "Phase Stability": Physics: *Edwin Mattison McMillan* (1907–1991), United States. Edwin McMillan shared the 1951 Nobel Prize for chemistry with Glenn T. Seaborg.

Using variable frequencies of electrical impulses in a cyclotron, it is possible to compensate for the increase in mass of accelerating subatomic particles.

Edwin McMillan was aware of the problem that Ernest Lawrence's cyclotron experienced with the ever-increasing "speeds" obtained during acceleration of subatomic particles. According to Einstein's theory of relativity, the greater a particle's acceleration, the greater its increase in mass. This is why particles with mass can never reach the speed of light, since the particle would become more massive than the entire universe—if there was enough energy to accelerate it to such a velocity. Lawrence's cyclotron used a fixed frequency of electrical stimulation to accelerate the beta, alpha, or other subatomic particles. As these particles spun, faster and faster in the cyclotron, they increased in mass and thus became out of phase with the frequency of the electrical impulse. McMillan's solution was to use a variable frequency that could change as the mass of the particles changed. This led to a new device, the synchrocyclotron (also known as the synchrotron). It was so named because it could synchronize the frequencies required to maintain the speed of the particles within the cyclotron and continue to accelerate them to even greater energies, thus enabling physicists to explore additional collision particles and radiation. McMillan and Seaborg shared the 1951 Nobel Prize for the isolation of the elements neptunium and plutonium. *See also* Einstein; Lawrence; Seaborg.

Meitner's Theory of Nuclear Fission: Physics: *Lise Meitner* (1878–1968), Austria. Lise Meitner received the 1966 Enrico Fermi Prize from the U.S. Atomic Energy Commission for her work on nuclear physics.

As the nuclei of uranium absorb neutrons, they become unstable and "fission" into two lighter nuclei; in addition, they produce extra neutrons and radiation.

Lise Meitner, an Austrian physicist who emigrated to Sweden in the 1930s, collaborated with Otto Hahn on nuclear physics research. In 1938 Hahn and Fritz Strassman (1902–1980) became perplexed when they "shot" neutrons into nuclei of uranium. To their amazement, the uranium nuclei became lighter rather than heavier when absorbing neutrons. Hahn contacted Meitner and requested assistance in solving this problem. In 1939 Meitner and her nephew, Otto Frisch, who had escaped Germany to live in the Netherlands, solved the problem. As the nuclei of uranium atoms absorb neutrons, they become unstable due to an excess of neutrons. Thus, they split into two smaller nuclei while at the same time ejecting two or three other neutrons as well as energy in the form of radiation. Frisch named this process *fission*, after the process that cells undergo when dividing. It was never understood why Hahn, Strassman, Meitner, or Frisch were never recognized for the importance of their discovery of a chain reaction of fissionable unstable uranium nuclei for the production of energy. It was not long after that this concept was used by the United States to develop the atomic bomb. *See also* Hahn.

Mendeleev's Theory for the Periodicity of the Elements: Chemistry: *Dmitri Ivanovich Mendeleev* (also spelled Mendeleyev) (1834–1907), Russia.

There is a definite repeating pattern of the properties of elements based on the elements' atomic weights and valences.

Dmitri Mendeleev was not the first to recognize some sort of pattern related to similar characteristics of elements based on either their atomic weights or atomic numbers. Several chemists recognized that elements seemed to be grouped in triads, or in repeating groups of seven, or with some evidence of "octave" periodicity of their properties. (See Figure N1 under Newlands.) In 1870, one year after Mendeleev published his **Periodic Table of the Chemical Elements**, Julius Meyer conceived a table similar to Mendeleev's. Meyer plotted a chart relating physical and chemical properties of elements with their atomic weights. His work, however, was overshadowed by Mendeleev's publication in 1869. Mendeleev classified the elements by their atomic weights as well as valences, even though some of the valence numbers for the elements conflicted with the arrangement of the atomic weights. (See Figure M3.) He realized after a row of seven there must be another column to complete that segment, so that a new row, based on continuing periodicity of atomic weights, could be recognized in the organized chart (the octet or rule of eight). Mendeleev exhibited great insight by "skipping places" in his periodic table for elements not yet discovered. He called the yet-to-be-discovered elements *eka* (meaning "first" in Sanskrit) elements, which he predicted would fit the blank spaces he provided in his table by predicting atomic weights and properties. It was later discovered that the few inconsistencies in Mendeleev's periodic table were due to the use of atomic weights instead of atomic numbers (the number of protons in the nucleus). Once this was corrected, the current periodic table proved to be not only one of the most useful but also one of the most elegant organization charts

PERIODIC TABLE OF THE ELEMENTS

GROUPS → PERIODS ↓	1 IA	2 IIA	3 IIIB	4 IVB	5 VB	6 VIB	7 VIIB	8	9 VIII	10	11 IB	12 IIB	13 IIIA	14 IVA	15 VA	16 VIA	17 VIIA	18 VIIIA
1	1 H 1.0079																	2 He 4.00260
2	3 Li 6.941	4 Be 9.01218											5 B 10.81	6 C 12.011	7 N 14.0067	8 O 15.9994	9 F 18.9984	10 Ne 20.179
3	11 Na 22.9898	12 Mg 24.305											13 Al 26.9815	14 Si 28.0855	15 P 30.9738	16 S 32.06	17 Cl 35.453	18 Ar 39.948
4	19 K 39.0983	20 Ca 40.08	21 Sc 44.9559	22 Ti 47.88	23 V 50.9415	24 Cr 51.996	25 Mn 54.9380	26 Fe 55.847	27 Co 58.9332	28 Ni 58.69	29 Cu 63.546	30 Zn 65.39	31 Ga 69.72	32 Ge 72.59	33 As 74.9216	34 Se 78.96	35 Br 79.904	36 Kr 83.80
5	37 Rb 85.4678	38 Sr 87.62	39 Y 88.9059	40 Zr 91.224	41 Nb 92.9064	42 Mo 95.94	43 Tc (98)	44 Ru 101.07	45 Rh 102.906	46 Pd 106.42	47 Ag 107.868	48 Cd 112.41	49 In 114.82	50 Sn 118.71	51 Sb 121.75	52 Te 127.60	53 I 126.905	54 Xe 131.29
6	55 Cs 132.905	56 Ba 137.33	57 La★ 138.906	72 Hf 178.49	73 Ta 180.948	74 W 183.85	75 Re 186.207	76 Os 190.2	77 Ir 192.22	78 Pt 195.08	79 Au 196.967	80 Hg 200.59	81 Tl 204.383	82 Pb 207.2	83 Bi 208.980	84 Po (209)	85 At (210)	86 Rn (222)
7	87 Fr (223)	88 Ra 226.025	89 Ac▲ 227.028	104 Und (261)	105 Unp (262)	106 Unh (263)	107 Uns (264)	108 Uno (265)	109 Une (266)	110 Uun (267)	111 Uuu (272)	112 Uub	113 Uut	114 Uuq	115 Uup	116 Uuh	117 Uus	118 Uuo

TRANSITION ELEMENTS

6 ★ Lanthanide Series (RARE EARTH)

58 Ce 140.12	59 Pr 140.908	60 Nd 144.24	61 Pm (145)	62 Sm 150.36	63 Eu 151.96	64 Gd 157.25	65 Tb 158.925	66 Dy 162.50	67 Ho 164.930	68 Er 167.26	69 Tm 168.934	70 Yb 173.04	71 Lu 174.967

7 ▲ Actinide Series (RARE EARTH)

90 Th 232.038	91 Pa 231.036	92 U 238.029	93 Np 237.048	94 Pu (244)	95 Am (243)	96 Cm (247)	97 Bk (247)	98 Cf (251)	99 Es (252)	100 Fm (257)	101 Md (258)	102 No (259)	103 Lr (260)

© 1996 ROBERT E. KREBS

Figure M3: A modern version of Mendeleev's Periodic Table of the Chemical Elements that includes his "missing" elements and the most recently discovered "heavy" elements.

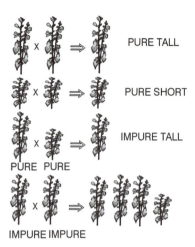

PURE TALL

PURE SHORT

IMPURE TALL

PURE   PURE

IMPURE IMPURE

Figure M4: Statistical representation of Mendel's crossbreeding of tall and short pea plants.

ever conceived by humans. *See also* Cannizzaro; Dobereiner; Frankland; Meyer; Newlands.

Mendel's Law of Inheritance: Biology: *Gregor Johann Mendel* (1822–1884), Austria.

Characteristics of offspring are determined by two factors, one from each parent.

Gregor Mendel, a meticulous experimenter, began by keeping records of as many as seven different characteristics of parent pea plants in succeeding generations. He was interested in the ratio of specific characteristics that passed from parent plants to offspring plants. (See Figure M4.) He calculated ratios for the inheritance of stem length, the position of the flower on the stem, the color of the unripe pods, the smoothness and roughness of the pea pods, color of seeds, forms of seeds, and cotyledon (seed coat) color. From his observations and calculations, he based his law of inheritance on three theories:

1. As the female parent's egg and male parent's sperm sex cells mature, the formerly paired inheritance factors divide, resulting in just one specific factor for each characteristic from each parent. These single factors are then combined into a new pair during fertilization and are responsible for the inherited characteristics of the offspring. This is now known as the principle of *segregation*.

2. Characteristics are inherited individually. This means that one factor or characteristic can be inherited along with another factor and may be either dominant or recessive (e.g., tall stems with wrinkled pea pods). This is known as the principle of *independent assortment*.

3. Each characteristic is the result of the connection of at least two genes, one from each parent. One of these two factors is always dominant over the other. This is now known as the *law of dominance*.

Mendel's law indicated there was not a blend of inherited characteristics, but rather "fractional" inheritance, which strengthened Darwin's concept of natural selection. *See also* Darwin; De Vries.

Mesmer's Theory of Animal Magnetism: Medicine: *Franz Anton Mesmer* (1734–1815), Austria.

The behavior of all things—all living organisms, as well as the heavens, earth, moon, and sun—is affected by a "universal fluid" that can be received, propagated, and communicated with each other through motion.

Franz Mesmer based his theory on the three laws of motion as explained by Sir Isaac Newton. He believed that since the moon and sun cause tidal movements, their motions will also cause any earthly object, including humans, to affect each other through some unexplained "universal fluid." As a physician, he applied this theory of fluids and motion to treating patients with magnets, with the idea that the magnets might influence the "fluids" as they do metal. He soon found that the magnets were not needed if the patient was open to his suggestions of how to be "cured." His ideas were not well accepted in France in the late 1700s and resulted in a report on his methods by several scientists who investigated Mesmer's claims. This report stated that magnetism had no medical effect, that Mesmer's "suggestions" seemed to produce nothing but odd behavior in his patients, and a cure with magnetism without imagination did not exist. Nevertheless, "mesmerism" later became recognized as hypnotism, which was separated from the original discredited concept of animal magnetism. Although Mesmer's theories are no longer considered valid, his legacy remains when someone claims to be "mesmerized" or hypnotized. Hypnotism may affect a person's behavior, but it has never been proven to cure a disease caused by a bacterium or virus. His concept of using magnets to cure all types of human ailments has been modernized and is considered by some as a form of alternative medicine, whose efficacy is yet to be proven.

Meyer's Theory for the Periodicity of the Elements: Chemistry: *Julius Lothar Meyer* (1830–1885), Germany.

There are step-wise changes in the valences of elements as related to their atomic volumes and weights.

Familiar with the work of Stanislao Cannizzaro who related Avogadro's number to the atomic weights of elements, Julius Meyer measured the volumes and atomic weights of elements and plotted the results on a graph. In 1864 Meyer recognized that his graphs indicated definite peaks and valleys, which related to the physical characteristics of different elements. Several examples of these peaks of plotted data were exhibited by the alkali metals, such as hydrogen, lithium, sodium, potassium, rubidium, and cesium, and one element not known at that time, francium. He also related the valences of elements that appeared at similar points on the graph to their chemical characteristics. This graph, referred to as *Lothar Meyer's curves*, was, in essence, the basis for the modern **Periodic Table of the Chemical Elements**. (See Figure M4 under Mendeleev.) Unfortunately, Julius Meyer did not publish his work until 1870, one year after Dmitri Mendeleev published his periodic table. *See also* Avogadro; Cannizzaro; Mendeleev; Newlands.

Michelson's Theory for the "Ether": Physics: *Albert Abraham Michelson* (1852–1931), United States. Albert Michelson was awarded the 1907 Nobel Prize in physics.

If there is an aether, then the speed of light from space traveling directly toward the earth should be less as the earth moves toward the light source. Also, the speed of light traveling at right angles to the earth's motion should be greater than the speed of light coming toward the earth.

The ancient concept of an **aether** or (ether) was used to explain the existence of some kind of matter in outer space because a pure vacuum was thought to be impossible. Since the ancient scientists believed that nature abhors a vacuum, there must be some type of "matter" in space rather than a vacuum. More recently, the ether was considered as something beyond earth's atmosphere that could carry electromagnetic waves (light, radio waves, etc.), similar to how air molecules carry sound waves. Albert Michelson invented an instrument called the interferometer, designed to split a beam of light by using a half-silvered mirror, which allows half of the light from a source to be transmitted and the other half to be reflected. Each split beam then proceeded to separate mirrors arranged on arms of the apparatus, so that the split beams were again combined at the point at which they would interfere with each other to produce a characteristic pattern of fringes. (See Figure M5.) The type of patterns formed depended on the time it took for each of the two beams to complete the trip from the source, through the split mirror, and return. The apparatus could be adjusted so that the light could approach at 90 degrees, which then should produce a different fringe pattern. Since it was believed that the ether had no motion, as Earth moved through the ether (as sound moves through air molecules), the light coming directly toward Earth would be slower than the light coming toward Earth at a 90-degree right angle. After many repeated experiments using his interferometer, Michelson found no difference in the speed of light with his instrument despite the direction from which the light's speed was measured. This was the death knell for the aether concept. *See also* Einstein.

Miller's Theory for the Origin of Life: Chemistry: *Stanley Lloyd Miller* (1930–   ), United States.

Under conditions of the primitive earth, the correct mixture of chemicals along with the input of energy could spontaneously form amino acids, the building blocks of life.

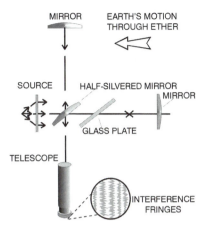

Figure M5: A diagram of Albert Michelson's interferometer designed to determine if an ether existed in space. If so, it might be detected by comparing a split beam of light as the earth moved through this hypothetical medium.

As a graduate student of Harold Urey, Stanley Miller conducted experiments at the University of Chicago to demonstrate how life could have started. He theorized that the primitive atmosphere on earth was the same as now exists on some of the other planets. For example, Jupiter and Saturn are very rich in methane gas (CH_4) and ammonia (NH_3), as well as possible water and lightning, and these conditions could be responsible for the formation of life. In his laboratory Miller attempted to recreate a primordial environment where an **autocatalytic** process might facilitate the formation of **prebiotic** life from organic molecules. Miller combined ammonia, methane, hydrogen, and water vapor and subjected the enclosed mixture to discharges of high-voltage electricity. Then he analyzed the mixture by the use of chromatography and detected several organic substances, of which two, glycine and alanine, were simple amino acids. This experiment was conducted many times with various mixtures and sources of energy. All produced some complex organic molecules, but none came anywhere near forming a substance that could reproduce and maintain its metabolism. The current thinking is that life was not formed by a random process (as is evolution) and that it may require the application of some new, possibly unknown, principles to form complex organic life from inorganic substances, or that life arrived on earth from some source in outer space. *See also* Darwin; Ponnamperuma; Urey.

Millikan's Theory for the Charge of Electrons: Physics: *Robert Andrews Millikan* (1868–1953), United States.

By indirectly measuring the effects on electrons by an electrical field whose intensity is known, the charge on the electron can be calculated.

Robert Millikan knew of James J. Thomson's discovery of the electron in

1896, as well as Thomson's use of the Wilson cloud chamber to compare the charge of an electron to its mass (e/m) and arrive at an approximate charge for the electron, which he stated as $1*10^{-19}$ coulombs. (*See also* C. Wilson.) It was a somewhat inaccurate measurement; today the constant for the electron charge is stated as 1.60219×10^{-19} C. Millikan conceived of a device similar to the Wilson cloud chamber but that used an electrostatic charge instead of a magnetic field to measure the charge of the electron. This was the classical "oil drop" experiment. Millikan atomized tiny oil droplets, which, due to the "friction" of falling, obtained a charge of static electricity as they fell through a small opening between two charged plates. He created a variable electric potential between the plates (+ and − charges) and exposed this area with a light beam enabling him, with the use of a microscope, to observe what occurred as the charged drops fell through the small opening between the charged plates. When the current was off, there was no charge on the plates, thus permitting the oil drops to fall at a constant rate due to gravity. As he adjusted the charges on the plates, the oil drops were deflected up or down according to several factors (the electrical potential between the charged plates, gravity, the electron's mass, and its electrical charge). He calculated the basic charge on an electron, a constant unit in physics, to be 4.774×10^{-10} ($\pm$ 0.0009) electrostatic units. With this information and using Thomson's data for the e/m formula, Millikan determined the electron has only about 1/1836 the mass of a hydrogen ion (a proton). *See also* Compton; Thomson; C. Wilson.

Minkowski's Space-Time Theory: Physics: *Hermann Minkowski* (1864–1909) Germany.

Any event occurring in both local space and time exists in the fourth dimension of space-time.

In 1907 Hermann Minkowski proposed his theory known as *Minkowski space*. His space-time concept provided the mathematics for local (measurable continuum) events occurring simultaneously that led to Einstein's general theory of relativity published in 1916. In addition to the three dimensions of space (x, y, and z, or width, height, and depth), Minkowski's theory included the fourth dimension of time; thus space-time represents the inertial frame of reference for all bodies in motion. This concept indicated that such a phenomenon would account for a curvature of space-time, which accounts for gravity. *See also* Einstein.

Misner's Theory for the Origin of the Universe: Astronomy: *Charles William Misner* (1932–), United States.

The universe began in a nonuniform state, which in time became uniform as it expanded due to natural forces and physical laws.

Charles Misner based his "mixmaster" model for the origin of the universe on the idea known as the *horizon paradox*. He theorized that in the beginning, all forms of matter were very much mixed up, with no or very little order. As the universe expanded, forces such as friction and gravity affected this diverse mixture and formed a more homogenous, isotropic, and uniform universe. His evidence was the horizon paradox, which states that the universe, when viewed from earth, is so huge that it appears, even to incoming microwave signals, as a very uniform structure—just as when viewing the horizon of earth, the distant landscape on earth appears more uniform than when viewing the same scene up close (e.g., viewing earth from airplanes). Misner's mixmaster concept and the horizon paradox provided support for the big bang inflationary theory for the origin of the universe. *See also* Gamow.

Mohorovicic's Theory of the Earth's Interior Structure: Geology: *Andrija Mohorovicic* (1857–1936), Croatia.
There are definite boundaries between earth's crust and mantle.
Based on his observation of an earthquake that occurred in 1909 in his native Croatia, Andrija Mohorovicic believed there was a boundary between the layers of earth. Using a seismograph, he recorded waves from an earthquake as they penetrated the deep areas of Earth and compared them with the waves that traveled on the surface. He discovered the waves from a deep layer traveled back to earth faster than did the surface waves, because the deep mantle layer was of greater density than was the crust at the surface. (The denser the medium, the greater the speed of sound and vibrations traveling through it.) He concluded that there must be a relatively abrupt separation between these two layers with the mantle starting about 20 to 25 miles below the surface. It was later determined the crust under the oceans is much thinner—only about 3 miles deep. The continental surface crust ranges in thickness from 22 miles in valleys to about 38 miles under mountains. The mantle is about 1800 miles thick, and the outer core is about 1400 or 1500 miles thick, with an inner core of about 1500 miles in diameter. This discontinuity between the crust and mantle was named after Mohorovicic and is sometimes referred to as the "Moho."

Montagnier's Theory for the HIV Virus: Medicine: *Luc Montagnier* (1932–), France.
A number of biomolecular mechanisms may be responsible for a depletion of lymphocytes in HIV-infected individuals.
In 1983 in Africa, Luc Montagnier and his colleagues discovered and isolated the human **retrovirus**, named HIV-1, which is related to AIDS. Later his team discovered a second retrovirus, HIV-2. (Animal retroviruses were known, but they were not generally associated with humans.) Montagnier's theory states that HIV exhibits characteristics of a retrovirus, which is the main mechanism that reduces the bacterial-viral-fighting lymphocytes in the human immune system, thus allowing **AIDS** to develop. He also investigated several other possible mechanisms that could relate the virus to AIDS. In the meantime, Robert Gallo of the United States, using a sample of Montagnier's retroviruses, discovered two viruses, one similar to HIV-1 and HIV-2, which Gallo named HTLV-1, and

later another variety named HTLV-3, both of which were found in T-4 cells (special lymphocytes of the immune system). The viruses Gallo identified were not exactly the same as Montagnier's but may have been mutations of HIV. A dispute arose over who discovered what and when, which eventually was settled by naming Montagnier's virus LAV and Gallo's as HTLV-3. In 1986 it was agreed to call all varieties of the retrovirus HIV. *See also* Baltimore; Delbruck; Gallo.

Moseley's Law: Physics: *Henry Gwyn Jeffreys Moseley* (1887–1915), England.
There is a distinct relationship between the x-ray spectrum and the proton number of chemical elements. When exposed to x rays, each element with its specific atomic number produces a unique spectrum of wavelengths.

Henry Moseley, using x-ray spectrometry, examined the lengths of electromagnetic waves emitted by different elements when exposed to x rays. He observed that each element produced its own specific wavelength, which he examined by using crystal diffraction. His data indicated that, each element may be considered a separate integer that is proportional to the square root of the frequency of its specific wavelength. We now refer to this integer as the *atomic number*, which is the number of positive protons in an atom's nucleus that determines a specific element's physical and chemical properties. Moseley's data improved Mendeleev's **Periodic Table of the Chemical Elements** by arranging the elements in the table according to their atomic numbers rather than their atomic weights. *See also* Mendeleev.

Muller's Theory of Mutation: Biology: *Hermann Joseph Muller* (1890–1967), United States. Hermann Muller was awarded the 1946 Noble Prize for physiology or medicine.
X rays and other ionizing radiation can induce chemical reactions that produce genetic mutations.

Familiar with Mendelian heredity and the concept of genes as the carriers of inherited characteristics, Hermann Muller experimented with the fruit fly *Drosophila*. His early research indicated that raising the temperatures of the eggs and sperm of fruit flies led to an increase in the rate of mutations. In 1926 he discovered that x rays would also cause mutations. Some mutations were recessive, but mostly the mutated genes were dominated and thus harmful and passed onto offspring. He concluded that mutation was a chemical reaction and could be caused by exposing the eggs and sperm to a variety of chemicals and forms of ionizing radiation. Mueller was concerned with the increased exposure of humans to all types of ionizing radiation (medical x rays, nuclear radiation, cosmic radiation, ultraviolet, etc.). He believed excessive amounts of exposure to radiation caused genetic mutations—some good, but mostly bad—that would be passed onto future generations. There is an ongoing debate as to the consequences of excessive mutations in the general population. *See also* Lysenko; Mendel.

Mulliken's Theory of Chemical Bonding: Chemistry (physical): *Robert Sanderson Mulliken* (1896–1986), United States. Robert Mulliken was awarded the 1966 Noble Prize for chemistry.

Nuclei of atoms produce fields that determine the movement of electrons in their orbits. Thus, the orbits of electrons for atoms combined in molecules may overlap and include two or more molecules.

Robert Mulliken formulated the concept of molecular orbits in which the valence electrons located in the outer orbits are not bound to any particular atom, but may be shared with several different atoms within a molecule. Familiar with Niels Bohr's quantum electron orbital model of the atom, Mulliken and his colleague, Friedrich Hund, applied quantum mechanics to explain how the valence electrons are delocalized in the molecular orbit where the bonding (combining) takes place. (See Figure S2 under Sidgwick.) In other words, the orbiting electrons of isolated atoms become molecular orbitals that may represent two or more atoms for each molecule. Thus, the energy of bonds could be determined by the amount of overlap of atomic orbitals within the molecular orbitals. This may be one reason that there are so few "free" atoms of elements in the universe. They are mostly joined in a great variety of molecular compounds. Related to chemical bonding is the concept of electronegativity that Mulliken devised. *Electronegativity* is the ability of a specific atom within a molecule to attract electrons to itself and thus enable the molecule to carry an extra negative charge. To explain this phenomenon, he developed the formula $\frac{1}{2}(I + E)$, where I is the ionization potential of the atom and E is the atom's affinity for electrons. *See also* Bohr.

N

Nambu's Theory for the "Standard Model": Physics: *Yichipo Nambu* (1921–), Japan.

Quarks can exist with three values, one of which is an extra quantum number referred to as color.

The "standard model" is one of several postulates of the quantum theory that is applicable to both sub-micro (quarks) and macro (supernovae) events. The other two deal with the quanta nature (tiny packets) of energy and the particle-wave nature of matter. Yichipo Nambu investigated "baryons," which are particles with three quarks exhibiting one-half spin as they interact with the strong force. The regular proton in the nucleus has only two "up" quarks and one "down" quark, with the symbol *uud* used to designate this structure. Baryons are composed of three identical quarks referred to as "strange" quarks, with the symbol *sss* designating their structure. Nambu's theory stated that quarks really exhibited a value of three, with an extra quantum number referred to as "color." The terms *up, down, strange*, and *color* are arbitrary and do not mean the same as the words are commonly used but refer to the particles orientations in space and spin directions. The three arbitrary colors are red, green, and blue, which enabled quantum rules to be followed by allowing three up quarks (uuu), three down quarks (ddd) or three strange quarks (sss) to exist as long as each "triple" quark has one of the three different "colors." This is now known as the **Standard Model** for elementary particles (quarks) obeying the dictates of quantum theory. Nambu's theory was expanded to explain the possibility for the absence of free quarks which, if detected, would exist as massless, one-dimensional entities. He suggested that the reason "free" quarks have yet to be detected is that they are located at the ends of "strings." Thus, this became known as the *string theory*, which had some problems when a string was "cut," which produced a quark and an antiquark pair but not a "free" quark. The string theory was later revised, and is now known as the *superstring theory. See also* De Broglie; Gell-Mann; Hawking; Schrödinger.

Nernst's Heat Theorem: Chemistry: *Walther Hermann Nernst* (1864–1941), Germany. Walther Nernst received the 1920 Noble Prize for chemistry.

When the temperature approaches absolute zero, so does entropy approach zero (kinetic energy and motion cease).

Walther Nernst's heat theorem was based on experimentation with ions in solution and attempting to determine the specific heat related to chemical reactions. He measured the heat absorbed in a chemical reaction, which fell along with the chemical's temperature as they both (heat and temperature) approached zero Kelvin in value (absolute zero $-273.16°C$). This theorem is called the *third law* and together with the first two laws of thermodynamics is the formulation of the science of thermodynamics. His theorem deals with the calculation of the *absolute* **entropy**, while the second law of thermodynamics only measures the *differences* in entropy. *See also* Carnot; Clausius; Fourier; Kelvin.

Newcomb's Theory for the Speed of Light: Physics: *Simon Newcomb* (1835–1909), United States.

The use of statistical methods can improve the accuracy of the constant for the speed of light.

In 1880, Newcomb, an employee of the U.S. Naval Observatory in Washington, D.C., was ordered by the U.S. secretary of the navy to measure the speed of light accurately. Newcomb applied statistical techniques to data gathered through repeated measurements of the speed of light. He placed a mirror at the base of the Washington Monument in Washington, D.C., and proceeded to shine a light from his laboratory onto the mirror, measuring the time it took to make a round trip. From these repeated events he recorded his data as a histogram (a graph representing the statistical means of his data in block form), from which he then considered the distribution and confidence interval for the data to arrive at an average figure for the speed of light. He noticed some so-called outliers, which were measurements way off from the average, which he eliminated. This technique of excluding spurious measurements, which are often artifacts of the measuring instruments or the observer, is frequently used by experimental scientists. *See also* Michelson.

Newlands' Law of Octaves: Chemistry: *John Alexander Reina Newlands* (1837–1898), England.

The chemical elements, when listed by their atomic weights, show a pattern of certain properties repeating themselves after each group of seven.

John Newlands first stated his law of octaves in 1864, but it was not accepted as anything more than an odd arrangement of the elements. In essence, his law states that any given element will have similar characteristics to and behave like

S I M I L A R	I	II	III	IV	V	VI	VII	VIII
	H							
	Li	Be	B	C	N	O	F	
	Na	Mg	Al	Si	P	S	Cl	
	K	Ca	□	□	As	S	Br	
	Cu	Zn	□	Ti	V	Cr	Mn	Fe Co Ni
	Rb	Sr	In	Sn	Sb	Te	I	
	Ag	Cd	Y	Zr	Nb	Mo	□	Rh Rh Pd

PERIODS

Figure N1: John Newlands' early version of Mendeleev's Periodic Table of the Chemical Elements, indicating the grouping of chemicals according to similar characteristics. Gaps in the table indicate yet-to-be-discovered elements.

another related element when organized in rows of seven according to their increasing atomic weights. (See Figure N1.) Many scientists used various methods to arrange and classify the fifty-five to sixty then known elements. The "noble" or inert gases found in group 8 of the modern **Periodic Table of the Chemical Elements** were not yet discovered. Newlands tried something different. He organized these elements into groups by atomic weights and noticed that similar elements repeated similar properties when listed by rows of seven. For instance, pairs in the second group of seven (by atomic weight) were similar to the pairs in the first row of seven elements. His elements did not include the unknown group 8, so this repeating of properties reminded him of the seven intervals of the musical scale, where the same seven notes are repeated several times and the eighth note in each octave (row) resembles the first note in the next higher octave. Therefore, he called his organization of the elements the *law of octaves*, which was later corrected to include the eighth group. In addition, Newlands' "table of the elements" did not allow blank spaces for yet-to-be discovered elements, so his pairing of elements by rows of seven was not always accurate. In 1869, Mendeleev published his periodic table, which provided blank spaces for undiscovered elements, and thus it represented a more realistic organization of the elements. (Later, the periodic table was improved by organizing the elements by their **atomic numbers** rather than their atomic weights. See Figure M4 under Mendeleev.) Following the success of Mendeleev's table, Newlands finally published his law of octaves, which he had withheld due to criticisms of his theory. Since that time, Newlands has been given credit for the original concept of the periodic arrangement of the elements by their atomic weights. *See also* Mendeleev.

Newton's Laws and Principles: Physics: *Sir Isaac Newton (1642–1727)*, England.

Newton's First Law: *An object at rest will remain at rest and an object in motion with constant velocity will remain in motion at that velocity until and unless an external force acts on the object.*

This first law refers to the concept of *inertia*, which is the tendency of a body to resist changing its velocity. Therefore, a body at rest remains at rest, and a moving body continues to move with a constant velocity unless acted upon by an external force. This was a major step in revising the ancient concept of motion, which presumed that for a force to move an object, something must be in contact with the object that was "pushing" it. In other words, before Newton's first law, people did not believe in "force at a distance." The accepted belief was that the movements of planets were caused by heavenly angels or an **aether** was pushing them.

Newton's Second Law: *The sum of all the forces (F) that act on an object is equal to the mass (m) of the object multiplied by the acceleration (a) of the object.*

Newton's second law explains the relationship between acceleration and force. Acceleration is the rate of change in the velocity of an object with respect to the time involved in the change of velocity. Velocity involves both speed of an object and its direction. The second law is expressed as $F = ma$. F is the force exerted on the mass (m) of the object, and a is the acceleration of the object. To determine the acceleration of an object, its mass would be inversely proportionate to the force acting on it: $a = F/m$. Both acceleration and force are considered *vectors* (arrows), since both have a direction and a magnitude. Vectors can be added and subtracted, so it is possible to arrive at a sum of several forces acting on an object by adding the magnitude of one vector arrow to the next.

Newton's second law also explains the concept of *momentum*, which is the product of an object's mass times its velocity: momentum $= mv$. The rate of change in momentum equals the strength of the force applied to the object. Momentum explains why, in an accident or crash, a lightweight car traveling at a high speed will sustain more damage than a heavy car involved in an accident traveling at a lower or the identical speed as the lighter car.

Both Newton's first and second laws of motion led to the concept of *inertial frames of reference*. An inertial frame occurs when an object that has no external forces acting on it continues to move with a constant velocity. This is why, once you start sliding on ice, it is difficult to stop unless there is some external force to impede your progress.

Newton's Third Law of Motion: *When two bodies interact, the force exerted on body 1 by body 2 will be equal to (but opposite) the force exerted on body 2 by body 1.*

Another way to say this that is when one object exerts a force on a second object, the second object will exert an equal but opposite force on the first object. It may be expressed as $F_{1*2} = -F_{2*1}$ and is commonly worded as, *For every action, there is an equal and opposite reaction.* This explains why a pot of soup will remain on a table. As the force of weight of the pot of soup "pushes" down, the table is also pushing up on the pot with an equal and opposite force. Sometimes this relationship can be exaggerated. For instance, when a person walks on the ground, he or she is exerting a backward force on earth while earth is exerting an equal and opposite force on the person. Thus, while the person moves forward, earth does not seem to move backward because it has a much greater mass than a human. If a person and the other object (earth) had identical mass, not much progress would be made by walking. Another example is the equal-and-opposite reaction in a rocket motor. Many people believe the exhaust fire and fumes "push" against the air and thus propel the rocket forward. But there is no air in space, and rockets surely operate in space. The greater the mass of the exhaust exiting the rear of the rocket with tremendous velocities, the greater the opposite (and equal) reaction inside the front end of the rocket. Therefore, air is not required for a rocket's exhaust to "push" against. Rather, the faster the gases (mass) are expelled, the greater is the opposite reaction to the direction of the gases.

Newton's Law of Gravity: *Two bodies of mass$_1$ and mass$_2$, separated by a distance r, will exert an attractive force on each other proportional to the square of the distance separating them.*

Galileo demonstrated that all bodies in free fall do so with an equal acceleration. In other words, he determined that, disregarding air and other sources of friction, two bodies of unequal weights will accelerate at the same rate when in free fall. He came very close to explaining gravity, but it was Newton who applied the mathematics of his three laws of motion (in particular the third law) to the concept of constant gravitational acceleration as applied to all bodies. The force of gravity is expressed as $F = Gm_1m_2/d^2$ where F is the force, G is the proportional constant for gravity, m_1 and m_2 are the masses of the two bodies, and d^2 is the square of the distance between them. Newton determined that the force of gravity is the cause of the acceleration of a body, and the force is not changed by the motion of that body. In other words, the rate of acceleration is independent of the force and thus is a property of the body. Weight is distinguished from mass in the sense that the weight of an object (on earth) is dependent on the attraction between the object and the Earth. In real life, earth is so much more massive than the object that it does most of the attracting. Thus, the more massive the object on Earth, the more it weighs. Weight is determined by the force of gravity on an object. Since the moon has less mass than earth, its gravity is about one-sixth that of earth. Therefore, a 180-pound person would weigh only 30 pounds on the moon. The mass of an object might be thought of as the amount of "stuff" in the object; it is the same anywhere the object is located in the universe.

Newton's Theories of Light: *(1) Light is composed of a great multitude of very tiny particles of different sizes, which are reflected by shiny polished surfaces. (2) White light is a heterogeneous composition of these different-sized particles, which form rays of different colors.*

Sir Isaac Newton did not accept the wave theory of light, which many other scientists compared with sound waves. His rationale was that light, unlike sound, could not travel around corners. Therefore, light cannot have wave properties and must be composed of a multitude of very tiny particles.

For the second part of his theory, Newton placed a large "plano-convex" lens on a shiny surface and exposed it to direct sunlight. A distinct pattern of rainbow-like concentric rings of colored light became visible near the edges of the lens. Although this was a good indication that light had wave properties, Newton still maintained his corpuscular (tiny particles) theory of light. He also believed white light was composed of light rays of various colors, each color of which was a different-size corpuscular particle. He was determined to prove sunlight was a heterogeneous blend of a variety of light rays (and particles) and that both reflection and refraction could be used to separate these individual-sized particles and rays. To do this, he covered the window of a room to shut out sunlight, leaving a small opening for a narrow shaft of sunlight, which he directed through a prism. Some of the light rays going through the prism were bent (refracted) more than other rays, and together they produced a color spectrum of the sunlight. (See Figure F4 under Fraunhofer.) He then placed another prism in contact with the first and discovered that the second prism caused the individual rays to converge back into the white light of the sun. This proved sunlight was indeed composed of different-colored light rays. However, rather than proving his corpuscular theory, it gave further evidence for the wave theory of light, which was further confirmed when new diffusion gratings were used instead of prisms to demonstrate the wave nature of light. Newton delayed publishing his results for several years fearing adverse criticism of his work. *See also* Maxwell.

Newton's Calculus: There is a mathematical relationship between the differentiation and integration of small changes for events.

Newton referred to this as the "fluxional method," which became known as *calculus*. Gottfried Leibniz independently developed what he called differential calculus. Calculus is the branch of mathematics that deals with infinitesimally small changes. It is used extensively in almost all areas of physics, as well as in many other fields of study.

Today, both Newton and Leibniz share the honor for the development of calculus. *See also* Euler; Galileo; Kepler; Leibniz.

Nicholas' Theory of an Incomplete Universe: Astronomy: *Nicholas of Cusa* (1401–1464), Italy.

Because he was a bishop and papal adviser, Nicholas of Cusa's theories assumed a religious and metaphysical leaning. Nevertheless, he proposed some rather revolutionary ideas for his time in history.

There is nothing fixed in the universe, yet it is not infinite.

Nicholas of Cusa did not accept the theories proposed by Aristotle and other astronomers that the universe was composed of a series of crystal domes over a flat earth, or the alternative—that the universe was one big sphere with earth at its center. Nicholas proposed a more flexible universe. It was in a state of becoming where nothing was fixed; there was no outer edge or circumference, and the center was not yet established. Another way to look at this is to consider an infinite circle whose infinite circumference is composed of an infinite straight line. He believed that the universe was extremely complex, with everything in a state of flux and in motion. Even so, God's universe was not infinite. He believed that things are finite, but God is infinite. Nicholas' view of the solar system was ahead of his time. He also believed Earth, as well as all the other planets, revolved around the sun. *See also* Aristotle; Copernicus; Eudoxus; Galileo.

Nicolle's Theory for the Cause of Typhus: Biology: *Charles Jules Henri Nicolle* (1866–1936), France. Charles Nicolle received the 1928 Nobel Prize for physiology or medicine.

The disease typhus is spread by a parasite or body louse that lives on the patient's clothing and body.

In the late 1800s, Charles Nicolle, director of the Pasteur Institute, was asked to determine the cause of typhus, an infectious disease that had been known to follow wars and plagues as early as 1400 B.C. As with measles, people who recovered from typhus were immune to it. Typhus was also confused with influenza because it was thought to be spread by "droplet" infection or direct contact with patients' clothes or dust. Its death rate was over 50 percent for adults contracting the disease. Nicolle visited homes where whole families were infected, exposing himself and his coworkers to the infection; several died while gathering information. He observed that while almost all members of a family contracted the disease, once they came to the hospital, it no longer spread. He noticed a relationship between patients brought to the hospital who were undressed, bathed, and whose clothes were either burned or laundered and the cessation of new infection. As new patients were admitted and went through this procedure, the disease did not spread to others in the hospital. He deduced there must be some type of insect in their clothes or on their bodies.

In 1909 he concluded from this evidence that the culprit was the body louse. Using this knowledge and information, Charles Nicolle experimented with animals, exposing them to the disease carried by the louse. He further established that the typhus germ was not transmitted to a new generation of the louse parasite. Rather, once the germ-carrying adult louse dies, the epidemic ends. He found the guinea pig and some monkeys were susceptible, but not many other animals. Using the blood from infected animals that recovered, he tried to develop a vaccine to prevent typhus and understand how it affected the immune system. Nicolle's work emphasized the need for better personal and public hygiene to prevent typhus. His discovery that typhus was carried by lice explained why, since ancient times, it was associated with wars. It was not until World War II that the insecticide DDT, now banned, was used to control typhus outbreaks in army troops stationed in Italy and other countries, ultimately saving many lives.

Noddack's Hypothesis for Producing Artificial Elements: Chemistry: *Ida Eva Tacke Noddack* (1896–1979), Germany.

When uranium is bombarded with slow neutrons, artificial elements and their isotopes should be produced.

All isotopes of elements above atomic number 81 are unstable and radioactive. Ida Eva Tacke Noddack was familiar with Enrico Fermi's work with radioactive elements. Fermi realized that fast neutrons that were slowed could more readily penetrate the nuclei of uranium, thus causing fission to occur. This gave Noddack the idea that artificial isotopes of elements could be produced by the same process of bombarding uranium with slow neutrons. This is exactly what occurred when Otto Frisch confirmed Noddack's hypothesis by bombarding the heavy nuclei of uranium with slow neutrons. This bombardment broke up the nuclei into a few large fragments, which proved to be isotopes of other elements. *See also* Fermi; Frisch.

Northrop's Hypothesis for the Protein Nature of Enzymes: Chemistry: *John Howard Northrop* (1891–1987), United States. John Northrop shared the 1946 Nobel Prize for chemistry with James Summer and Wendell Stanley.

If enzymes can be crystallized, their composition must be of a protein nature.

Several other chemists claimed that enzymes did not have the characteristics of proteins. In the late 1920s James Summer (1877–1955) claimed to have crystallized the common enzyme urease, which manifested the characteristics of a protein. In the early 1930s John Northrop and his colleagues were successful in crystallizing several more important enzymes, including trypsin, perpsin, chymotrypsin, and more important, ribonuclease and deoxyribonuclease (related to RNA and DNA). The exact nature of proteins was difficult to determine because of their very long molecular structures. The crystallization provided a means for identifying the structures of the enzyme and confirmed the theory that they were of a protein nature, enabling scientists to study and understand their chemical composition. Later, Northrop proved that bacteria-type viruses (bacteriophage) also consist of proteins and cause diseases by infecting specific species of bacteria. (See Figure D2 under Delbruck.) *See also* Delbruck.

Noyce's Concept for the Integrated Circuit: Physics: *Robert Norton Noyce* (1927–1989). United States.

A series of transistors can be combined on a small single piece of semiconducting material by etching microscopic transistors onto the surface of a chip to form circuits that can integrate the individual transistors.

Rectifying crystals were in use before the vacuum radio tube was used to control the unidirectional flow of alternating current. They were not very effective, but the concept of using crystals to control the flow of electricity and electromagnetic radiation was not lost on a number of scientists. In the late 1940s, William Shockley (1910–1989), Walter Brattain (1902–1987), and John Bardeen (1908–1991) used a different substance, a germanium crystal. It was an ineffective conductor of electricity but a good insulator, making it what is now known as a *semiconductor*. Silicon crystals were less expensive and soon replaced germanium. It was discovered that if tiny amounts of certain impurities were placed in the semiconductor material, its characteristics could be controlled. These were referred to as solid-state devices, which act as vacuum tubes without having the tubes' large size, generation of heat, or possibility of breakage, and they used very little electricity. This unique device was developed by Shockley, but John Robinson Pierce (1910–) gave it the name *transistor*, for its property of transmitting current over a specified resistance. (See Figure S1 under Shockley.) Robert Noyce is credited with the concept of combining a series of transistors onto a small silicon semiconductor chip (about 1/4 square inch or less) to form an integrated circuit. In 1959 Noyce received funding from the Fairchild Corporation to form Fairchild Semiconductor, the first major semiconductor electronics plant in the Silicon Valley to exploit the use of the new chip. Later, he formed the INTEL Corporation. In 1959 Noyce filed for a patent for the new chip, even though a few months previously, Jack S. Kirby of Texas Instruments also filed for a patent. Noyce's patent was approved because the Patent Office said Kirby's application was not specifically clear. *See also* Shockley.

O

Odling's Valence Theory: Chemistry: *William Odling* (1829–1921), England.

Elements can be grouped according to their analogous properties based on their representative values of replacement.

The concept of valence, proposed by Edward Frankland in 1854, was unknown to William Odling. Prior to Frankland's valence concept, Odling proposed that during chemical reactions, a distinct ratio existed when one element replaced other elements in a chemical reaction. This was a forerunner of the theory of atomic valences, which is the ability of elements to combine with other elements, it can be expressed as the number of univalent atoms with which they are capable of uniting. (See Figure S2 under Sidgwick.) Some elements may be univalent (1); others are divalent (2), trivalent (3), or tetravalent (4) with respect to the number of univalent atoms with which they can combine. Still others may possess variable valences (e.g., nitrogen and phosphorus). At first Odling, as well as others of his time, rejected the existence of atoms. At one time it was incorrectly assumed that just one atom of hydrogen combined with one of oxygen to form water. But after conducting experiments with oxygen, Odling came to believe in the valence theory. He was the first to realize that oxygen had an atomic weight of 16, not 8. This convinced him that oxygen gas had to be diatomic—a molecule composed of two oxygen atoms. He also speculated about a triatomic molecular form of oxygen (ozone, O_3). *See also* Frankland; Sidgwick.

Oersted's Theory of Electromagnetism: Physics: *Hans Christian Oersted* (1777–1851), Denmark.

A magnetized compass needle will move at right angles to the direction of an electric current flowing through a wire suspended over the compass.

Hans Christian Oersted's discovery of the relationship between electricity and magnetism (electromagnetism) was accidental. Aware of the experiments dealing with static electricity and the new form of "flowing" electricity described by Alessandro Volta, Oersted performed various "galvanic" experiments using Volta's cells to produce current electricity. He knew of others who demonstrated that by passing an electric current through water it could be separated into oxygen and hydrogen gases. He believed that this established the connection between electrical forces and chemical reactions and that water must be a compound, not an element, as had been believed for centuries. Oersted then conducted experiments with this new electricity. He attempted to demonstrate that when a wire was heated by carrying electricity, it would act as a magnet and attract a compass needle. He noticed at once that the needle was not attracted to the wire but rather moved 90 degrees from the direction in which the current was flowing in the wire. Saying nothing about this observation to his students, he continued to turn the current in the wire on and off. Each time he did so, the needle of the compass moved at a right angle to the wire. In 1820 he published his results describing the existence of a circular magnetic field around a wire carrying an electric current, thus establishing the connection between current electricity and magnetism, now known as *electromagnetism*. The unit for the strength of a magnetic field is named after Oersted and is defined as the intensity of a magnetic field's strength expressed in the centimeter-gram-second (cgs) electromagnetic system of physical units. Oersted's concepts sparked many experiments and theories related to electromagnetism, the end result being our modern "electric" oriented society. *See also* Ampère; Faraday; Henry; Maxwell; Volta.

Ohm's Law: Physics: *Georg Simon Ohm* (1787–1854), Germany.

A unit of electrical resistance is equal to that of a conductor in which a current of 1 ampere is produced by a potential of 1 volt.

Georg Ohm related electrical resistance to Joseph Fourier's concept of heat resistance, which states the flow of heat between two points of a conductor depends on two factors: (1) the temperature difference between the origin of the point of heat and the end point of the conductor and (2) the physical nature of the conducting material being heated. Ohm speculated how this information related to electricity and this led to his experimenting with wires of different thicknesses (cross sections). He demonstrated that electrical resistance to current passing through these different wires was directly proportional to the cross section of the wires and inversely proportional to the length of the wires. Almost all effective conductors of heat are also excellent conductors of electricity. The law can be applied to both direct and alternating currents. Ohms's law is very versatile and can be used to measure conductance, current density, voltage, resistors, inductors, capacitors, and impedance. It is stated as $R = V/A$, where R is the natural resistance of the wire (conductor) to the flow of electricity, V is voltage, or strength of the electric current, and A is ampere, which is the amount of the electricity (also maybe expressed as I). There is an expression for conductance (G), which is referred to as the reverse ohm (mho), $I = GV$, where I

is the current (same as amps), G is the conductance factor of the wire (how well it conducts electricity), and V is the voltage. The unit (Ω) of electrical resistance is named after Ohm; it is the amount of resistance to voltage of 1 ampere of current. The importance of Ohm's law was not recognized by other physicists for some time, but by the early 1840s, the Royal Society in England accepted both his law and Ohm as a member. *See also* Ampère; Faraday; Fourier; Volta.

Oken's Cell Theory: Biology: *Lorenz Oken* (1779–1851), Germany.
Living organisms were not created but rather originated from vesicles (cells), which are the basic units of life.

In 1805 Lorenz Oken theorized that humans and animals not only originated from but were also composed of cells that he called *vesicles*. Until this point, the source of life—its origin and composition—had been the subject of speculation by scientists and philosophers over many centuries. Some believed life began within a "primeval soup" or was carried to earth from outer space, or was derived from self-organizing, self-replicating inorganic molecules, or, more acceptable to most people, was created by a supreme being. The discovery of fossils and the use of a microscope led to further concepts of living tissue. Robert Hooke viewed tiny enclosures in cork bounded by walls that reminded him of the rooms occupied by monks in monasteries; thus, he named them *cellulae*, meaning "small rooms" in Latin. Oken further speculated that these "cells" were the basic units of life, from which all complex organisms were derived and developed, and he theorized that cellular structure was basic for all organic substances. Oken was one of the first of many scientists to contribute to and expand the concept we now know as the cell theory. *See also* Schleiden; Schwann; Virchow.

Olbers' Paradox: Astronomy: *Heinrich Olbers* (1758–1840), Germany.
If the universe is old, eternal, unchanging, infinite, and uniformly filled with stars, why is the night sky dark instead of bright?

Heinrich Olbers' paradox has intrigued astronomers, physicists, and mathematicians for decades. It is based on several question that are still being investigated by scientists: Is the universe finite or infinite? Is it an evolving and expanding universe or a steady-state universe? Are galaxies (groups of stars) evenly distributed in the heavens, or is space nonhomologous? We know that light follows the basic inverse square law. An appreciation of the inverse square law as related to light can be demonstrated at night by shining a flashlight at a 1-foot-square white sheet of cardboard held away from the light at several different distances. The illumination on the cardboard is greater at 10 feet than at 25 feet and greater in intensity at 25 feet than at 100 feet. It will be obvious that the light intensity diminishes as the distance between the flashlight and white cardboard increases. (The intensity of the light at different distances can be measured with a light meter.) Light, over distance, is dispersed and becomes less focused, but it will travel in a straight line forever if not absorbed or distorted (affected) by gravity in space. Therefore, the intensity of light received on earth from stars is much reduced from its brightness at the source. But at the same time, the average number of stars at any given distance increases in number by the square of the distance to earth. This is the basic distribution of stars in the universe. Therefore, according to one part of the paradox, the night sky should be as bright as the sun. On the other hand, Olbers claimed that the reason the night sky is not as bright as the sun is that interstellar "dust" absorbs the starlight. Today, this is an unacceptable solution to his paradox because the universe is assumed to have come into existence at a finite time, even though it might be infinite in space. It has a beginning, it has history, it seems to continue to expand, and for the most distant and possibly oldest galaxies, light has not had time to get to earth. Light does disperse over long distances, and galaxies and their stars are not evenly distributed throughout space. Currently, the question concerning the universe's being finite or infinite is being examined seriously by astronomers using the Hubble Space Telescope and the Gemini North telescope installed at Mauna Kea, Hawaii. The mirror for the Gemini is 8.1 meters in diameter and only 20 centimeters thick. It is difficult to cast a single piece of glass this size without it cracking. The mirror is the largest single-piece glass mirror ever cast for a reflector telescope. (See Figure H1 under Hale.) It is expected that both of these instruments will locate galaxies at the limit of the speed of light, which means these distant galaxies are receding faster than the speed of light, which may be considered a boundary formed by the limits of just how far we can see into the past using electromagnetic radiation (light, radio, microwaves, x rays, etc.) originating from the edge of the universe. Or it might mean that we may never be able to "see" the edge of the universe, even if it is finite. Or, if the universe is forever expanding, it may be too young for light to have reached us. Therefore, Olbers' paradox addresses several phenomena of physics and is a puzzle of unknowns.

Oliphant's Concepts of Isotopes for Light Elements: Physics: *Marcus Laurence Elwin Oliphant* (1901–   ), Australia.

By "shooting" an ion beam at targets of lithium, beryllium, and related elements, new atoms of hydrogen and helium can be created by a process of atomic transformation.

Marcus Oliphant, known as Sir Mark Oliphant, was influenced by Ernest Rutherford's work with the nuclei of atoms. In 1932, Sir James Chadwick bombarded nuclei of atoms and discovered a third basic particle in the atom, the neutron. (The basic particles of atoms are electrons, protons, and neutrons.) Also in 1932, Harold Urey discovered a heavy form of hydrogen called *deuterium*. It was known for some time that the hydrogen atom contained only one particle in its nucleus, the positively charged proton, which made it the lightest of all elements. Urey discovered that the nucleus of hydrogen could also contain a neutron; thus this isotope of hydrogen had an atomic weight of 2 instead of 1,

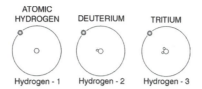

ATOMIC
HYDROGEN   DEUTERIUM   TRITIUM

Hydrogen - 1   Hydrogen - 2   Hydrogen - 3

Figure O1: Harold Urey produced heavy hydrogen atoms with one proton and one neutron (deuterons). Marcus Oliphant bombarded deuterons with other deuterons to produce a third isotope of hydrogen-tritium, with two neutrons and one proton. The three isotopes of the hydrogen atom exhibit minor differences in chemical properties due to differences in atomic weights. These differences are minor for chemical reactions involving isotopes of heavier elements. Heavy hydrogen atoms are used to facilitate nuclear fusion reactions.

as for ordinary hydrogen. The nuclei of this isotope were referred to as *deuterons*. Oliphant and a colleague bombarded these heavy hydrogen deuterons with other deuterons, producing a new isotope of hydrogen that contained two neutrons plus the proton. Thus, now there were three forms of hydrogen: H-1, H-2, and H-3. (See Figure O1.) This third form, composed of one proton and two neutrons, is named *tritium*. It is unstable due to its radioactive nature, with a half-life of about 12 years. At about the same time, it was discovered that the nuclei of these forms of hydrogen could react with each other, producing other new elements (particularly helium) and releasing tremendous energy. This was the beginning of our understanding of the thermonuclear reactions that take place in the sun, as well as the development of the hydrogen bomb. Oliphant moved to the United States and perfected the electromagnetic method for separating rare fissionable uranium (U-235) from the more common form of uranium (U-238). U-235 was used to produce the first atomic bomb. Oliphant was also the first to realize that nuclear reactors used to generate electricity can also produce plutonium. Thus, any country with such a reactor could develop atomic (nuclear) weapons. He became a firm critic of nuclear weapons and an advocate for the peaceful uses of atomic energy. *See also* Chadwick; Rutherford; Urey.

Oort's Galaxy and Comet Cloud Theories: Astronomy: *Jan Hendrik Oort* (1900–1992), Netherlands.

Oort's Theory for the Structure and Motion of the Milky Way: *Composed of billions of stars, our Milky Way rotates as an entire giant disk, but the rotation is differential, not uniform, since the outer stars rotate at a slower speed than do the inner stars.*

Jan Oort determined several facts about galaxies, including our Milky Way galaxy. He theorized that the rotating stars in a disk galaxy follow Newton's laws of motion, in particular, the concepts of conservation of energy and angular momentum. Therefore, the outer stars move more slowly than do the inner ones in the gigantic cluster of stars forming a disk galaxy. In addition to his theory of the motions of galaxies, he determined the sun is only 30,000 light-years

from the center of the Milky Way, located about one-third from the outer edge
of one of its arms. He ascertained too that the sun makes one complete revolution around the Milky Way's axis once every 225 million years.

Oort's Comet Cloud: *Comets originate in an area beyond the solar system.*

In 1950 Jan Oort identified about twenty comets with orbits so large that it
requires many months, or years, for them to complete each orbit around the sun.
He believed they originated from a great cloud or reservoir of over 1 trillion
comets, comet material, and assorted objects that swarm far beyond the edge of
our solar system. This cloud was still close enough to be affected by the sun's
gravity even though it was about 100,000 astronomical units (AU) in diameter.
In comparison, the planet Pluto's orbit is only 40 AU in diameter. The theory
is these objects in the Oort cloud are "leftover" matter remaining at the edge of
the *solar nebula*, the swirling mass of dust and gas that formed the solar system,
planets, meteors, and comets. He suggested there were two kinds of comets,
both of which were disturbed from their paths within this cloud by perturbations
caused by the gravity of a passing (not too close) star. The path of one type of
comet follows a hyperbolic orbit or path through the solar system, meaning it
makes a wide sweep but not a closed orbit. Therefore, it appears only once as
a comet visible to the naked eye or low-powered telescopes from earth, never
to return again. The paths of other comets become very eccentric, but their orbits
are closed. They do not follow a circular path through the solar system but
rather an eccentric ellipse. This is the most familiar type, such as Halley's comet,
and the more recent Shoemaker-Levy comet, which smashed into Jupiter on
July 11, 1994. More recently, there is evidence that an inner Oort comet cloud
exists beyond the planet Neptune. It is believed it contains about 100 times the
number of comets as does the outer Oort cloud, and these inner cloud comets
have a much more circular orbit, similar to the planets, than do the comets in
the outer Oort cloud. (See Figure K3 under Kirkwood for a similar asteroid
belt.)

Oparin's Theory for the Origin of Life: Biology: *Alexsandr Ivanovich Oparin*
(1894–1980), Russia.

*The first living organisms subsisted on organic substances, not inorganic
matter.*

How life started is one of the oldest philosophical and biological questions.
Charles Darwin's theory of evolution did not include an explanation for the
origin of life. The concept that life just "arrived" from nonliving substances was
known as *spontaneous generation*. It was disproved by Redi, Pasteur, and others.
In 1922 Alexsandr Oparin was the first to theorize there was a slow accumulation in the oceans of simple organic compounds formed from inorganic compounds. He conjectured that the original living organisms were *heterotrophic*,
because they did not synthesize their food from inorganic materials as would
autotrophic organisms, such as some bacteria and green plants. A more recent
theory states that energy-producing elements and compounds self-organized into

microcomponents of plant and animal cells (mitochondria and chloroplasts) combined through the process of **autopoiesis** to form simple cells. *See also* Miller; Oken; Pasteur; Redi; Virchow.

Ostwald's Theories and Concept of Chemistry: Chemistry: *Friedrich Wilhelm Ostwald* (1853–1932), Germany. Friedrich Ostwald received the 1909 Nobel Prize for chemistry.

Ostwald's Theory of Catalysts: *Nonreacting foreign substances can alter the rate of chemical reactions.*

Cognizant of the kinetics of chemical reactions, their speeds, and equilibrium states, Ostwald theorized that certain substances (catalysts), when added to a chemical reaction, can either speed up or slow down the rate of that reaction; but at the same time, the catalyst will not alter the energy relationship within the reaction nor will the catalyst itself be changed. This concept became extremely important in the development of modern technology for controlling a great variety of chemical reactions, including the platinum beads used in the modern catalytic converter in automobile exhaust systems to reduce harmful exhaust gases to less toxic fumes.

Ostwald's Law of Dilution: *The extent to which a dilute solution can become ionized can be measured with a high degree of accuracy.*

Ostwald's law of dilution was a means for determining the degree of **ionization** in a dilute solution with some degree of accuracy. A dilute solution is one with a small amount of solute (the substance dissolved) compared to the amount of solvent (the substance dissolving of the solute). He patented his process, now known as the Ostwald process, which is used worldwide to produce nitric acid by oxidizing ammonia.

P

Paracelsus' Concepts of Medicine: Chemistry: *Philippus Aureolus Theophrastus Bombastus von Hohenheim (Paracelsus)* (1493–1541). Germany.

Iatro-chemistry is superior to herb chemistry for treating diseases and illnesses.

Early in his career, Philippus Aureolus Theophrastus Bombastus von Hohenheim changed his name to Paracelsus. The Roman Galen, a herbalist physician, subscribed to the **humoral** theory of disease; his teaching persisted for 1500 years and was considered the authority in medicine during Paracelsus' time. But Paracelsus continually challenged Galen's doctrines as accepted by his contemporary physicians in the early sixteenth century. Paracelsus was, in a sense, a compassionate patient advocate. For most of his life and for many reasons, he was an outcast of the medical community. One reason was that he considered the human body from a chemical point of view, not just a spiritual vessel. Paracelsus' iatro-chemistry (the use of chemicals for medical treatments) was based on the *Tria prima*, which was predicated on three basic types of matter: mercury was the spirit, sulphur is the soul, and salt is the body, with the inflammable sulphur combining the body and spirit into one unit. This, and more, was the basis of medicinal alchemy in the Middle Ages. Alchemy was the study and practice of combining a few basic elements to form the philosophers' stone, considered the key to transmuting base metals (e.g., lead) into gold or to produce the "elixir of life" which was sought as the cure for all illnesses.

Paracelsus was the first link between medieval medical practices and modern scientific medicine and pharmacy. He cured a few powerful men who became his sponsors, which enabled him to continue teaching and to use his unique methods of curing the ill. Paracelsus still believed in astrology, magic, and alchemy but used combinations of chemicals, such as mercury, iron, arsenic, sulphur, antimony, and laudanum (opium), rather than herbs. He was the first physician to try specific remedies for specific diseases and to connect heredity and other patterns to certain diseases and physical conditions, such as goiter, cretinism, and patterns of syphilis. He criticized his colleagues for their practice of "torturing" the ill by bleeding them and other inhumane treatments. He was respected by the common people, but after his sponsors died, he was driven out of the country by his many enemies. For the remainder of his short life, he was a physician to miners, from whom he learned about metals and minerals, as well as their unique lung diseases. Paracelsus introduced pharmacology, antiseptics, modern surgical techniques, and microchemistry (homeopathy). His work with the miners also qualifies him as the first physician to develop the field of occupational and industrial medicine.

Pardee's Theory for Cell Enzyme Synthesis: Biology: *Arthur Beck Pardee* (1921–), United States.

A mutant gene can induce dominance on a molecule to suppress production of the enzyme beta-galactosidase.

Arthur Pardee and his staff crossed a mutant bacteria cell with normal bacteria, which then became capable of synthetically producing a metabolic enzyme crucial to living cells. Their process produced synthetic beta-galactosidase without requiring outside stimulation. This led to the production of purines and a better understanding of nucleic acids. Purines are double-ring nitrogenous organic molecules such as adenine (A), guanine (G), thymine (T), and cytosine (C), which form the base pairs of the nucleic acids of DNA and RNA. A new pathway was provided for newly formed proteins that can be stimulated by growth factors to duplicate their own DNA, and thus continue to divide. The human immune system T-cells, which fight not only infections but cancer cells, do not normally reproduce, but with this new technique, it became possible to produce numerous T-cells. Pardee's research advanced the understanding of the immune system, enzymes, T-cells, and the HIV virus infections. *See also* Crick; R. Franklin; Watson.

Pascal's Concepts, Laws, and Theorems: Physics: *Blaise Pascal* (1623–1662), France.

Pascal's Concept of a Barometer: *The height of a column of mercury decreases as altitude increases due to a decrease in air pressure.*

Blaise Pascal pursued the concept of atmospheric pressure proposed by Evangelista Torricelli's experiment that demonstrated that a 30-centimeter vertical column of mercury could be suspended in a closed tube. Torricelli was the first to theorize that the mercury was not suspended because a vacuum formed inside the closed tube, but rather by the weight of the air outside the tube. (See Figure T2 under Torricelli.) Pascal set out to prove that the height of this column of mercury was dependent on the weight of the air above the mercury in the dish that contained the column of mercury, and that the weight of air pushing down on earth varied with altitude. He and his brother first measured the exact height of the suspended mercury in the column at the altitude of Paris, France. They proceeded to move the experiment to the top of a high mountain and demonstrated the column fell (was less than 30 cm) as the altitude became greater, proving air above a mountain exerts less pressure on earth than at sea level. Thus, he not only confirmed Torricelli's concept of air pressure but also discovered that this phenomenon could be used as a crude altimeter. The concept of the barometer can be used to predict weather conditions based on the fact that warm, stormy air is less dense, and thus weighs less per square centimeter than does cold, clear, denser air. The altimeter's basis is the fact that air becomes less dense as the altitude increases. Thus, the greater the altitude, the less air weighs per square centimeter. The barometer is vital for weather forecasting, and the altimeter is important for determining the altitude of airplanes. *See also* Galileo; Torricelli.

Pascal's Law of Hydraulics: *Pressure applied to a contained fluid is transmitted throughout the fluid in all directions regardless of the area to which the pressure is applied or the shape of the container.*

Blaise Pascal based this law on his work with atmospheric pressure and Simon Stevin's (1548–1620) demonstration that the pressure on a fluid of a given surface depends on the height of the fluid above that surface but not on the surface area of the fluid or the shape of the fluid's container. (Today we know this as the science of hydrostatics.) For example, if you fill two 20-foot vertical pipes with water, one 2 inches in diameter and the other 12 inches in diameter, and with a valve tap at the base of each pipe, the pressure of the water coming out of each tap will be equal since the height of the columns of water was equal in each pipe (but not their volumes or surface areas). Pascal's law of hydraulics basically states that fluids transmit pressure equally in all directions. For instance, the pressure exerted on a confined liquid with a plunger having a small 2 cm cross section will exert a greater pressure on a surface area having a much larger cross section. If the second surface's area is twice as great as the first plunger's surface area, so will be the pressure exerted on the second area by the plunger. The original pressure is multiplied by the differences in the areas of the cross sections of the two surfaces. He used this law to develop the hydraulic press, which converts a small force into a much larger force. When a person steps on the brake pedal to stop a car, the hydraulic system converts the small force of the foot into the much larger force required to stop the car. This force is transmitted through the hydraulic fluid and tubing system to the brake's mechanism on the car's wheels to create adequate friction on the wheel's brake pads or discs, which then stops the car. In this case, a special oil-like fluid is used instead of water, since water would freeze and prevent the hydraulic system from working. *See also* Archimedes.

Pascal's Theorem: *If a hexagon is inscribed in any conic section, the points related to where opposite sides meet will be collinear.*

At a young age Blaise Pascal had already formulated several theorems of projective geometry that later became known as Pascal's theorem. Viewing a diagram of his theorem is easier to understand than a written statement. (See

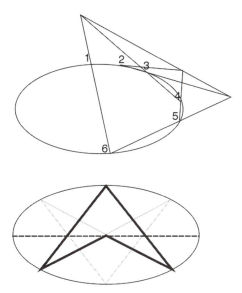

Figure P1: Two versions of Pascal's hexagram theorem. (Top) The points of a hexagon when inscribed in a conic section (ellipse) will meet on opposite sides. (Bottom) The opposite vertices of a hexagon inside an ellipse will be connected by three lines that meet in a point.

Figure P1 for two aspects of Pascal's hexagram theorem.) In essence, the theorem states if a hexagon (six-sided figure) is circumscribed onto a conic figure (touches the outside of the circle or oval), then the vertices of this hexagon (lines connecting the opposite points where the sides of the hexagon meet) can be connected by three lines that meet at a central point inside the conic figure.

Pasteur's Germ and Vaccination Theories: Biology: *Louis Pasteur* (1822–1895), France.

Pasteur's Germ Theory of Fermentation: *Fermentation can be prevented in fermentable (organic) substances by not exposing them to airborne dust particles.*

Louis Pasteur speculated that only living organisms could differentiate between the various shapes of molecules (*molecular dissymmetry*). He concluded that a definite distinction existed between nonliving (inorganic) and living (organic) chemistry. He then studied fermentation and the effects of yeasts on living substances, which he claimed was a chemical reaction involving microorganisms. This prefaced his famous experiment that disproved the ancient concept of the spontaneous generation of life. About 200 years earlier, Francesco Redi had conducted a similar experiment to disprove spontaneous generation, but without the knowledge of the germ theory. Pasteur placed a cooked broth into a sterilized flask structured with a curved neck, which prevented air, and thus dust, from entering the flask. The broth did not ferment nor develop any bacterial growth. Then he broke off the curved neck, allowing air and dust to enter, which soon caused the broth to ferment. His interest in fermentation was inspired by France's wine and brewery industries, which were unable to control the quality of their wine and beer. Pasteur was asked by these industries to study the problem, which he did, using the results of his famous sterilized flask experiment. He realized "germs" from the air contaminated the wine, causing it to ferment. He also found microorganisms in the lactic acid that result from the fermentation of milk. As a solution, he devised a process whereby the wine was heated but not boiled, and then cooled in closed containers to prevent airborne dust from contaminating the product. He also applied this process of *pasteurization* to milk, which was heated to about 108°F and then quickly cooled. Pasteur also knew that some infections are caused by airborne microorganisms; thus he was one of the first to recommend the use of carbolic acid as an antiseptic, the boiling of surgical instruments, and maintaining proper hygiene in hospitals. *See also* Lister.

Pasteur's Vaccination Theory: *Attenuated microorganisms from animals with anthrax can be used to inoculate healthy animals, which will prevent the disease.*

This aspect of Louis Pasteur's work is an excellent example of serendipity— an occasion when an unexpected development or insight presents itself to a knowledgeable observer. It seems that after identifying a batch of chicken cholera bacilli under his microscope, he neglected it for several weeks during a hot summer. (Chicken cholera is similar to but not the same as the waterborne cholera contracted by humans.) Even though this was not a fresh batch of his bacteria, he injected it into healthy chickens. These chickens contracted only a mild case of cholera. He then proceeded to infect both the vaccinated group of chickens and a control group (nonvaccinated) of chickens with fresh cholera bacilli. The chickens that received the attenuated (weakened) bacteria contracted only mild cases of the cholera, while the noninoculated chickens died of the disease. Thus, Pasteur discovered how to produce vaccines that could immunize animals against diseases. He proceeded to heat gently, to about 75°F, the anthrax bacilli that cause diseases in sheep, cattle, and horses, as well as humans. In one experiment he injected twenty-four sheep with the virulent form of anthrax and twenty-four sheep that had been immunized with his attenuated bacilli. All twenty-four nonimmunized sheep died; all of the immunized sheep lived. His most famous accomplishment was his treatment of a young boy who had contracted rabies from the bite of a rabid dog. Pasteur developed the antirabies vaccine by injecting the virus into the spines of rabbits and then using these attenuated viruses to inoculate the boy who, after several treatments, survived. The concept of vaccination was improved and expanded by others to include inoculations for diphtheria, typhoid, cholera, plague, poliomyelitis, smallpox, measles, and other diseases of humans and animals. Louis Pasteur is known as the Father of Microbiology. *See also* Koch; Lister; Redi.

Pauling's Theory of Chemical Bonding: Chemistry: *Linus Carl Pauling* (1901–1994), United States. Linus Pauling was awarded the 1954 Nobel Prize for chemistry.

The nature of chemical bonding of elements and molecular structures can be determined by the application of quantum mechanics.

Linus Pauling's career covered not only his early work in determining the nature of chemical bonding of atoms, the complex structure of organic molecules and crystals, the nature of oxygen binding to hemoglobin (sickle cell anemia), and vitamin C therapy, but also for his activities to halt nuclear testing, for which he received the 1962 Nobel Peace Prize. Pauling used many techniques to study the structure and bonding properties of atoms, including electron and x-ray diffraction of large protein molecules and electromagnetic instruments to assist in determining molecular structure. His pioneering approach was the use of quantum mechanics to describe how electrons are arranged in orbits, how they bond, at what angles they bond, their bonding energies, and the distances between electrons in different atoms that are combined. (See Figure S2 under Sidgwick.) This was important because up to this time, quantum mechanics was usually limited to explanations of phenomena at the larger atomic level rather than the subatomic and energy levels. Pauling found that some elements and compounds do not follow the classical valence bonding of single electrons, but rather exist in two or more forms through the process of resonance. He made two unique observations. One was the concept of hybrid molecules, which accounted for various shapes of molecules. The other was the concept of resonance for molecules, which explained how a molecule could appear to be somewhat like similar molecules yet have a slightly different structure and thus different characteristics. This is how he conceived the idea that oxygen molecules do not bond in a normal way with some types of hemoglobin cells that are "sickle" shaped. This disease was first discovered by James Bryan Herrick, but Pauling identified its cause as a genetic malformation of blood cells. People who inherit this *molecular disease* often die at an early age since it affects their blood's capacity to carry oxygen to tissue cells. In 1950 Pauling described the structure of the complex protein molecules involved in the chromosomes of cells as an alpha-helix structure, which almost described the double-helix structure of DNA announced in 1954 by James Watson and Francis Crick. Pauling was the first to use quantum mechanics to explain how atoms bond to form molecules, including the concept of electronegativity. His concepts that describe the types of atomic bonding are depicted as complex three-dimensional figures. *See also* Bohr; Crick; Ingram; Pauli; Watson.

Pauli's Exclusion Principle: Physics: *Wolfgang Pauli* (1900–1958), Switzerland. Wolfgang Pauli was awarded the 1945 Nobel Prize for physics.

Only two electrons of opposite spin can occupy the same energy level (same quantum number) simultaneously in an atom's orbit.

Wolfgang Pauli was aware of Niels Bohr's application of quantum theory to the electrons orbiting the nuclei of atoms. At the time, Bohr based his "planetary" structure of electrons orbiting the nuclei on the new concept of quantum mechanics. In cooperation with Arnold Sommerfeld (1868–1951), Bohr expanded his "solar system" atom to include the energy associated with the electrons in different orbits. The quantum aspect stated that each orbiting electron could have only one of three quantum numbers, referred to as n, l, and m. Pauli proposed the concept that required each electron to be one member of a pair of electrons, one of which "spun" around its axis in one direction, while the other spun in the opposite direction. For this situation to exist, he introduced a fourth quantum number for electrons, s, which has the value of $+1/2$ or $-1/2$. These numbers correspond to the spin of each member of the pair. This is the point at which his famous 1924 *exclusionary principle* enters the picture (also referred to as the *Pauli principle of exclusion*). The principle states that no two electrons in an atom may have the same four quantum numbers—n, l, m, or s. The spin and exclusion principle, which explained many mysteries related to the structure of the atom, was confirmed a few years later. In other words, if an electron is at a specific energy level in a particular orbit, no other electron can be at that same exact level; thus, other electrons are "excluded." In 1930, Pauli identified a particle in the atom's nucleus that he believed was another type of neutron that had no charge and was emitted along with an electron during beta decay of a neutron. The problem was that the new particle was too light to be a neutron. The actual neutron that exists in the nucleus was discovered later by Sir James Chadwick. Pauli's "neutron" was later verified and renamed by Enrico Fermi as the *neutrino. See also* Bohr; Chadwick; Fermi; Steinberger.

Peano's Axioms and Curve Theorem: Mathematics: *Giuseppe Peano* (1858–1932), Italy.

Peano's Axioms: *The use of symbolic logic and the axiomatic method will provide rigor to the theorems of mathematics.*

Mathematical logic is the use of symbols instead of words to explain mathematical statements. This form of logic is sometimes referred to as a *universal language* because universal symbols make working with equations easy for anyone to understand, no matter what language is spoken. Giuseppe Peano conceived nine different logical mathematical axioms. Five of his most famous axioms involved the logic of developing numbers:

1. The figure 1 is a number.

2. Any number that follows 1 in sequence is also a number.

3. No two numbers can have the same successor number.

4. 1 cannot be a successor to any other number.

5. If 1 has a property and any successor also has that property, so do all numbers.

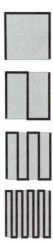

Figure P2: Peano's curve is related to Hilbert's curve, and Koch's curve. All versions of continuous curves are based on the concept that many different types of spaces can be completely filled by continuous curves (of many kinds) if the space is contained, the curves are connected, and the curve (lines) does not cross itself.

These axioms are in the form of one of Peano's syllogisms, which describe numbers in terms of a set of elements. They are a logical series of statements used to define what a "number" is and is not. Peano claimed that the natural (real) number system can be derived from his axioms.

Peano's Curve Theorem: *There are continuous curves that completely fill all the space inside squares or cubes.*

Peano's continuous curve theorem is based on his idea of the cluster point of a function, which is the basic element for geometric calculus. He defined a curved surface as the length of an arc as compared to the area within a curved surface. He named this geometric function a *space-filled curve.* Peano's curve theorem was revised to include a great variety of types of space that may be completely filled up by a continuously drawn curve, if and only if that space is connected, compact, has a continuous border, and is measurable. For the theorem to apply to both enclosed circles and squares, the "curve" or continuous line must not cross itself. For a square, the space-filling curve must have at least three points of multiplicity. (See Figure P2.) Peano's curve theory is based on geometric calculus, which can be used to measure space within a curve. *See also* Leibniz.

Pearson's Statistical Theories: Mathematics: *Karl Pearson* (1857–1936), England.

It is possible to measure statistically the continuous variations responsible for natural selection.

Karl Pearson developed several important statistical methods for treating data related to evolution. The concept of natural selection as a series of continuous variations was, according to Pearson, a problem dealing with biometrics (using information from living systems to develop synthetic systems). His statistical approach was in direct conflict with that of other scientists, who believed that evolution (natural selection) was a process of discontinuous variations based on breeding as the main mechanism. Among the important statistical concepts that grew out of Pearson's work with evolution are these:

- *regression analysis*, used to describe the relationship between two or more variables. In controlled experiments, it is used to estimate the value of the dependent variable (the experimental factor) on the basis of independent variables (controlled factors).

- *correlation coefficient*, the statistical measurement between two variables that are quantitative or qualitative in nature. The measurements are unchanged by the addition or multiplication of random variables. This statistical method expresses measurable (and probable) differences between two events.

- *chi-square test* of statistical significance, a means for determining the "goodness of fit" between two binomial populations (two different but related groups), each having a normal distribution. It is a test to determine how far the event is from the statistical mean or how factors for two different events "match."

- *standard deviation* (a term originated by Pearson), the statistical treatment used to determine the difference between a random variable and its mean. It is expressed as the square root of the expected value of the square of this difference.

Many of Pearson's concepts deal with averages. *Mean* is the term we usually associate with "average." It is determined by adding all the values or a set of numbers and dividing the sum by the total number of values—for example, $2+4+9+3 = 18 \div 4 = 4.5$ as the mean (average).

Median is the central point in a series (set) of numbers when arranged in order of numbers or value. For the median, there is an equal number of greater (larger) and lesser (smaller) numbers above and below the median value for example, in the sequence 1,2,3,4,5,7,9, 4 is the number halfway between those numbers above and below it, and thus the median. The median may equal the mean, or it may not.

The *mode* is the number most frequently occurring in a set of numbers or values. In the sequence, 1,1,2,2,5,5,5,6,8,9,12,16,25, the mode is 5; it occurs most often in the set. Using the mode instead of the mean can influence the meaning of the data. For instance, if the salaries of 100 workers and 5 executives in a company are listed and the mode is used to express an average salary for company employees, the figure will certainly be lower than if the mean (average) salary is used for the calculation. *See also* Darwin; Galton; Mendel.

Peierls' Concept for Separating U-235 From U-238: Physics: *Sir Rudolph Ernst Peierls* (1907–1995), England.

Rare uranium-235 can be separated from uranium-238 by the process of gaseous diffusion.

Rudolph Peierls was familiar with Henri Becquerel's accidental discovery of radioactivity, the Curies' work in separating radium from uranium ore, and Otto Frisch's and Lisa Meitner's theory of fission of unstable nuclei of uranium. In 1940 Peierls and Frisch collaborated on forming the rare isotope of U-235 into a small mass that would spontaneously fission, resulting in the production of tremendous energy that might be used to construct a giant bomb. In 1913, Frederick Soddy theorized that when a radioactive element gives off alpha particles (helium nuclei), it changes from one element into a different element with a different atomic number. With the loss of a beta particle (electron), there is also a loss of a negative charge. Thus, the nuclei gains one positive charge, which also makes it a different element. Soddy realized that if neutrons were added to or removed from the nuclei of atoms, the charge would not change (proton/atomic number), so the element would still fit its same place in the **Periodic Table of the Chemical Elements**, although it would have a different atomic weight. Soddy called these variations of elements by weight *isotopes*. Based on the information that uranium contained at least two different isotopes (common uranium-238 has 92 protons and 146 neutrons, while uranium-235 consists of 92 protons and 143 neutrons), Peierls and Frisch attempted to separate the two isotopes of uranium. This proved to be difficult since only 1 out of 140 atoms of uranium is the U-235 isotope. Even so, they made some calculations for the energy output that could result from the fission of U-235. In addition, they determined it would take only 1 pound of U-235 to sustain a chain reaction. For example, once a few atoms started to fission, they would, in very rapid geometric progression, cause all the other nuclei also to fission almost instantly, thus causing a huge explosion. The next problem was how to separate the fissionable U-235 from the stable U-238. This was accomplished by a massive effort using gaseous diffusion. This process converted purified uranium ore into a gas that allows the heavier U-238 to be separated through diffusion filters from the lighter U-235. The first atomic bombs were produced by this method. A much simpler and less expensive method of producing a fissionable material was to force uranium-238 nuclei to absorb neutrons inside a nuclear reactor. This nuclear reaction formed neptunium-239 (93 protons and 146 neutrons), which decays into plutonium-239 (94 protons and 145 neutrons). Although plutonium is stable, if it is forced to absorb slow neutrons, it becomes fissionable. This procedure became the basis for producing modern nuclear (atomic) weapons. *See also* Becquerel; Curies; Frisch; Meitner; Soddy.

Penrose's Theories for the Black Hole, "Twistors," and "Tiling": Physics: *Roger Penrose* (1931–   ), England.

Penrose's Hypothesis: *Black holes are singularities with "event horizons."*

Roger Penrose, along with Stephen Hawking, applied Einstein's theory of general relativity to prove that black holes are *space-time singularities* (a single event in space-time within a trapped surface). They proposed this concept even though such phenomena have no volume, are infinitely dense, and evolve as space-time events. Penrose's hypothesis is based on the fact that singularities are not "naked" but rather have an "event horizon," which is the outer limit boundary where all mass, including light once it enters this border area, will be sucked into the black hole by tremendous gravity. The outer rims of black holes (accretion disks) emit visible light but much less than do neutron stars. Penrose stated that things do happen in a black hole. Only a massive, dense object in deep space with an event horizon creates the energy adequate to cause energy (light) to disappear. Particles break down into two new particles, one of which would be trapped in the black hole while the other, containing more mass and energy than the original particle, might escape out the bottom. Such an arrangement of matter being changed and compressed on entering one hole and then exiting might be responsible for the birth of new galaxies or universes. If this proves to be correct, the physical concept of conservation of energy will be preserved. Orbiting space craft carrying x-ray instruments are exploring black holes to learn more about them.

Penrose's Twistor Theory: *There are massless objects in space existing in "twistor space."*

Penrose's "twistor" theory is a new, complex geometric construct that he developed to explain a synthesis of quantum theory and relativity in space-time. Penrose refers to his theory as "twistor space," where particles have no mass at rest but do exhibit properties of linear and angular momentum when the particles change position from their point of origin. The particle in space-time may have spin as well as a vector of four dimensions. He developed his theory to replace the Einsteinian theory of relativity and a four dimensional space-time. Twistor theory is a complicated and a not-well-understood or accepted area of research in mathematics.

Penrose's Tile Theory: *"Tiles" can intersect at their boundaries while never overlapping.*

Tilings, also known as *tessellations*, occur when geometric figures continue to repeat themselves. In other words, tessellations happen when an arbitrarily large plane surface is arranged with nonoverlapping tiles (i.e, the tiles connect only at their boundaries). The tile figures can be constructed by triangles, squares, and hexagons (with three, four, and six possible symmetries). (See Figure P3.) At the same time a "tiling" figure can never have five sides, such as a pentagon, because it cannot be "folded" so that the edges meet perfectly (i.e., it is not symmetrical). Penrose's tiling also explains the structure of crystals that can have three-, four-, or six-folded rotation symmetry, but not a five-fold rotational symmetry. In 1984 this belief about a five-fold crystal structure being impossible seemed to be disproved when a crystal of an alloy composed of aluminum and manganese was rapidly cooled to form such a crystal. The possible distinction could be that the crystal is three-dimensional while the three, four, and six symmetries were two-dimensional. Tiling may be thought of as a periodic pattern that carries the design into itself. **Fractals** are similar to tiling in that they are self-contained repeating patterns of decreasing size. (See Figure

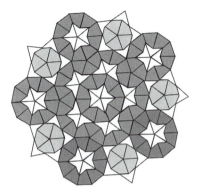

Figure P3: Penrose tiles use the same graphic figure repeatedly, but not overlapping, to cover an arbitarily large large two-dimensional area with a continuous design. Tessellation is the placing of tiles as congruent polygons in a plane.

W1 under Wolfram.) Generating Penrose's tiling and fractals on personal computers has become a popular exercise for creating geometric designs.

Penzias' Theory for the Big Bang: Astronomy: *Arno Allan Penzias* (1933–), United States. Arno Penzias shared the 1978 Nobel Prize for physics with Pyotr L. Kapitza and Robert W. Wilson.

Background radiation received on earth from all directions in space is the leftover microwave radiation following the big bang.

In the early 1960s Penzias and his colleague at Bell Laboratories, Robert W. Wilson, were working with a special radio antenna they designed to detect signals from communication satellites, when they received some unexplained background noise. They continually picked up all kinds of signals—some generated by the internal electronics of their instruments, some from earth sources, and some unexplained. By 1964 they had eliminated much of the other noises but still continued to receive signals near the wavelength of 10^{-3} meters, indicating that the signal was at about 3.5 kelvin. This type of signal proved to be on a timescale much older than earth itself. Therefore, they concluded it was leftover microwave emission from the big bang origin of the universe about 14 or 15 billion years ago. This theory concerning cosmic background microwave radiation is considered an important breakthrough in modern astrophysics. *See also* Gamow; Hale; Hawking.

Perl's Theory for a New Lepton: Physics: *Martin Lewis Perl* (1927–), United States. Martin Perl shared the 1995 Nobel Prize with Fredrick Reines.

A new lepton, the tau particle, generated by a particle accelerator will decay into either an electron plus a neutrino/antineutrino or a muon.

There are two classes of subatomic particles: (1) the *hadrons*, which include the proton and neutron (protons and neutrons along with their "binding" quarks are also considered **fermions**), which have half-integer spin and are quantum individualistic in the sense that they obey the Pauli exclusion principle), and (2) *leptons*, which include four types of particles: the *electron* and the *muon*, along with their two related *neutrinos*. Perl proceeded to use a powerful accelerator to generate a new lepton. It is necessary to use some form of detection device to record the existence of the subatomic particles before they decay. He generated a record of over 10,000 events, of which only a few of the new particles were detected as he predicted. This new particle turned out to be a new, heavy lepton with more mass than a proton. This caused some problems since the four known leptons are strongly related by symmetry to their four quarks. Therefore, to maintain symmetry, one of the basic constants of physics, a new quark was predicted to match the new lepton. In 1977 this new quark was discovered and named the upsilon particle by Leon Lederman. Thus, both a new lepton and a new quark joined the growing multitude of subatomic particles. *See also* Lederman.

Perrin's Theory of Molecular Motion: Physics: *Jean Baptiste Perrin* (1870–1942), France. Jean Perrin was awarded the 1926 Nobel Prize for physics.

Einstein's formula for molecular motion can be confirmed by determining the size of molecules.

Jean Perrin was familiar with Robert Brown's (1773–1858) concept that the motion of molecules of water caused the motion of tiny pollen grains suspended in water (Brownian motion) and Einstein's theory that the average distance the pollen particle traveled in the water increased with the square of the time elapsed for the motion. After controlling for conditions of temperature and the type of liquid in which the pollen was suspended, Einstein was able to predict, on the average, how far a particle would travel. However, at the time Einstein made this prediction, there was no way to confirm it. Perrin, who was known for his theories concerning the discontinuous structure of matter and how sediments obtain equilibrium, related these ideas to the importance of knowing the size and energy of molecules. In 1908 Perrin, who had access to a "super" microscope, experimentally controlled and measured the size of molecules as related to molecular movement, thus confirming Einstein's theory. This formula for the size of water molecules was important for the confirmation of both the kinetic theory of matter (motion) and Avogadro's number. *See also* Avogadro; Einstein.

Planck's Formula and Quantum Theory: Physics: *Max Karl Ernst Ludwig Planck* (1858–1947), Germany. Max Planck was awarded the 1918 Nobel Prize for physics.

Planck's Formula: $E = \hbar v$; where E = the energy involved, $\hbar$ = Planck's constant of proportionality, and v = the frequency of the radiation.

For centuries, physicists were puzzled by the two theories of light. During the nineteenth century, some thought the corpuscular (particle) theory and the wave (electromagnetic radiation) theory were inconsistent with the then current theory of molecules and thermodynamics. In the 1860s, Gustav Robert Kirchhoff and other scientists experimented with black body radiation, an ideal surface, such as a hollow metal ball with a small hole that absorbs all light, that does not reflect back any light but rather emits radiant energy of all wavelengths. They found that a body at "red heat" emitted radiation at low frequencies for the spectrum of light waves (infrared and red), and that "white heat" emitted radiation at the higher frequencies at the yellow, blue, green end of the light spectrum. From these data, scientists projected curves on a graph to explain their theory. Max Planck plotted a new set of curves representing these data and advanced a different formula. His formula $E = \hbar v$ explained that energy radiated from the black body specifically in quanta (small bits) of energy, not continuously as previously believed. These quanta were represented by the $\hbar v$ in the formula, where v is the frequency of the radiation and the $\hbar$ is the action of the quanta of energy, which is a proportionality that can only assume integral multiples of specific quantities (quantum theory). The $\hbar$ is now known as *Planck's constant* and is one of the major constants in physics. In other words, the energy of a quantum of light is equal to the frequency of the light multiplied by Planck's constant.

Planck's Elementary Quantum Action Theory: *Energy does not flow in an unbroken stream but rather proceeds or jumps in discrete packets or quanta.*

The science of quantum mechanics is based on Planck's theory that energy can only be emitted or absorbed by substances in small, discrete packets he called quanta. This theory has been used and expanded by many scientists. After the 1950s, Planck's quantum theory was used extensively in producing and identifying numerous subatomic and subnuclear particles and energy quanta. There are dozens of other examples in science that make use of Planck's quantum theory. Today many people mistakenly consider a "quantum leap" to be a great stride or large advancement of events or accomplishments. Originally it referred to a very small packet of energy or mass, which may be thought of as a tiny "particle wave" of light or an electron's tiny gain or loss of energy when it moves from one orbit to another. (See Figure D1 under Dehmelt.) *See also* Chadwick; Compton; Einstein; Heisenberg; Kirchhoff; Pauli; Rutherford; Schrödinger.

Pogson's Theory for Star Brightness: Astronomy: *Norman Robert Pogson* (1829–1891), England.

Pogson's ratio is the interval of star magnitudes that might be represented by a multiple of five magnitudes.

From ancient times, the magnitude (brightness) of stars was based on what could be judged from Earth. The brightness of stars was ranked in just six magnitudes, the first being the brightest stars (excluding our sun) and the sixth faintest were those just barely visible. Norman Pogson, an Englishman who spent his life in India as an official astronomer, proposed a more rational and useful system to determine the magnitude of stars. He realized the first-magnitude stars were about 100 times brighter than those in the sixth-magnitude category. From these data he devised a *ratio of brightness* of 2.512. This means a fifth-magnitude star is 2.512 (about two and half) times as bright as is a sixth-magnitude star. It was soon evident his "ratio" was not adequate for the actual range of brightness to cover all luminosities, so negative magnitudes were introduced. For instance, the sun is a −26.7 magnitude star (as viewed from earth), and the brightest star beyond the sun is the −1.5 magnitude star, Sirius. The moon has a luminosity of −11.0. Pogson's ratio is still used today, but it is augmented by using the spectrum and colors of stars as recorded on photographic plates, which, through timed exposures, can record stars beyond the 20 magnitude level. (See Figure H2 under Hertzsprung.)

Ponnamperuma's Chemical Theory for the Origin of Life: Chemistry: *Cyril Andrew Ponnamperuma* (1923–1994), United States.

It was possible for chemicals and energy existing in the primordial atmosphere to synthesize protein molecules and nucleic acids required for life.

Cyril Ponnamperuma's theory is based on three processes that must proceed in sequence for life to form from chemicals and energy. First, the necessary atoms must form into the required molecules. Second, these molecules must combine into self-replicating polymers. And, third, these polymers (large organic molecules) must unite into living cells, tissues, organs, systems, and finally organisms. Ponnamperuma and several other scientists attempted to achieve this process actinically in the laboratory. One attempt exposed a mixture of water, methane, and ammonia to beta radiation, expecting to produce adenine (a purine found in RNA), but no success in the second and third stages of this process was achieved. Another attempt exposed formaldehyde to ultraviolet light to produce a polymer, again without success. Several other scientists have synthetically produced a variety of organic molecules, but none of these experiments met the three stages required in the process of producing life as described by Ponnamperuma. *See also* Chambers; Miller.

Poseidonius' Concept of the Earth's Circumference: Astronomy: *Poseidonius of Apamea* (c. 135–51 B.C.), Greece.

The circumference of the earth can be calculated by measuring distances between two locations, both on the same meridian circle.

About 200 years before Poseidonius conceived his method for measuring the circumference of earth, Eratosthenes of Cyrene measured the distance between two distant cities at the same time of the summer solstice and used this figure to calculate that earth's circumference was 25,054 miles (the current average circumference of earth is 24,857 miles). Poseidonius used the figure of 5000 stadia as the distance between two cities located on the same meridian. (*Stadia* is an ancient Greek measurement of distance based on the length of the course in a stadium. It is equal to approximately 607 feet, or 185 meters.) This meridian encompasses 1/48 of the circle of earth's circumference. In other words, he projected that the distance between the cities equaled about 1/48 of the distance around the globe at that particular meridian. From this he multiplied 48 times 5000 to arrive at a circumference of 240,000 stadia, which compared favorably with Eratosthenes' figure of 250,000 stadia. We now assume there are 8.75 stadia to the mile, so the circumference comes out to about 27,000 miles. Poseidonius thought 240,000 stadia was much too large, so he reduced his figure to only 180,000 stadia. About 1000 years later this had unexpected consequences when Christopher Columbus used Poseidonius' figure for a much smaller earth rather than the one provided by the ancient astronomer Eratosthenes. Therefore, Columbus believed Asia was only 3000 miles west of the European coast, which made his trip to the New World, which he thought was India, much longer than expected. *See also* Eratosthenes.

Priestley's Theories of Electrical Force and Dephlogisticated Air: Chemistry: *Joseph Priestley* (1733–1804) England.

Priestley's Theory of Electrical Force: *The force between two charged bodies decreases as the square of the distance separates the charged bodies.*

Joseph Priestley was a friend of Benjamin Franklin, who encouraged him to investigate the new phenomenon of electricity. He was the first to measure the electrical force between two charged bodies as related to the distance between them. He calculated that if the distance between the two bodies is increased by a factor of 2, the electrical force is decreased by a factor of 4. This follows the well-documented general square law of physics, which was confirmed by other scientists. He also was the first to determine that charcoal (carbon) could conduct electricity. This became an important concept when applied to the new uses of electricity, such as the carbon arc light, the arch furnace, and electric motors.

Priestley's Dephlogisticated Air: *When the oxides of certain metals are heated, they produce an air that has lost its phlogiston.*

Although not the first to experiment with the heating of materials to drive off gases, Joseph Priestley was one of the first chemists to make careful observations of what happened to the materials he used. He was also the first to try different experiments to help him understand respiration and combustion. In 1771, he hypothesized that when a candle is burned in a closed jar, it consumes much of the "pure" air. Therefore, there must be some way for nature to replenish the air dissipated by burning objects, or else it would all be used up in the atmosphere and none left for respiration. Then he placed a small green plant in the same jar with the candle and found that after several days, the air again would support combustion. Using similar techniques, Priestley isolated several other gases by heating different substances. He produced sulfur dioxide, ammonia, nitrous oxide, hydrogen chloride, and carbon monoxide. He collected a gas, which was known as "fixed air" (CO_2), given off from the vats in a brewery. He then bubbled it through water; the result was carbonated water. Priestley then heated a small amount of mercury in a closed container and noticed that it formed a red "calx" on the surface similar to rust. He proceed to place a candle, and then a mouse, in this air given off by the heating of the mercury oxide. The candle burned much brighter and the mouse lived much longer in this new air that he called "dephlogisticated" air because he believed it lost its "phlogiston." Although Joseph Priestley is credited with the discovery of oxygen, Antoine Lavoisier named *oxygen*, from the Greek word meaning "sharp" because, at one time, scientists mistakenly thought that all acids contained oxygen. *See also* B. Franklin; Lavoisier; Scheele.

Prigogine's Theories of Dissipative Structures and Complex Systems: Chemistry: Ilya Prigogine (1917–   ), Russia. Ilya Prigogine received the 1977 Nobel Prize for chemistry.

Prigogine's Dissipative Structures: *States of thermodynamic equilibrium for systems are rare. More common states exist where there is a flow or exchange of energy between systems.*

One example Ilya Prigogine used to explain his "dissipative structures" was the solar system. Without the sun's continual bathing of the earth with energy, earth's atmosphere would soon reach thermal equilibrium, meaning it would reach a sustained very cold temperature since heat always flows to cold, not the reverse. Since the sun provides a steady flow of energy to earth, this might be thought of as *negative entropy* or, as Prigogine believes, a process that reverses irreversible equilibrium states. His work on irreversible processes is credited with forming a bond between the physical sciences and biology that deals with systems that over time have not obtained equilibrium—life and growth. An example is what happens in living cells as they constantly exchange substances and energy with their surroundings in tissues. The process of entropy is irreversible only if there is no exchange of energy between or among complex systems. Theoretically, some billions of years in the future of the universe, entropy (the complete disorganization of matter) and the irreversible attainment of thermal equilibrium will win out unless a new source of universal energy is forthcoming.

Prigogine's Theory of Complex Systems: *Simple molecules can spontaneously self-organize themselves into more complex structures.*

Ilya Prigogine is known as the grandfather of *chaos theory*, which, in the science of complex structures, has a more specialized meaning than the one used in ancient as well as modern times. Before the development of Prigogine's theory dealing with dissipative and irreversible processes, chaos theory was thought of as a mathematical curiosity. More recently, chaos theory, as related to complex systems, has had a widespread impact on several science disciplines, particularly biology, but also economics. Chaos deals with initial conditions and how these conditions alter the causes that create problems when trying to predict effects. A classic example is that the knowledge of initial conditions of a weather system does not provide adequate information, down the line, to be able to predict the weather with any degree of accuracy. Weather is a classic complex system where the chaos theory is applicable. (See Figure W1 under Wolfram.)

Prigogine believed that very simple inanimate and inorganic molecules, at least at one time, had the ability to organize themselves spontaneously into higher, more complex organic molecules and organisms. This process must have involved some exchange of energy for the self-organizing molecules to reverse entropy (or, as Prigogine would say, "nonequilibrium thermodynamics"). This is not exactly the same as the old idea of spontaneous generation, but it might be thought of as a modern version of that idea, which he also related to evolution. Ilya Prigogine received a Nobel Prize for his work for nonequilibrium thermodynamics (dissipative structures), which relates to concepts in chemistry, physics, and biology.

Ptolemy's Theory of a Geocentric Universe: Astronomy: *Claudius Ptolemaeus* (Ptolemy of Alexandria), (c. 90–170) Egyptian.
Earth being the heaviest of all bodies in the universe finds its natural place at the center of all the cosmos.
Ptolemy collected and compiled a great deal of information from other astronomers. From Aristotle, he gleaned there were two parts to the universe—earth and the heavens and that earth's natural place was at the center of all the universe. He considered earth the sublunary region where all things are born, grow, and die, while the heavens are composed of compact concentric crystal spheres surrounding earth. (See Figure P4.) Each shell was the home of a heavenly body arranged in the order of the Moon, Mercury, Venus, Sun, Mars, Jupiter, and Saturn, followed by the fixed stars and the "prime mover," who kept the whole system moving. The other person who influenced Ptolemy was Hipparchus (c. 190–120 B.C.), who formulated positions and motions for the

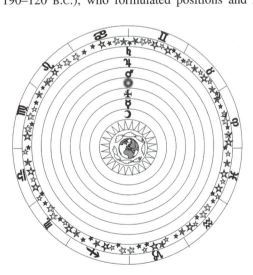

Figure P4: Artist's conception of Ptolemy's geocentric universe. The original contained as many as eighty epicycles, which he devised in order to match observations.

planets and moon. From this background Ptolemy not only claimed the universe was geocentric, but that all bodies that revolve in orbits do so in perfect circles and at constant velocities, while the stars move in elliptical orbits at inconsistent velocities. This required the application of complicated geometry, which Ptolemy used to describe these motions. These three kinds of motions traced the following geometric paths: eccentric, epicycle, and the equant. Ptolemy combined these to form his *Ptolemaic system* for planetary motion. His system was not accurate enough to determine all the motions of heavenly objects, but it was used by other astronomers for over 1300 years, until in 1514 Nicholas Copernicus developed his first heliocentric model of the universe, which he continued to refine for the next thirty years. *See also* Aristotle; Copernicus; Galileo.

Pythagoras' Theorem: Mathematics: *Pythagoras of Samos* (c.580–c.500 B.C.), Greece.

The square of the hypotenuse of a right triangle is equal to the sum of the squares of the other two sides of the triangle.

Pythagoras believed that whole numbers, as well as fractions expressed as ratios of whole numbers, were not only "rational numbers" but also explained the basis of the universe. However, when he compared the sides of a right triangle with the ratio of 1 to 2, the opposite side of the 90 degree angle (hypotenuse) was an "irrational number." In other words, the diagonal of a square cannot be related to the sides expressed in whole numbers and thus the ambiguity of the square root of 2. The original concept for the Pythagorean theorem goes back 1000 years before Pythagoras to Babylon when the idea was first conceived that any three-sided figure with sides containing the ratio of 3:4:5 would form a 90 degree right angle triangle. Proof for the theorem was derived by the Pythagoreans ($a^2 + b^2 = c^2$) with the credit for the proof given to Pythagoras who was the leader of an academic "cult" of mathematicians who believed their work was sacred and should be kept secret. They believed all events and all things can be reduced to mathematical relationships. Their motto was stated as, "All things are numbers," and their secrecy is one reason it is difficult to determine which writings were by Pythagoras and which by his fellow Pythagoreans.

R

Rabi's Theory of Magnetic Moment of Particles: Physics: *Isidor Isaac Rabi* (1898–1988), United States. Isidor Rabi was awarded the 1944 Nobel Prize for physics.

Neutron beams can be used to determine the magnetic moments of fundamental particles such as the electron.

Isidor Rabi advanced the work of Otto Stern who in 1922 used a beam of molecules to determine the spacing of atomic particles referred to as *space quantization*. Rabi proceeded to develop a beam combined of various atoms and molecules that he used to produce magnetic resonance (oscillations), which could accurately determine the magnetic moments of fundamental particles. His theory and experiments resulted in the development of nuclear magnetic resonance (NMR), making it possible to measure the energies both absorbed and given off by the resonating atoms and molecules, which then can be used to identify substances. The process was revised and improved to produce a better image of human tissue than x rays, and its name was changed from NMR to magnetic resonance imaging (MRI) because of the mistaken belief that "nuclear" referred to nuclear radiation, rather than the oscillating nuclei of the atoms in the tissue cells of human bodies. *See also* Ramsey; Stern; Tyndall.

Raman's Theory of Light Scattering: Physics: *Sir Chandrasekhara Venkata Raman* (1888–1970), India. Chandrasekhara Raman was awarded the 1930 Nobel Prize in physics.

A small amount of light of specific frequencies will be reflected from a substance exposed to a direct beam of light of a single frequency.

Chandrasekhara Raman determined that a beam of light of a single frequency, when striking a substance at right angles, would produce some frequencies different from the original one. He further discovered that these new frequencies were specific to the type of material from which the beam was reflected. This became known as the *Raman effect*, which is the exchange of infrared frequency of the light and the material reflecting the light. Although the Raman effect is very weak—about 1/100,000 times less intense than the light of the incident beam—this scattered light of different frequencies can measure the exchange of energy between the light and the substance being examined. The characteristics of the molecules of the examined substance exhibit an intensity proportional to the number of scattering molecules that happen to be in the beam of light. This technique can identify specific gases, liquids, and solids. Gases have a low molecular concentration and thus produce a very weak Raman effect. Even so, the Raman effect is a very accurate and effective tool for both quantitative and qualitative analysis. *See also* Tyndall.

Ramsay's Hypothesis for Inert Gases: Chemistry: *Sir William Ramsay* (1852–1916), England. William Ramsay was awarded the 1904 Nobel Prize for chemistry.

The placement of the inert gas argon in the Periodic Table of the Chemical Elements indicates there will be other similar inert gases, still to be detected, with greater atomic weights.

Sir William Ramsay followed the work of Lord Rayleigh and Henry Cavendish, both of whom experimented with air and discovered that, after removal of all the nitrogen and oxygen, there appeared to be some "leftover" gas. Rayleigh and Cavendish believed this small amount of unidentified gas was the result of contamination. Ramsay collected this small amount of gas and used spectroanalysis to examine its properties. In 1898 he determined this gas was a new element, naming it *argon* from the Greek word *argos*, meaning "inert." Based on his theory of the placement of argon in the **Periodic Table of the Chemical Elements**, Ramsay predicted there were more inert gases yet to be found. He and several colleagues proceeded to identify helium as the gas emitted from the radioactive decay of radium. Other inert gases are neon (new), krypton (hidden), and xenon (stranger). These gases have many uses, including the gas in light bulbs, neon tubing, lasers, photographic speed lights, the decarbonizing of iron during smelting, and as nonoxidizing gas for welding. *See also* Cavendish; Rayleigh.

Ramsey's Chemical Shift Theory for Improved MRI: Physics: *Norman Foster Ramsey* (1915–   ), United States. Norman Ramsey shared the 1989 Noble Prize for physics with Hans G. Dehmelt and Wolgang Pauli.

The chemical shift of molecules affected by nuclear magnetic resonance can be identified by using magnetic shielding.

Norman Ramsey improved Isidor Rabi's nuclear magnetic resonance (NMR) technique by using two different radio frequencies, which result in more accurate measurement of the magnetic effects on atoms. Magnetic shielding enclosed the magnetic field within a specified area, preventing external static charges from interfering with the process. The NMR process causes nuclei to resonate, thus revealing the magnetic properties of their atoms and molecules. It can be used to analyze the structure of molecules and their interactions with other nuclei in close proximity. The modern NMR, now called magnetic resonance imaging (MRI), can detect a variety of conditions in the human body. Ramsey also utilized his concept of separate oscillating fields to produce molecular beams in a **maser** to run a very accurate atomic clock. In addition, he worked out a statistical model for negative thermodynamic temperature systems, theorizing the possibility of temperatures below absolute zero (below 0 kelvin or $-273.16°C$). *See also* Dehmelt; Pauli; Rabi.

Raoult's Law: Physics: *François-Marie Raoult* (1830–1901), France.
The amount of decrease in the freezing point of a dissolved substance (as compared to just the solvent) is related to the amount of the dissolved solute as well as to the molecular mass of the solute.

Raoult's law was based on Jacobus Van't Hoff's work on solutions previously done with the optical activity of organic compounds when in solution. Raoult's observations were founded on how and why the freezing point of salt changed when dissolved in water as compared to its being dissolved in an organic solvent. Many organic compounds are optically active in the sense that they rotate the plane of polarized light. Raoult's law is important for understanding the structure and determining the molecular weights of organic compounds. *See also* Van't Hoff.

Raup's Theory of Cyclic Extinction of Animals: Biology: *David Malcolm Raup* (1933–), United States.
The cyclic extinction rates for animals peaks every 26 million years.

David Raup based his theory of a 26-million-year cycle for the mass extinction of life on earth on data gathered in cooperation with his collaborator John Sepkoski (1948–1999). Raup believed that fossil evidence of a smooth, evolutionary transition from one species to another, as Darwin claimed, is not convincing. Darwin assumed these gaps in the fossil records would be filled in time and with more exploration. Over a hundred years after Darwin, Raup maintained that the fossil record was still too incomplete to account for a gradual evolution and suggested that general extinctions were caused by extraterrestrial catastrophic phenomena (e.g., asteroids, comets, meteors), not terrestrial disasters, such as earthquakes and volcanic eruptions. Raup's theory is somewhat related to the catastrophic theory proposed by Eldredge and Gould for the extinction of the dinosaurs 65 million years ago by an asteroid. Raup proposed that earth has a "companion" star with a 26-million-year orbital period, meaning it returns to the region of the solar system on a periodic basis, bringing with it showers of asteroids that impact Earth. He called this companion sun Nemesis. Most astronomers reject the Nemesis theory. *See also* Agassiz; Cuvier; Darwin; Eldredge; Gould.

Rayleigh's Light Scattering Law: Physics: *Third Baron Rayleigh* (1842–1919), England. (Rayleigh was born John William Strutt.)

When energy is removed from a beam of electromagnetic radiation (light), the change in the direction (angle) and wavelength of the emitted radiation is dependent on the scattering nature of the medium through which it passes (i.e., gases or liquids).

Third Baron John Rayleigh confirmed John Tyndall's theory that light passing through the atmosphere is scattered by small particles suspended in the air. The Tyndall effect explains that since water droplets in clouds are larger than the wavelength of light, the clouds appear white. Rayleigh applied mathematics to this concept of scattering to explain why the sky is blue in color. He claimed that light from overhead (midday) is more direct and thus is less scattered (fewer particles to travel though) than light coming from the sun when it is near the horizon. Since there is less scattering of overhead light, the wavelengths of this visible light of the electromagnetic spectrum are shorter, and thus appear blue. The same reasoning can explain red sunsets, which scatter more light toward the red end of the spectrum. Rayleigh accomplished this by determining that the amount of scattering was dependent on the wavelength of light. There are two kinds of scattering: instantaneous, considered "true" scattering, which occurs rapidly when electromagnetic energy is absorbed from the incident beams and then re-radiated; and "delayed" scattering, during which a time lapse between the absorption of the energy and its re-radiation takes place. Delayed scattering causes luminescence. This led to an expansion of the scattering law to include how longer AM radio (amplitude modulation) waves are scattered by the atmosphere and thus can travel around corners, while waves of shorter lengths such as FM radio (frequency modulation) and TV (television) waves cannot. Scattering experiments that cause beams of electrons, alpha particles, or other subatomic particles to collide with atomic nuclei have uncovered much about atomic structure and the fundamental nature of matter. These experiments use high-energy particle accelerators designed to scatter the particles and record the resulting paths of collisions. Rayleigh also explained there was another type of wave that followed along a surface whose motion decayed exponentially with the depth of the source from the surface. This type of surface wave is now called a *Rayleigh wave* and is the basis for the development of the science of earthquake detection. *See also* Maxwell; Raman; Tyndall.

Ray's Theories of Fossils and Plant Classification: Biology: *John Ray* (1627–1705), England.

Ray's Theories for the Origin of Fossils: *Fossils were formed by natural processes, not by God.*

John Ray's religion was based on his concept of "natural theology" in which he claimed that if one wants to understand God, one must study his creation of the natural world. Ray proposed several theories about fossils that were considered controversial at that time. Other scientists claimed fossils were formed by a creative force on Earth, or that God was making "models" of life, or fossils had satanic origins whose purpose was to confuse. Ray's proposed theory was that some organisms possibly washed into big cracks in the earth during the biblical flood. However, he did not believe this was a major cause since most fossils are found in beds, meaning they most likely would have been washed away by the flood. His major theory stated that at the time of creation, earth was covered by oceans where these organisms were created and lived. As the water receded, the living organisms were left on dry land, covered with mud and silt, later to become fossils. His theories that fossils were at one time "natural" living organisms laid the groundwork for future scientists, including Charles Darwin, to explore evolutionary adaptation. *See also* Darwin.

Ray's Classification System: *Plants and animals can be classified by differences in structure of species rather than by individuals.*

Ray is best known for his classification systems of plants and his later attempts to classify animals by structural similarities and differences. A major contribution was his division of the plant world by distinguishing between *monocotyledons* (based on seeds with a single opening leaf, e.g., grass, corn) and *dicotyledons* (based on seeds with two opening leaves, e.g., trees, beans). Ray also established the basis of classification systems on species, not individuals, and used this system to classify about 19,000 different plant species. Ray's classification system influenced Carolus Linnaeus and other taxonomists for several centuries and led others to explore the concept of biological evolution. *See also* Agassiz; Darwin; Linnaeus; Lyell.

Redi's Theory of Spontaneous Generation: Biology: *Francesco Redi* (1626–1697), Italy.

Flies do not generate spontaneously but rather develop from eggs, while some other worms and types of flies may appear by spontaneous generation.

From ancient times through the Renaissance period, it was accepted that some forms of lower life formed spontaneously from nonliving matter. It seemed obvious to most people that garbage generated rats, and food and manure sooner or later spontaneously generated flies. William Harvey was one of the first to contend that vermin, such as flies and rats, do not appear spontaneously but rather come from eggs and breeding. Francesco Redi decided to investigate Harvey's idea and conducted one of the first examples of a controlled experiment. First, he placed cooked meat in eight jars, covering four of them while leaving four uncovered. Maggots and flies developed in the uncovered jars but not the covered ones. He wondered if the air had something to do with the appearance of flies. Next, he placed more meat in another eight jars, covering four with gauze but leaving them open to air. He left the other four jars uncovered and exposed to the air. Redi concluded that maggots do not develop in covered jars that allow air in and keep out flies; therefore spontaneous generation is not a reality, at least for flies. He also concluded that flies must lay eggs too small to be seen in the open jars, and these eggs develop into maggots, which hatch into flies. However, Redi still believed that spontaneous generation was possible for some living organisms, but his controlled experiments did encourage others to perform more definitive experiments. *See also* Harvey; Pasteur; Spallanzani.

Reichstein's Theory of the Chemical Role of the Adrenal Gland: Chemistry: *Tadeus Reichstein* (1897–1996), Switzerland. Tadeus Reichstein shared the 1950 Nobel Prize for physiology or medicine with Philip Hench and Edward Kendall.

Six of the twenty nine identified chemical steroids are essential to prolong life in animals with damaged adrenal glands.

In 1946 Tadeus Reichstein isolated and identified twenty nine **steroid** hormones in adrenal glands. He synthesized aldosterone, corticosterone, and hydrocortisone and synthetically produced, on an industrial scale, the steroid deoxycorticosterone, which is used to treat Addison's disease. Earlier, he isolated what is known as ACTH (adrenocorticotropic hormone) or, more commonly, cortisone, used in the treatment of arthritis, skin rashes, and joint diseases. In 1933 Reichstein also artificially synthesized ascorbic acid (vitamin C), the first vitamin that could be mass produced.

Reines' Theory of Natural Neutrinos: Physics: *Frederick Reines* (1918–1988), United States. Frederick Reines shared the 1995 Nobel Prize for physics with Martin L. Perl.

If neutrinos exist in the high levels of radiation inside nuclear reactors, they should exist in the cosmic radiation.

In 1930 Wolfgang Pauli proposed the theoretical existence of what was called the **neutrino**, a fundamental physical particle that seemed to have no charge and much less mass than the neutron. Pauli claimed that such a particle was necessary to comply with the law of conservation of matter. (See Figure F2 under Fermi.) The problem was that it existed for only a very short period and was no longer detectable when it weakly interacted with other particles. Frederick Reines and Clyde Cowan (1919–1974) were the first to investigate the neutrino's properties, interactions, and role. First, they confirmed the neutrino's existence as being produced by the high radiation in nuclear reactors. The neutrino is difficult to detect because it travels only a very short distance before weakly interacting with matter and then disappearing. Confirming the neutrino's existence was an extremely difficult task, which Reines and Cowan accomplished in a deep pit near a nuclear reactor. It was necessary to shield out other high-energy particles and to use tanks of water to slow the neutrinos produced by the reactor so their instruments could record the neutrinos' interactions with other particles. At first, they detected only about three or four events per hour, but this was adequate to prove the existence of neutrinos. Reines and other collaborators were the first to discover neutrinos being emitted from the stellar supernova SN1987A, confirming his theory that neutrinos can be generated from outer space, most likely from the collapse of stars. Reines also found that neutrinos from outer space enter the ground (earth), which then produce **muons**;

neutrinos scatter electrons, which produce **antineutrinos**; and oscillating neutrinos can be transformed into different types. Reines' theory of cosmic neutrinos was the forerunner to neutrino physics and neutrino astronomy, which study the interactions of cosmic neutrinos with particles in the atmosphere and their sources in the cosmos. The research related to neutrinos continues as a study of particle physics, which may someday lead to a better understanding of the fundamental law of conservation. *See also* Fermi; Pauli.

Ricciolo's Theory of Falling Bodies: Physics: *Giovanni Battista Ricciolo* (1598–1671), Italy.

A pendulum that beats once per second can be used to confirm Galileo's theory of falling bodies.

Giovanni Ricciolo, an observational astronomer, disagreed with many of Copernicus' theories. Even so, he mapped mountains and craters on the moon and was the first to identify Mizar as a double star. He attempted to confirm Galileo's theory that the period of a swinging pendulum is the square of its length. He, and others who assisted him, tried to count the number of swings each day. If the number of swings per day could be adjusted to 86,400 (60 sec/min × 60 min./hr. × 24 hr./day = 86,400 seconds per day), they would succeed in developing a pendulum that could count seconds accurately. They tired of counting all day and night and so abandoned the project. However, Ricciolo used his pendulum to measure falling bodies. It is assumed that it was either Ricciolo or Simon Stevin, not Galileo, who dropped two balls of different sizes (weights) from the leaning tower of Pisa. Galileo used inclined planes to slow the descent of the balls and thus timed them with his heart pulse beat. Ricciolo used a pendulum as an accurate timekeeper. They both came up with the figure for g (the gravity constant) of approximately 9.144 meters per second squared which compares with today's figure of 9.807 (about 30 feet per second squared). *See also* Galileo.

Richardson's Law of Thermionic Emission: Physics: *Sir Owen Willans Richardson* (1879–1959), England. Sir Owen Richardson was awarded the 1928 Nobel Prize for physics.

The kinetic energy of electrons emitted from the surface of a solid is exponentially related to the increase in the emitter's temperature.

Sir Owen Richardson proposed an explanation for Thomas Edison's observation of the emission of electrons from hot surfaces (*Edison effect*). The electrons came from inside the solid, which was heated, and escaped from this material when the electrons achieved enough kinetic energy to overcome the "grasp" of the surface of the solid. Richardson's law states: *The electron's temperature increases exponentially with the increase of the emitter's temperature.* He related **thermionic emission** of metals to molecules that achieve adequate kinetic energy to escape from the surface of a liquid during the process of evaporation and boiling. This law became important in the development of electron tubes used in early radio, TV, and radar prior to the days of **transistors** and computer chips. *See also* Edison.

Richter's Theory of Earthquake Magnitude: Geology: *Charles Francis Richter* (1900–1985), United States.

Earthquakes can be measured on an absolute scale based on the amplitude of the waves produced.

Several earthquake scales existed before Charles Richter developed his absolute log scale. In 1902 Giuseppe Mercalli devised a descriptive scale based on the extent of devastation caused by an earthquake, as well as descriptions of the aftereffects. This was a very subjective means for determining the actual strength of earthquakes. In 1935 Charles Richter created a scale based on the maximum magnitude of the waves as log_{10} (logarithm base 10 or a tenfold increase in power for each numerical increase in the scale), as measured in microns. His scale has values of 1 to 9 on this scale, where 1 is the least damaging and 9 the most damaging. Using logarithms for this scale can be confusing since each increase in number represents a tenfold increase in the power or severity of the earthquake. In other words, an earthquake measured at 5 on the Richter scale is ten times stronger than one with a 4 reading, and a 6 earthquake is 10 times stronger than a 5 and so on. This scale also means an earthquake with a magnitude of 9 is 1 billion times more powerful than one at the 1 level. It is estimated that for every 50,000 earthquakes of 3 or 4, only one of 8 or 9 will occur. One of the strongest ever measured was in 1899 with the magnitude of 8.6 in Yakutat Bay, Alaska.

Robbins' Theory for the Polio Virus: Biology: *Frederick Chapman Robbins* (1916–   ), United States. Frederick Robbins shared the 1954 Nobel Prize for physiology or medicine with John Enders and Thomas Weller.

Since the polio virus can multiply outside nerve tissue, it can exist in other tissue as well.

Frederick Robbins' medical background included collaborating with the U.S. Army to find cures for diseases caused by viruses and parasitic microorganisms. In 1952, Robbins and his colleagues grew the virus that causes poliomyelitis in cultures produced outside a living organism. Up to this time, it was thought the polio virus could exist only in nerve cells of the central nervous system. They proved this particular virus could live in tissue other than nerve tissue, leading to the theory that the virus survives in body tissue and later attacks the central nervous system. This research resulted in the development of vaccines and new techniques for culturing and detecting the polio virus that may be dormant in body tissue and later attack the central nervous system. *See also* Delbruck, Sabin.

Roberts' Theory of Split Genes: Biology: *Richard J. Roberts* (1943–), England. Richard Roberts shared the 1993 Nobel Prize for physiology or medicine with Phillip A. Sharp.

The DNA of prokaryotic cells becomes messenger RNA, which acts as templates to assemble amino acids into proteins.

Prokaryotic cells have very primitive, poorly defined nuclei, and their DNA has no membrane surrounding them. Some examples are blue-green algae and some bacteria, such as *Escherichia coli.* Prokaryotic primitive nuclei have a single chromosome with only about 3 million DNA base pairs. Since amino acids require about 900 DNA base pairs to form proteins, this type of cell should be able to produce about 3000 different proteins. This is in comparison with cells of mammals, called *eukaryotic cells*, which contain about 4 billion DNA base pairs and can produce over 3 million proteins—many more than mammals need. Roberts found that part of the DNA of the prokaryotic cells with no nuclei can split into separate messenger RNA capable of producing proteins. Robert's simple explanation of the structure of primitive prokaryotic nuclei made the study of the formation of RNA as genetic messengers for the DNA much easier than simply studying the very complex RNA and DNA of mammals. Roberts' theory advanced a better understanding of how amino acids form proteins within the human body. *See also* Crick; Sharp; Watson.

Roche's "Limit" Theory: Astronomy: *Edouard Albert Roche* (1820–1883), France.

A satellite of a planet cannot be closer than 2.44 radii of the larger body without disintegrating.

In 1850 Edouard Roche proposed what is now known as the *Roche limit*, based on the concept that if both a satellite and the planet it is orbiting are the same density, there is a limit to their proximity to each other without the satellite, or both, breaking up under the force of gravity. The Roche limit explains the existence of the rings of the planet Saturn. Since the outer ring of Saturn is only 2.3 times the radius of Saturn, it might have been a solid satellite that came too close and broke into fragments. The Roche limit also explains why these many small chunks of matter did not re-form into a solid body orbiting Saturn. The orbit of Earth's satellite (moon) is many times the 2.44 radii of the Roche limit; thus, there is little chance for it to be affected by gravity to the extent it would break into fragments. *See also* Cassini; Schiaparelli.

Roentgen's Theory of X Rays: Physics: *Wilhelm Conrad Roentgen* (1845–1923), Germany. Wilhelm Roentgen received the first Nobel Prize for physics in 1901.

Cathode rays are capable of sending unknown rays to screens to which they are not directed, thus causing fluorescence.

In 1895 Wilhelm Roentgen experimented with a Crookes' tube (a high-voltage gaseous-discharge tube), which produces cathode rays, which produce fluorescence when focused onto a sensitive screen, and detected an unknown form of radiation. (See Figure C3 under Crookes.) He noticed that a cardboard coated with a yellow-green crystal fluorescent material, $BaPt(CN)_4$ (barium cyanoplatinite), located in another part of the room, also was fluorescing when the tube was in operation, even though no rays were directed toward it. He concluded that since **cathode** rays can travel only a very short distance, they must originate from some unknown radiation. Thus, he called them x rays (also known as Roentgen rays in his honor). He continued to study x rays, recording accurate descriptions of their characteristics as listed below:

- They had a much greater range than did cathode rays.
- They traveled in straight lines, but may also be scattered in straight lines from their source.
- They were not affected by magnetic fields or electrical charges.
- They could pass through cardboard and thin metal sheets. Most materials, except lead, are transparent to them to some degree.
- They could expose photographic materials.
- They passed through the human hand and outlined the bone structure.
- They are longitudinal vibrations (waves), while light consists of transverse vibrations.

The discovery of x rays did not solve the issue of the particle-wave duality nature of light, which was being explored at that time. Rather, it complicated the dilemma because some characteristics of x rays are similar to light rays and some are not. The use of x rays became important in the study of crystal structures, as well as in medical diagnosis, and later led to the discovery of radioactivity. Roentgen and his assistant were subjected to excessive exposure to x rays; both died from radiation poisoning. *See also* Becquerel, Curies.

Romer's Theory for the Speed of Light: Physics: *Olaus Christensen Romer* (1644–1710), Denmark.

The motion of Earth to or away from Jupiter can be used to establish the speed of light.

In the 1670s while examining the records of Giovanni Cassini who had determined Jupiter's rotational period and its distance from Earth, Olaus Romer noticed the figures varied depending on whether Earth and Jupiter were approaching each other in their orbits or receding from each other. There was a difference of 10 minutes from the time Jupiter's four then-known moons went behind the planet (were eclipsed), while at the same time Earth's path was *proceeding* in the direction of Jupiter, and when Jupiter's major moons were next eclipsed as Earth's path was *receding* from Jupiter. This 10-minute difference was the amount of time it took the light from Jupiter to reach Earth from these two different distances between Jupiter and Earth. This provided the necessary data for Romer to calculate the speed of light since Cassini had previously determined the distance of Jupiter from Earth. In 1676, Romer announced his theory for establishing the fundamental constant of the speed of light as 140,000 miles per second, which is only about 75 percent of today's figure of 25,000 kilometers per second (about 186,000 miles/sec.). This was the first proof that light has a finite speed. *See also* Cassini, Michelson.

Rossi's Theory for Cosmic Radiation: Physics: *Bruno Benedetti Rossi* (1905– 1994), Italy.

The charge on cosmic rays can be detected by the influence of the earth's magnetic field.

Cosmic rays were first detected in the early 1900s, but little was known about them except they were a form of high-energy, penetrating radiation. In 1930 Bruno Rossi tested his cosmic ray theory using the east-west symmetry concept. Earth's eastward and westward magnetic fields would act differently on incoming cosmic rays due to the direction of the fields' motions. Rossi set up several Geiger counters (radiation detectors) on a high mountain, facing some east and facing several west, so they could detect and count the cosmic rays coming from outer space from different directions as Earth rotated on its axis. He found an excess of 26 percent of cosmic rays coming eastward toward Earth. Thus, he concluded they were mainly composed of positive protons and other positive particles, along with some electrons. These were all high-energy particles coming from both the sun and possibly supernovae (stars). Later, Rossi believed x rays must also originate from astronomical bodies in outer space but were not detectible on Earth because they were all absorbed by the atmostphere. In the 1960s, Rossi was one of the pioneers in the use of rockets that carried instruments to detect cosmic x rays above Earth's atmosphere. He found some x rays originating from the Crab Nebula, the Scorpio constellation, and many other sources beyond the solar system. Currently a special telescope is orbiting Earth, detecting x rays leftover from the big bang.

Rowland's Theory of Chlorofluorocarbons' Effects on the Ozone: Chemistry: *F. Sherwood (Sherry) Rowland* (1927–   ).

Chlorofluorocarbons will decompose in the upper atmosphere, releasing chlorine, which reacts with and breaks down ozone molecules.

In the early 1970s, Sherry Rowland and his student began investigating the possible effects of chlorofluorocarbons (CFCs) on the ozone. In the laboratory they worked out the reaction as $Cl + O_3 \rightarrow ClO + O_2$, and $ClO + O \rightarrow Cl + O_2$, where Cl are free chlorine atoms, O_3 is a form of oxygen molecule called ozone, O are oxygen atoms, and O_2 are oxygen molecules. The chlorine atom combines with the ozone molecule to form chlorine monoxide molecules and a regular oxygen molecule. In the second reaction, the chlorine is regenerated to start the process all over again, but some of the oxygen atoms in the second reaction can also combine with the oxygen molecule to re-form the ozone molecule. It takes only a relatively small amount of the CFCs to start the reaction. The question that has not yet been settled is whether this laboratory reaction is the same as what actually happens in the 15–30-mile-high ozone layer. As the amount of CFCs entering the atmosphere increased after the 1970s, there was a detectable decrease in the **ozone layer** over Antarctica, but not much of a hole over the North Pole. The thickness of the ozone layer has always been cyclic and is always thinner over the equator because this is the area where it is generated and then spreads out to the polar regions. Since refrigeration and air-conditioning used most of the CFCs, these industries have eliminated their use and are substituting less reactive substances, such as hydrochlorofluorocarbons (HCFCs) and hydrofluorocarbons (HFCs). Experts claim that these and about eighteen other possible hydrofluorocarbon substitutes will not cause **global warming** and ozone depletion, since they do not contain free chlorine and decompose in the lower atmosphere. Research continues for even better substitute fluids for refrigeration and air-conditioning use. *See also* Arrhenius.

Rubin's Theory of Dark Matter: Astronomy: *Vera Cooper Rubin* (1928–   ).

Galactic rotation indicates there is more mass in galaxies than is visible from Earth.

Vera Rubin studied spiral galaxies by measuring the rotational velocity of their arms. This can be done by applying the Doppler shift, which indicates that the light from a body moving away from the viewer will appear redder, and when moving toward the viewer it appears bluer. In addition, Kepler's law of rotation of bodies in space states that the velocity of a rotating body decreases with the distance. When the gravitational constant is applied to a revolving mass, the following equation should apply: $v^2 = GM/r^2$, where v is the velocity, r is the radius of the orbiting mass, M is the mass, and G is the gravitational constant. Rubin found this equation did not apply to some spiral galaxies because they increased their speed with distance and their mass seemed much too low. She interpreted this to mean the mass had to be there but that it was not visible from earth. She called this unseen mass **dark matter**. Rubin also concluded that over 90 percent of all the matter in the universe does not emit much radiation and thus is dark and relatively "cold" in the sense that no light or infrared (heat) radiation is detected. Finally, she concluded there are more dark galaxies than luminous ones. In addition, she believes there has to be much more matter than can be seen since it is required to provide the gravity to hold galaxies together so they do not "fly apart." Solving the puzzle of dark matter may lead to an understanding of the fundamental nature of the universe. Most astronomers accept her concept of dark matter. *See also* Doppler; Kepler.

Rumford's Theory Relating Work to Heat: Physics: *Count Benjamin Thomson Rumford* (1753–1814), England.

A specific amount of work can be converted into a measurable amount of heat.

Benjamin Rumford was impressed by the amount of heat generated by the process of boring out holes in metal cannon barrels even when water was used to cool the operation. Rumford was familiar with the old concept of caloric as being the property within substances that was released by friction or by forcing it out of solids in some way. Some scientists said the boring process "wrung out" the caloric from the metal; others said all those fine shavings created the heat. Rumford had a different theory. He believed heat was generated by the mechanical work performed and proposed there was a conservation of work (friction) and heat (motion or energy). This was one of the first concepts of the conservation of matter and energy and that heat involves motion of some sort (kinetic energy). Several other developments furthered Rumford's theory. These and other theories led to the laws of conservation of mass, energy, and momentum. Rumford invented the calorimeter, which measures the amount of heat generated by mechanical work. *See also* Joule; Lavoisier.

Russell's Theory of Stellar Evolution: Astronomy: *Henry Norris Russell* (1877–1957), United States.

Based on the correlation of the magnitude of stars to their types, stars evolve through stages of contraction from hot giants, to smaller stars, and finally into cold dwarfs.

In 1913 Henry Russell published the results of his research relating the classification of stars by type to their brightness (magnitude). At about the same time another astronomer, Ejnar Hertzsprung, produced similar data. Their combined data were placed in graph form, known as the Hertzsprung-Russell diagram, which depicts a *main sequence* of stars as distinct from sugergiants, giants, and white dwarfs (See Figure H2 under Hertzsprung for a depiction of their graph.) Russell was the first to use the terms *giant* and *dwarf* to describe groups of stars. He was also the first to use photographic plates to record stellar **parallax** and measure a star's luminosity. The diagram depicts the concentration of the supergiants and giants (located in the upper right of the graph) that in time become hot stars in the main sequence, then collapse by gravity to form cool, white dwarfs (located in the lower left of the diagram). Russell developed a method for measuring the size and orbital period for stars as well as their spectra. His work enabled other astronomers to determine galactic distances for stars that were beyond the parallax technique for making measurements. His work also led to new theories for stellar evolution. *See also* Hertzsprung.

Rutherford's Theories of Radioactivity/Transmutation and Atomic Structure: Astronomy: *Baron Ernest Rutherford* (1871–1937), New Zealand.

Rutherford's Theory of Radioactivity and Transmutation: *Radioactive substances emit three different types of radiation by which one element is changed into a different element.*

Baron Ernest Rutherford was one of the first to explore the emissions of polonium and thorium, in addition to radium. In 1899 he discovered there were two different types of radiation emissions from these elements which he referred to as *radioactivity* to describe this emission. He named one type of radiation,

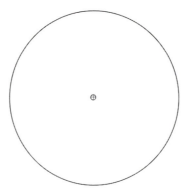

Figure R1: Rutherford's experiment indicated that the positively charged atom consisted almost entirely of a massive but very small positive nucleus. The atom consisted of a vast area around this tiny nucleus. The outer reaches of this area are occupied by negatively charged electrons, which weigh much less than the positively charged central nucleus. His conclusion was that the atom is mostly empty space.

alpha, which would cause ionization but could be stopped by a piece of paper (helium nuclei). The other he named *beta*, later known as high energy electron emission, which was not ionizing but somewhat more penetrating than beta radiation. He then determined there was a third type of radioactivity, which was characterized by high energy, deep penetration, and high ionization, but was not affected by a magnetic field, which he named *gamma rays*. Rutherford used this knowledge to devise an unusual theory, called *atomic transmutation*, which almost sounded like the old alchemists' dream of the philosophers' stone that could change lead into gold. Rutherford's idea stated that as some these radioactive particles were emitted from their source element, the mass and charge (number of protons) of the original atoms were changed to become a different element. Rutherford and Frederick Soddy confirmed this theory with experiments using radioactive thorium that decayed into another active form, which they called *thorium X*, which resulted when a series of chemical and physical changes converted one type of atom into another. This was known as *transmutation*. As a result, Rutherford became most interested in the alpha particles and their effect on substances.

Rutherford's Theory of Atomic Structure: *The atom is composed mostly of "empty space" whose mass is concentrated in a very small, dense, central particle with a charge.*

Ernest Rutherford knew alpha particles could expose photographic plates and could be beamed through very thin pieces of material to produce a fuzzy image. Two of his students conducted an experiment where alpha particles were beamed through a very thin piece of gold foil (about 0.00004 centimeter, which is only a few atoms thick) to determine what type of pattern the particles would form on the other side of the foil. Rutherford noticed that most of the alpha particles went straight through the foil and were recorded by the detecting instrument directly behind the foil. But a detector off to the side at about 45 degrees also picked up some signals, indicating that something in the foil was deflecting a few of the alpha particles. Rutherford noticed that although most particles went through the foil as if nothing was there, a few actually seemed to bounce back. He said, "It was almost as incredible as if you fired a 15-inch shell at a piece of tissue paper, and it came back to hit you." After making some calculations, he concluded that this backward scattering of the alpha particles is evidence of a few collisions with something where almost all the mass is concentrated in a central, very small "nucleus." It was at this point that he realized this central nucleus had a positive charge. (See Figure R1.) *See also* Bohr; Curies; Soddy.

Rydberg's Theory of Periodicity for Atomic Structure: Physics: *Johannes Robert Rydberg* (1854–1919), Sweden.

Elements can be organized by the structure of their atoms based on their spectra rather than according to their mass.

In the 1880s Johannes Rydberg was aware that Johann Jakob Balmers (1825–1898) first discovered the relevance of the spectral lines of the hydrogen atom. Balmers found there was a simple relationship between the wavelength of the lines and the spaces between them when expressed on a graph. Rydberg's theory and experiments provided the explanation for this relationship. He examined the spectra of hydrogen atoms and discovered that the frequencies of the excited atoms produced a spectrum that can be stated as a constant, relating the wavelength to a series of lines in the spectrum. The Rydberg equation can be stated as: $\lambda = R(1/m^2 - 1/n^2)$, where λ is the wavelength, R is the Rydberg constant, and m and n are whole numbers. It can also be expressed as $1/\lambda \ R(1/1^2 - 1/m^2)$, where m must be an integer larger than 1. The Balmers spectra series for atoms represented only the shorter ultraviolet range. Rydberg proceeded to re-organize Dmitri Mendeleev's **Periodic Table of the Chemical Elements** according to the structure of atoms based on their spectral lines. (See Figure M4 under Mendeleev.) After applying his equations to the patterns of atomic structure, Rydberg developed a spiral form of the periodic table. Soon after Rydberg developed his equations, Henry Moseley determined that the nuclei of atoms had positive charges, which confirmed Rydberg's and Rutherford's theories. *See also* Bohr; Newlands; Mendeleev; Moseley; Rutherford.

S

Sabin's Theory for Attenuated Live Polio Vaccine: Biology: *Albert Bruce Sabin* (1906–1993), United States.

If live polio virus can be grown in tissue cultures, it can be attenuated (weakened) to vaccinate against poliomyelitis (infantile paralysis).

During World War II Albert Sabin developed vaccines for diseases such as dengue fever and encephalitis and was familiar with the work of other microbiologists who experimented with the growth of viruses in the brains of mice. In 1954 Jonas Salk (1914–1995) used polio virus "killed" by formaldehyde, which was then injected to stimulate the human immune system, thus developing antibodies against the disease. It was not completely successful because the vaccine had to be injected several times over a period of years. Also, the dead virus's effects on the immune system did not last a lifetime. Albert Sabin developed a live but weakened version of the polio virus in the kidney tissue of monkeys. His version for providing immunity to the virus could be taken orally and had a lasting effect for producing antibodies as a preventative against the virus. The oral attenuated virus is still being improved. After testing his attenuated live virus on animals, Sabin tried it on himself and several prisoners who volunteered to test its efficacy. The U.S. public was skeptical of Sabin's vaccine due to the problems experienced with the Salk vaccine. Finally, after successful use in Russia and England, it was accepted in the early 1960s, and extensively used in the United States as an oral vaccine that prevented the outbreak of polio epidemics. Today it is estimated that the Sabin vaccine saves over 600,000 lives annually, and the disease has become almost nonexistent.

Sachs' Theory of Photosynthesis: Biology: *Julius von Sachs* (1832–1897), Germany.

The green pigment in plant cells is confined in discrete bodies he called chromoplasts.

Over the centuries a number of scientists theorized about how plants grow, as well as the nature of the green material in their leaves. Until 1862 scientists believed the green material in plants was distributed more or less evenly throughout individual plants. Julius von Sachs was the first to theorize that the green matter was contained in small, discrete bodies he named *chromoplasts* (a colored cell, later given the name *chloroplast*). He coated several leaves of a plant with wax and left others unwaxed. After exposure to sunlight, the unwaxed leaves produced starch, while the waxed leaves did not. Sachs concluded that the unwaxed leaves were able to absorb carbon dioxide, while the coated ones could not let this gas enter, even in sunlight. Thus, photosynthesis (from the Greek *photo*, which means "light," and *synthesis*, which means "put together") is the process whereby in the presence of light, chlorophyll in green plants converts carbon dioxide and water into starch: $6CO_2 + 6H_2O + $ light energy $\rightarrow C_6H_{12}O_6 + 6O_2$.

Saha's Theory of Thermal Ionization: Astronomy: *Meghnad N. Saha* (1894–1956), India.

The composition of a star's spectrum varies with the temperature of the light source.

Meghnad Saha theorized that the degree of ionization (electrons stripped from atoms to form ions) was dependent on the temperature of the atoms. He applied his concept to the spectrum of the light from stars. It was known that the light spectrum from some stars pointed to the presence of only hydrogen or helium, the two lightest elements. His examination of the spectra of stars indicated there were heavier elements (metals) that were being ionized in some stars. He developed a system that suggested spectral lines of stars could be represented by the degree of ionization, and thus the stars' temperatures. In other words, as the temperatures of stars increase, so does the degree of the ionization of the nuclei of the stars' atoms. Saha's theory led to the linking of gas thermodynamics with the kinetics of plasmas, which aided in interpreting the spectral lines of stars. His theory also enabled astronomers to determine the chemical makeup of different stars and confirmed the idea that heavy elements originated in stars, including the sun's gases.

Sakharov's Nuclear Fusion Theory: Physics: *Andrei Dmitriyevich Sakharov* (1921–1989), Russia. Andrei Sakharov was awarded the 1975 Nobel Peace prize.

Controlled nuclear fusion can be achieved by containing the plasma in a magnetic "bottle."

Andrei Sakharov's theory described how by confining a deuterium plasma (a highly ionized gas) within a strong magnetic field, the temperature could be raised to the point where the heavy hydrogen (deuterium) gas would be forced to fuse to form helium. The resulting reaction would release a tremendous amount of energy that could be used to produce electricity, much like the controlled nuclear fission reaction in nuclear power plants. In the early 1940s the United States developed the first nuclear fission bombs. In 1954 Sakharov was involved in the explosion of Russia's first atomic (**fission**) bomb, as well as its first nuclear (**fusion**) H-bomb. After realizing the tremendous destruction that would result from a nuclear war, Sakharov became an advocate for nuclear disarmament, which led to his demotion and exile, and eventually the Nobel Peace prize. *See also* Teller.

Salam's Theory for the Properties of Elementary Particles: Physics: *Abdus Salam* (1926–1996), Pakistan. Abdus Salam shared the 1979 Nobel Prize for physics with Sheldon Glashow and Steven Weinberg.

Both the electromagnetic and weak interacting forces act as a single interacting force for elementary particles at high temperatures.

Four basic forces account for the interactions of elementary particles. (1) Electromagnetic forces and (2) gravity are observed to interact over long distances throughout the universe. The electromagnetic force both attracts and repels, thus counteracting its strength. Gravity is a very weak force that only attracts and never repels. Even so, it is the most dominant force in the universe. (3) The strong force interacts with hadrons (nuclei of atoms), while the (4) weak interacting force, which is much less strong, interacts with leptons (similar to electrons and neutrinos). The forces that interact with hadrons and leptons are evident only at the very small atomic and subnuclear distances of matter. Albert Einstein attempted to devise a mathematical solution to combine the electromagnetic and gravity interacting forces. He was unsuccessful, and at that time the other two interactions with elementary particles were unknown. In 1968 Abdus Salam and his colleagues successfully combined both the electromagnetic and weak forces, which behaved as a single interacting force, but only at high temperatures. The analogy used is that at high temperature, water turns into steam. As the temperature drops, it again becomes liquid, and if it drops further, it becomes solid (ice). In other words, for electromagnetic and weak interactions, there are two different physical states of being, depending on the temperatures involved (combined at high temperatures and separate at low temperatures). Abdus Salam and the other Nobel Prize winners referred to the high-temperature combined state as the electroweak interaction. Salam's theory is a great step in the development of a grand unification theory (GUT) and Einstein's unified field theory. *See also* Weinberg.

Sandage's Theories of Quasars and the Age of the Universe: Astronomy: *Allan Rex Sandage* (1926–   ), United States.

Sandage's Theory of Quasars: *Quasars can be identified by their emitted radio signals, ultraviolet radiation, and blue light.*

In the early 1960s Allan Sandage detected radio signals from a small area in the distant universe. These radio signals seemed too strong to be originating from such a distant dim star. He and other astronomers referred to these objects that produced ultraviolet and blue light radiation as blue star objects (BSOs). Sandage determined they were not really radio stars. Therefore, he called them quasi-stellar or *starlike* bodies (*quasi* means "apparently" and *stellar* means "star"). This term was changed to *quasar*. Sandage realized these objects exhibited a great Doppler redshift, which overcame the ultraviolet and blue light. He concluded this could mean only, that the quasars were located at tremendous distances within the universe and that what could be seen from earth was really the center of a huge galaxy. More recently, Allan Sandage and his team estimated the distance of such quasar galaxies to be over 12 billion light-years away. The first quasar Sandage discovered was named 3C 48, which had the brightness of a sixteenth-magnitude star (See Pogson). More recently, the Hubble Space Telescope photographed a stellar object named NGC 4639, an excellent example of a quasar.

Sandage's Theory for the Age of the Universe: *The universe has an 80-billion-year cycle of growth: 40 billion years of expansion and 40 billion years of contraction.*

Allan Sandage does not believe the universe is static, regenerating itself, or expanding indefinitely. His contention is that the universe oscillates in cycles of expansion (birth and growth) and contraction (shrinking, death, and rebirth). His theory states that after a 40-billion-year period of expansion (it still has about 25 billion years left to reach maximum growth since the current universe is about 15 billion years old), the universe will cease to expand, reverse itself, and start contracting for another 40 billion years of the cycle. At the end of this contraction period, he contends it will form back into its original, tiny, very dense particle that will again create another big bang which will start the process all over again. His theory, one of many dealing with the nature of the universe, is partly based on the estimation of the density of matter in the universe, which has yet to be accurately determined. More recently, it has been estimated that the vast majority of matter in the universe is **dark matter**, which cannot be seen but implies an infinite universe. *See also* Schmidt; Rubin.

Sanger's Theories of the Structure of Proteins and Gene Splitting: Chemistry: *Frederick Sanger* (1918–), England. Frederick Sanger was awarded two Nobel Prizes in chemistry: one in 1958 and the other, which he shared with Paul Berg and Walter Gilbert, in 1980.

Sanger's Theory of Protein Structure: *The four amino acids that make up proteins connect in groupings and can be identified in sequences inside the protein molecule.*

Frederick Sanger used the process called paper **chromatography** to separate and count the number of amino acids in specific protein molecules. Once these groupings were broken into segments of two, three, or four amino acids, he determined the structure of larger and complete protein molecules. In 1953 he used this procedure to outline the complete structure of the molecule for the protein hormone insulin. It consists of fifty amino acids combined in two connecting chains. Since his identification of the exact order of the amino acid groups in the chains, it became possible for other scientists to produce synthetic insulin, used in the treatment of diabetes.

Sanger's Theory of Gene Splitting: *DNA can be split into fragments of various sizes, isolating a few cases of genes within genes.*

Frederick Sanger developed a new technique of splitting DNA into fragments in order to determine the base sequences of the nucleotides. In 1977 he was the first to describe the entire sequence of nucleotides in the DNA of a bacteriophage called Phi-X 174. To accomplish this, he needed to ascertain the order of about 5500 nucleotides in just one strand of the **phage**'s DNA. While examining his results, he unexpectedly discovered several situations where genes were located within other genes. Today this phenomenon is used to explain traits of genetic expression. Sanger's theory and research contributed to the foundations for the science of genetic engineering. *See also* Sharp.

Sarich's Theory of Utilizing Protein to Date Man/Ape Divergence Genetically: Biology: *Vincent Sarich* (1934–), United States.

When species split into two branches, future mutations for each branch are accumulated in a linear manner, and the greater the number of mutations, the greater the time of divergence.

Vincent Sarich used the albumin found in blood protein as a determinant for the divergence of humans and apes from a common ancestor. The concept is based on the facts that there is only about 1 percent difference between humans and apes in the DNA protein molecules and that mutations of individual genes not only differ for individual genes but genes mutate on a random basis at a measurable rate over time. Thus, it should be possible to determine the rate of changes in the albumin of humans and apes over a long time span. He called this technique the *molecular clock*. Sarich and his colleague Allan Wilson began with the base of 30 million years ago as the estimated time the species of humans and Old World apes evolved separately from a common ancestor (the "missing link"). After analyzing the data that compared antigens from humans and other anthropoids, which have a common genetic base, they proposed a time factor of about 5 to 7 million years ago when the two species evolved in their own directions. Other scientists disputed this short time period and claimed that fossil evidence places the division of **hominoids** and **hominids** from a common ancestor at about 15 million years. Sarich responded that his molecular data were more accurate than the estimations of the age of the oldest human fossils. His 5-million-year figure is now more or less accepted. Also there is now some evidence suggesting that chimpanzees are more genetically similar to humans (over 98 percent of the same DNA) than they are to gorillas and that chimps and humans should be classed in the same genetic family. *See also* A. C. Wilson.

Scheele's Theory of the Chemical Composition of Air: Chemistry: *Karl Wilhelm Scheele* (1742–1786), Sweden.

Air is composed of two gases, one of which supports combustion, while the other does not.

One of the gases that Karl Scheele isolated in air he named *fire air* (oxygen) because it supported combustion. The other, which prevented combustion, he named, *vitiated air* (nitrogen). In 1772, about two years before Joseph Priestley produced oxygen (which Lavoisier named), Scheele actually isolated oxygen and described it in a paper. However, his findings were not published until after Priestley's discovery of the same gas was published, and thus Priestley was given the credit. *See also* Lavoisier; Priestley.

Schiaparelli's Theory of Regularity in the Solar System: Astronomy: *Giovanni Virginio Schiaparelli* (1835–1910), Italy.

Meteors, planets, and the rings of Saturn follow regular patterns.

In 1877 the planet Mars was in conjunction to Earth—only about 35 million miles away. This proximity provided Giovanni Schiaparelli and other astronomers an opportunity to view Mars' surface for details. Despite the fact the atmospheres of both Earth and Mars hindered a clear view, Schiaparelli's record of his observations was confirmed by others. He reported both narrow and larger dark markings on the surface of Mars, which he concluded were bodies of water connected by narrow "channels" he called *canali*. He claimed these markings represented geometric patterns, which indicated some degree of regularity. It was speculated these patterns were the result of "constructions" by Martians. Others later expanded his concepts of regular structures to propose that Mars was a dying planet and these channels were the work of a desperate race attempting to bring water from the Martian ice caps to the tropical areas, where it could be used to grow plants. Myths about life on Mars existed for many decades until modern exploratory spacecrafts were sent to Mars to examine its atmosphere and surface.

Another example of Schiaparelli's theory of regularity is his claim that the rings of Saturn were formed by some natural physical process. Another is that meteor showers are caused by the breakup of comets. Thus, meteors must follow regular orbits similar to comets. One of his major theories of regularity states that the rotations of Venus and Mercury on their axes are synchronized with their sidereal periods. Therefore, these two planets always keep their same side facing Earth. He based this concept on his viewing of the same markings on these planets' surfaces when they were in a specific position. Schiaparelli's theory that these two planets keep their same side facing the sun was disproved only in the 1960s when radar signals bouncing off their surfaces indicated that Venus rotates on its axis every 243 days, while its sidereal period is 225 days; Mercury rotates on its axis about once every 59 days, while its sidereal period is 88 days. *See also* Cassini; Huygens; Lowell; Roche.

Schleiden's Cell Theory for Plants: Biology: *Matthias Jakob Schleiden* (1804–1881), Germany.

Plant structures are composed of small, distinct "walled" units known as "cells."

In 1838 Matthias Schleiden first recognized and reported on the "cellular" basis of plants, which he referred to as "units" of plant life. He was also first to note the importance of the nuclei in the possible reproduction of plant cells. He mistakenly thought that cells reproduced by the nuclei "budding" from the "mother" cell's nuclei. After Schleiden's discovery of plant cells, Theodor Schwann, also in about 1838, announced that animal tissues were also composed of cells, but with much less-well-defined cell walls. This led to the biological concept that cells are a basic unit of all living organic things. From this fundamental idea and the research of several other biologists, Schleiden and Schwann have been credited with the formulation of the cell theory, which states:

• All plants and animals are composed of cells or substances derived from cells.

• Cells are living matter, with membrane walls and internal components.

• All living cells originate from other cells; cells reproduce themselves.

• For multicellular organisms, the individual cells are subordinate to the whole organism.

Some of Schleiden's and Schwann's original observations and ideas were incorrect due to the very limited power of the microscopes available to them. But over time and with additional research by others, their concept that cells are the basic unit of life became an important step in understanding living organisms. *See also* Hooke; Leeuwenhoek; Schwann; Strasburger; Virchow.

Schmidt's Theory of the Evolution and Distribution of Quasars: Astronomy: *Maarten Schmidt (1929–   ), United States.*

Quasars exhibit a greater redshift than do stars. Therefore, they are younger, more distant, and more abundant stellar-like objects than are stars.

Maarten Schmidt expanded his research of our Milky Way galaxy to include the very dim and distant objects discovered in 1960 by Alan Sandage and Thomas Matthews. Sandage called these objects *quasars*, meaning "starlike." The first quasar, designated 3C 48, was identified by the radio signals it emitted. Schmidt studied its light spectrum. Even though it had the luminosity of only a sixteenth-magnitude star (see Pogson), he found it exhibited the spectral lines of the element hydrogen. When viewing other quasars, their spectra became more confusing, until Schmidt realized the hydrogen spectra lines shifted in wavelengths toward the red end of the spectrum. This Doppler redshift of light, which was greater than expected from a star, indicated that quasars are emitting great amounts of energy and light as they recede from Earth at fantastic velocities. The greater the redshift, the greater the speed at which they recede. Schmidt examined the quasar, named 3C 273, and determined that as the universe continued to expand after the big bang, the extreme redshift could mean only that this quasar was not only at least 1 billion light-years away, but its brightness was that of hundreds, or possibly thousands, of galaxies in a cluster. His theory asserts that quasars are among some of the earliest types of matter formed when the universe was young, and since they continually recede from us (and each other), they become more abundant as the universe ages. Schmidt proceeded to map the quasars in the universe, leading him to conclude that the so-called steady-state universe cannot exist. He and others interpreted their red-shift data as indicating that the number of quasars increases with distance and that no objects have been located at greater distances. The possible demise of the steady-state universe concept generated theories about an infinite universe. More recently, the discovery of **black holes** has advanced the theory that these huge dark masses of matter, from which matter or light cannot escape, are the source of the tremendous energy of quasars, or possibly the source of new quasars or even new universes. *See also* Doppler; Gold; Hawking; Hubble; Sandage; Schwarzschild.

Schrödinger's Theory of Wave Mechanics: Physics: *Erwin Schrödinger* (1887–1961), Austria. Erwin Schrödinger shared the 1933 Nobel Prize for physics with Paul Dirac.

An electron's position in an atom can be mathematically described by a wave function. The wave function can be determined by the solution of a differential equation that has been named after Schrödinger.

In 1925 the quantum theory was developed through efforts of Erwin Schrödinger, Niels Bohr, Werner Heisenberg, and others. Schrödinger was aware of Niels Bohr's application of the quantum theory to describe the nature of electrons orbiting the nuclei of atoms and Louis de Broglie's equation describing the wavelength nature of particles ($\lambda = \hbar/mv$, where λ is the wavelength, $\hbar$ is Planck's constant, and mv is the particle's momentum, i.e., mass times velocity). Schrödinger thought de Broglie's equation, which applied to only one electron orbiting the hydrogen nuclei, was too simplistic to describe the state and nature of the electrons in the inner orbits of more complex atoms. Schrödinger laid the foundation of wave mechanics as an approach to quantum theory, resulting in his famous complex wave differential equation, a mathematical nonrelativistic explanation of quantum mechanics characterized by wave functions. Quantum theory is based on two postulates: (1) energy is not continuous but exists in discrete bundles called "quanta" (e.g., the photon is an example of a discrete bundle of light), and (2) subatomic particles have both wave (frequency) and particle-like (momentum) characteristics. His equation proved more useful in describing the quantum energy states of electrons in terms of wave functions than Bohr's quantum mechanical theory of particles orbiting around the nuclei of atoms. Schrödinger's theory for the wave nature of particles was an advancement for the acceptance of the wave-particle duality of quantum mechanics. *See also* Bohr; de Broglie; Dirac; Heisenberg; Nambu.

Schwann's Theory of Animal Cells: Biology: *Theodor Schwann* (1810–1882), Germany.

The formation of cells is a universal principle for living organisms.

In 1838 Mathias Schleiden proposed a cellular theory for plants. At about the same time, Theodor Schwann made microscopic examinations of various animal tissues. He had already formulated the concept that animal tissues, particularly muscle tissues, were mechanistic rather than vitalistic (based on matter and energy rather than some other agent). Schwann suggested the substances that compose animal tissues do not directly evolve from molecules but rather from cells, and the material contained in animal "cells" is similar to plant cells. He stated that just as plant cells are derived from other plant cells, so are animal cells derived from other animal cells. However, Schwann mistakenly assumed that the material inside cells did not have any structure of its own except for what he called a "primordial blastema." Schwann's theory stated that animal cells represented fundamental units of life. Although he did not believe in spontaneous generation, he at first claimed that cells arose from nonliving matter. This proved to be a paradox until it was determined that all cells originate from other cells. Schleiden and Schwann are both credited for developing the cell theory that states all organisms are composed of cells, which are the basic structural and functional units of life. *See* Schleiden for details of the cell theory. *See also* Hooke; Virchow.

Schwarzschild's "Black Hole" Theory: Astronomy: *Karl Schwarzschild* (1873–1916), Germany.

Once a star collapses below a specific radius, its gravity becomes so great that not even light can escape from the star's surface, thus resulting in a black hole.

Karl Schwarzschild was an astronomer who, in addition to providing information about the curvature of space and orbital mechanics, studied the surface of the sun. His theoretical research indicated that when a star is reduced in size to what is now called the *Schwarzchild radius* (SR), its gravity will become infinite. In other words, if a star with a specific mass is reduced in size to the critical Schwarzchild radius, its gravity becomes so great that anything entering its gravitational field will not escape. The edge of the **black hole**, which is referred to as its *horizon*, is the zone where the hole's escape velocity exceeds the speed of light. The critical spherical surface region of a black hole where all mass and light are captured is called the *event horizon*. Schwarzschild used the sun to determine this critical radii for stars. The SR for the sun, when it does collapse, will be about 3 kilometers. (Because the sun's current radius is about 700,000 kilometers, it will be many billions of years before it "shrinks" to the SR of 3 km.) To determine the SR for other stars, divide the object's (star) mass by the mass of the sun and multiply by 3 kilometers. This equation might be expressed as: $M_o \div M_s \times 3$ km $= M_o$'s SR (M_o is the mass of the object for which SR is to be determined, M_s is the mass of the sun). Therefore, the critical radius for a black hole is proportional to its mass. The current theory suggests that a black hole may be open at the bottom of its "funnel shape," where the mass that was captured by the black hole may re-emerge as new stars or a new universe. Although the concept of black holes is demonstrated by mathematics, the idea that the "lost" mass will exit or escape to form another universe has not been proven. *See also* Hawking; Penrose.

Seaborg's Hypothesis for Transuranium Elements: Chemistry: *Glenn Theodore Seaborg* (1912–1999), United States. Glenn Seaborg shared the 1951 Nobel Prize for chemistry with Edwin Mattison.

The elements beyond uranium, atomic number 92, are similar in chemical and physical characteristics.

In 1940 Glenn Seaborg and his colleagues discovered the first two elements beyond uranium (92): neptunium (93), a beta-decay element somewhat similar to uranium, and plutonium (94), a radioactive fissionable element used for nuclear reactors and bombs. The discovery of these new elements resulted in Seaborg's hypothesis that elements beyond uranium (92) formed a group of elements with similar characteristics that represented a new and unique series of elements. He compared this new series, named the *actinide transition series*, to the lanthanide series of rare earths [lanthanum (57) to lutetium (71)], which also have very unique characteristics. Seaborg used his hypothesis to predict the existence of many more "heavy" elements in his proposed transuranic actinide series. Still later in his career, he speculated there was a "superactinide" series of elements ranging in atomic number from about 119 to as high as 168 or even 184. All of these super-heavy elements, if discovered, will be radioactive, very short lived, and difficult to detect. In 1944, Seaborg and his colleagues discovered two new elements beyond uranium (92) in the **Periodic Table of the Chemical Elements**: americium (95) and curium (96). Still later, Seaborg is credited with discovering berkelium (97), californium (98), mendelevium (101), nobelium (102), and lawrencium (103). [Elements einsteinium (99) and fermium (100) were discovered after the detonation of the 1952 H-bomb and are artificially produced in nuclear reactors.] Element 106, discovered in 1974, is currently named seaborgium (Sg), in honor of Glenn Seaborg. In 1999 Seaborg's Berkeley, California, laboratory announced the discovery of element 118, which has a half-life of about 0.00005 second.

Seebeck's Theory of Thermoelectricity: Physics: *Thomas Johann Seebeck* (1770–1831), Germany.

If electricity can produce heat when flowing through a wire, then a reverse effect should be possible; that is, heating a circuit of metal conductors should produce electricity.

Thomas Seebeck was familiar with Joule's law, which states: "A conductor (wire) carrying an electric current generates heat at a rate proportional to the product of the resistance (R) of the conductor (to the flow of electric current) and the square of the amount of current (I or A), (the currents amperage)." Using this as a basis, in 1820 Seebeck joined the ends of two different types of metals to form a loop or circuit. When a temperature differential was maintained between the two junctions, an electric force (voltage) proportional to the temperature differences between the junction was produced. This phenomenon is known as the *Seebeck effect*, where electricity is produced by temperature differences in the circuit. This device is now referred to as a *thermocouple*. If several thermocouples consisting of junctions between two dissimilar metals are connected in a series, a "thermopile" is formed, which can increase the voltage output equal to the number of junctions. It was later discovered that if the temperature of one junction increases and at another junction in the same circuit the temperature decreases, the heat is transferred from one junction to the other. The rate of transfer is proportional to the current, and if the direction of the current is reversed, so is the heat (it is absorbed). In 1854 Lord Kelvin demonstrated that if there is a temperature difference between any two points on a conductor carrying a current, heat will be either generated or absorbed, depending on the nature of the material. It was later discovered that magnetism can also affect this process. This principle has been applied to generate small amounts of electricity, as a "thermometer" to measure temperatures, and as a means of heating or cooling. Small heating and cooling devices in manned spacecrafts and portable refrigerators use the Seebeck effect. *See also* Joule; Kelvin.

Segre's Hypothesis for the Antiproton: Physics: *Emilio Gino Segre* (1905–1989), United States. Emilio Segre shared the 1959 Nobel Prize for physics with Owen Chamberlain.

If antielectrons (positrons) exist and can be produced in particle accelerators, antiprotons should also exist.

In 1932, Carl Anderson, building on Paul Dirac's idea that antiparticles are similar to elementary particles and the existence of the positron, established the existence of the antiparticle positron, which, unlike the negative electron, has a positive charge but is similar to the electron in all other characteristics. Emilio Segre, along with Owen Chamberlain, hypothesized that if antiparticles, such as positrons, can be generated in particle accelerators, antiprotons will also be generated if the accelerator is powerful enough. In 1955 they used the Berkeley Bevatron accelerator to generate 6 billion electron volts (BeV) to bombard copper with high-energy protons, which produced only one antiproton for about 40,000 or 50,000 other kinds of particles. They detected these few high-speed antiprotons by the unique radiation they emitted, which was later confirmed by exposing photographic plates to **antiproton** tracks. At the time antiparticles were discovered, it was also theorized these antiparticles annihilated regular particles when they met—for example: $e^- + e^+ \rightarrow$ energy. The question was, Why is not all the matter in the universe obliterated into energy? The answer is that at the time the universe was "created," more regular particles (e.g., electrons) were formed than antiparticles (e.g., positrons); thus they now dominate. *See also* Anderson; Dirac.

Sharp's Theory for the "Splicing" of DNA: Biology: *Phillip Allen Sharp* (1944–   ), United States. Phillip Sharp shared the 1993 Nobel Prize with Richard Roberts.

Messenger RNA found in eukaryotic cells hybridizes into four sections of DNA that loop from the hybrid regions of the DNA.

Phillip Sharp, as did other molecular biologists in the 1970s, believed **eukaryotic** cells (with nuclei) would act similar to **prokaryotic** cells (without nuclei), where the DNA would form triplets with RNA to provide the codes to form amino acids. After examining the results of his hybrid double strands of DNA/RNA in the adenovirus (the virus that causes common colds), he noted that small sections of the loops that formed from the hybrids broke off and became spliced with the messenger RNA. These then escaped from the cells to become templates for protein production. Sharp and his colleague Richard Roberts determined the "split genes" identified in the adenovirus were common to all eukaryotic cells (which includes the cells in the human body). They concluded that over 90 percent of the DNA was "snipped" out of the strands and became "junk" DNA. This result promised to provide some answers for the problems of genetic splicing related to some hereditary diseases. If this splicing or segmentation of the DNA molecules is better understood and can be controlled, it may be possible to find a cure for some hereditary cancers. *See also* Crick; Roberts; Sanger.

Shockley's Theory of Semiconductors: Physics: *William Bradford Shockley* (1910–1989), United States. William Shockley shared the 1970 Nobel Prize for physics with John Bardeen and Walter Brattain.

In crystal form, the element germanium will carry a current less well than a metal but much more efficiently than an insulator, which enables it to rectify and amplify electric currents.

In 1948 William Shockley and his collegues John Bardeen and Walter Brattain discovered that small impurities within the germanium crystal can determine the degree of its conductivity or capacity to carry electricity. This type of material, which allows some electricity to pass through, it, is called a *semiconductor*. Later they realized that other crystals, such as silicon, were even better and less expensive semiconductors. Since these devices are solid, they are also known as *solid-state semiconductors*. Shockley soon learned how to vary the small amount of impurities in the crystal's structure, enabling it to be used as a "switch," or as a **rectifier** or **amplifier**. These semiconductor elements are composed of atoms with four or five electrons in their outer shells (orbits), which act as a negative conductor and thus are named *n*-type conductors (*n* = negative). When elements with fewer than four outer electrons, such as boron which has just three outer electrons, are introduced as an impurity, they act like "holes" where electrons are missing. Thus, this is referred to a *p*-type conductivity (*p* = positive, or lack of a negative). This occurs when one electron from a close atom is transferred to fill up this "hole." In doing so, it creates another "hole,"

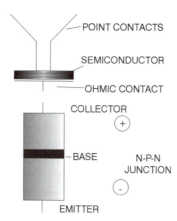

Figure S1: An artist's diagram of Shockley's transistor with an *n-p-n* junction. The *n*-type materials carry electrons, while the *p*-type conductivity occurs when the holes in the material left by moving electrons are filled, thus allowing the junction to act as rectifiers and transistors to alter and amplify current.

which results in these successive positive holes' being filled with electrons, forming a flow of electricity than can be regulated. Shockley made a "sandwich" of *p*-type material with *n*-type material to form a junction, which is known as an *n-p-n junction*, capable of amplifying electrical impulses (radio and TV). This *n-p-n* "sandwich" junction uses very little electricity as it transmits current across a resistor—thus the name *transistor*. (See Figure S1.) Transistors replaced glass vacuum tubes in radio and television receivers, resulting in the tremendous miniaturization of electronic equipment, and formed the basis for the current electronics industry.

Sidgwick's Theory of Coordinate Bonds: Chemistry: *Nevil Vincent Sidgwick* (1873–1952), England.

Two electrons from one atom can both provide "shared" electrons to form "coordinated" organic compounds.

Nevil Sidgwick became interested in the concept of valence as the sharing of electrons in shells of atoms as proposed by Richard Abegg, Gilbert Lewis, and Irving Langmuir. The valence concept is built on Niels Bohr's quantized atom, where the electrons orbit in specific shells based on their level of energy and was first proposed to explain how atoms combined to form molecules during inorganic chemical reactions. For example, each of two chlorine atoms shares an electron so that each can have eight electrons in its outer shell (orbit). This means each had seven electrons plus one shared with its close neighbor chlorine atom, thus forming the diatomic molecule of chlorine gas. (See Figure S2.) Sidgwick's theory stated that a similar "sharing" of electrons also occurred in the formation of both complex metal and organic compounds, but this sharing

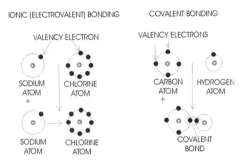

Figure S2: Chemical bonding uses electrostatic forces to form atoms into molecular compounds. Chemical reactions change, break, or re-form these bonds. There are two types of bonding: (Left) Ionic bonds between atoms occur when atoms with a dearth of negative charges in their outer orbits (valence) naturally attract electrons from other atoms to form ions and molecules. (Right) Covalent bonding occurs when atoms share electrons and each atom contributes one or more electrons to form the covalent bond. In both cases, the end results indicate the atoms have achieved an outer orbit electron configuration similar to the noble gases (Group 18, VIIIA of the Periodic Table of the Chemical Elements).

was different from the inorganic "covalent" electron bond, where each atom contributed one electron to the other atom. For this new type of bonding, one single atom could provide both electrons to combine with another atom, forming a new complex molecule. Therefore, he called these complex molecules *co-ordinated compounds*. The concept of coordinated bonds provided a better understanding of organic chemical reactions. *See also* Abegg, Bohr, Langmuir; Lewis.

Simon's Third Law of Thermodynamics: Physics: *Sir Francis Eugen Simon* (1893–1956), England.

The degrees of freedom for random paramagnetic molecules, which absorb heat from liquid helium, will become zero at the temperature of absolute zero.

Walther Nernst claimed that for thermodynamic reasons, absolute zero ($-273.16°C$ or $-459.69°F$) can never be reached since all materials at the absolute zero point would have no entropy, which he believed was impossible. Sir Francis Simon established the third law of thermodynamics by using a magnetic method of cooling as well as the use of liquid helium. Simon was able to reach the temperature of 0.0000016 kelvin, within about 1/200,000 of one degree above absolute zero by surrounding liquid helium with a magnetic field that, when removed, causes the paramagnetic molecules to orient themselves in a random fashion while absorbing the small amount of remaining heat from the helium. Although Nernst received the 1920 Nobel Prize for his concept that one could approach but never achieve absolute zero, it was Simon who established the third law of thermodynamics as the point where all molecular motion ceases—there is absolutely no heat (i.e., no kinetic energy). This is the point where material substances have no degrees of freedom (molecular motion), which is absolute zero (0 kelvin). *See also* Nernst.

Slipher's Theories of Interstellar Gases and Andromeda: Astronomy: *Vesto Melvin Slipher* (1875–1969), United States.

Slipher's Interstellar Gas Theory: *There are enormous amounts of dust and gaseous material dispersed between and among the stars and galaxies.*

In the early 1900s Vesto Slipher was the first to make telescopic observations of the great clouds of interstellar material located between the stars, that reflects the stars' light. Up to that time these clouds appeared to the unaided eye as "dust" or "gas." Slipher proposed that these observable but diffuse nebulae (clouds of gas and dust) become luminous due to light from nearby stars that is reflected off the dust and gases in space. He determined this radiation varies, thus altering the brightness of the night sky. Slipher also discovered the existence of sodium and calcium dispersed in interstellar space. *See also* Rubin.

Slipher's Theory for the Speed of the Andromeda Nebula: *The dark lines of the light spectrum of Andromeda indicate it is approaching us at a tremendous speed.*

For several hundred years, astronomers thought Andromeda was simply a large accumulation of gas concentrated at one location in the sky. In 1612, Simon Marius (1570–1624) was the first to view and describe this fuzzy luminous cloud he called the Andromeda nebula (*nebula* means "cloud" in Latin.) At one time it was also believed that Andromeda might be located in Earth's Milky Way galaxy. Slipher devised a technique using the Doppler effect to measure the radial velocity of spiral nebulae to determine the shift of their spectral lines. For instance, when an object moves away from us, its light's wavelegth lengthens toward the red end of the spectrum—thus, the redshift. Conversely, if the object moves toward us, its light's wavelength is shortened to the blue end of the spectrum. In 1912 Slipher determined the Andromeda nebula is not part of our galaxy but is rather a large galaxy moving toward Earth at a speed of more than 300 kilometers per second (about 200 miles per second). This theory, at first disputed by other astronomers, resulted in a better understanding of the nature of the universe. Slipher used the Doppler shift method to examine the spectra of several dozen extragalactic objects, predating Edwin Hubble's use of the redshift to measure the distance of far objects in the vast universe. Hubble estimated that Andromeda was 750,000 light-years from Earth. Since then this estimate has been increased to over 1 million light-years, meaning the light from the Andromeda nebula now viewed by astronomers started its trip over 1 million years ago. Today, it is estimated there are over 100 billion galaxies similar to Andromeda in the universe, each containing billions of individual stars. *See also* Doppler; Hoyle; Hubble.

Smoot's Theory of a Nonuniform Universe: Astronomy: *George Fitzgerald Smoot* (1945–   ), United States.

There are "spots" in the universe that are slightly warmer than the average temperature of the universe. Therefore, the universe is not absolutely isotropic.

Research from the early 1960s suggested that the universe must be isotropic (exactly the same in all locations). However, at the same time, the concept of an inflationary (ever-expanding) universe, proposed by Alan Guth (1947–   ), required the existence of areas throughout the universe that are less dense and/or with slightly different temperatures. In addition, Guth claimed that the observable inflationary universe could have originated from an infinitesimal "nothing." In 1989 a satellite carrying instruments designed to measure differences in radiation at different locations in the universe, as well as the absolute brightness in the sky, provided reams of data that Smoot analyzed. His theory of island structures in the universe is at odds with the isotropic concept of space, but it does agree with recently discovered clusters of galaxies and even superclusters of galaxies. He determined some areas of space are slightly warmer (by 1/30,000,000 of 1°C) than other regions of the universe. Thus, this difference in radiation indicates the universe is not isotropic. His data supported the now widely held theory of an inflationary universe.

Soddy's Displacement Law for Radioactive Decay and Theory of Isotopes:
Chemistry: *Frederick Soddy* (1877–1966), England. Frederick Soddy received the 1921 Nobel Prize for chemistry.

Soddy's Radioactive Displacement Law: *A radioactive element that emits an alpha particle (a helium nucleus) is transformed into another element with a lower atomic weight, while a radioactive element that emits a beta particle (electron) will raise that element's atomic number.*

Frederick Soddy was aware of the particles and radiation that were emitted by radioactive elements and that as their atoms lost either positively or negatively charged particles, these radioactive elements were changed from the original. At this time, the existence of the neutron was unknown. The discovery of the neutron in 1932 by Sir James Chadwick explained the difference in atomic weights for isotopes. But up to this time, some confusion persisted relative to what took place during the decay of radioactive elements. Soddy's displacement law of radioactive decay states that heavy radioactive elements, which emit **alpha particles**, will reduce that element's atomic weight by four. Conversely, if the radioactive element emits an electron, it will have a higher atomic number. Soddy's displacement law provided the information needed for his original theory of isotopes.

Soddy's Theory of Isotopes: *Some forms of the same elements have similar chemical characteristics (same number of protons) but exist with more than one atomic weight.*

Frederick Soddy believed that many elements were "homogeneous mixtures" of similar elements with similar atomic numbers but for some reason had slightly different atomic weights. He referred to these elements as isotopes, meaning the "same place" in Greek. At that time it was not known that the variances in weights of similar atoms were due to the different numbers of neutrons in atomic nuclei. Soddy demonstrated that two elements, uranium and thorium, are radioactive and will decay into isotopes of lead. Although these two radioactive elements decay in different sequences, their end products are both isotopes of lead (lead with different atomic weights). By subjecting uranium and thorium to different chemical reactions, Soddy and his mentor, Sir Ernest Rutherford, arrived at what is known as the *radioactive series*, which explains in detail the decay sequence taken by these elements before becoming stable isotopes of lead. Later in his career, Soddy also believed there were limited sources of hydrocarbon energy on earth, which culminated in his proposal to use the energy of radioactive elements as a solution for our energy problems. *See also* Chadwick; Rutherford.

Sorensen's Negative Logarithms Representing Hydrogen Ion Concentration: Chemistry: *Soren Peter Lauritz Sorensen* (1868–1939), Denmark.

The negative logarithm of the concentration of hydrogen ions in a solution can be used to measure the acidic or basic (alkaline) properties of solutions. This measure is called pH.

Soren Sorensen's system was one of the first attempts to measure the extent to which a solution was either acidic or basic. To make his concept work, he used negative logarithms to determine the concentration of the hydrogen ions (H^+). If the solution has a greater H^+ than OH^-, the solution is acidic. Conversely, if the solution has a higher concentration of OH^- ions than H^+ ions, it is basic (caustic or alkaline). The pH scale ranges from 0 to 14, with pH 7 being neutral (the H^+ and OH^- ions are equal). Solutions decreasing from pH 6 to pH 0 indicate increasing H^+ concentration or greater acidity, while a reading ranging from pH 8 to pH 14 indicates solutions of increasing OH^- or alkalinity. Since the pH scale is logarithmic, each unit increase in pH represents a tenfold increase in the concentration of either the H^+ or OH^- ions for each mole per liter (mol/L). For example, pH $0 = 1 \times 10^0$ H^+ concentration (mol/L), while pH $6 = 1 \times 10^{-6}$ concentration, and for pH 14 the H^+ concentration would only be 1×10^{-14}. Conversely, for pH 14, the concentration of the OH^+ ion would be 1×10^0, and for pH 0 the OH^- ion would be only 1×10^{-14} (mol/L). A simpler interpretation of the scale states that if a solution has a pH lower than 7, it is acidic; a pH higher than 7 indicates a basic; and a reading of 7 on the scale means the solution is neutral. This system measures the pH of all solutions in a number of ways, the simplest of which is the use of indicators, such as paper strips, that change color when wetted by the solution. The color change is matched with a color chart to determine the pH value of the solution. A more accurate and less subjective method is the use of a pH meter consisting of a glass electrode sensitive to the H^+ ions, where the reading can be compared to a reference electrode. The pH meter also permits continuous readings, which are not possible with the paper method. Determining the pH level of the acidity or alkalinity properties of solutions or other substances is important to many industrial processes dependent on the exact degree of pH required for some chemical reactions.

Spallanzani's Theory Refuting Spontaneous Generation: Biology: *Lazarro Spallanzani* (1729–1799), Italy.

If solutions containing microorganisms are boiled over a long period of time and not exposed to the air, all living organisms will be destroyed.

The concept of spontaneous generation dates back to the earliest humans and their curiosity about how living things just seemed to grow. In 1668 Francesco Redi, who studied insect reproduction, investigated William Harvey's concept that flies do not just spontaneously appear but are produced from eggs. (*See* Redi for details on his classic experiment.) In 1745 John Needham (1713–1781) proposed that a "life force" was present in all inorganic matter, including air, which could cause life to occur spontaneously. Therefore, after boiling his chicken broth but leaving it exposed to air, microorganisms grew, "proving" his belief that life could be generated from this "life force." Needham later repeated this experiment, but he sealed off the glass spout of the flask containing the boiled broth. The organisms still grew. Again he claimed to have proved the viability of spontaneous generation. Lazzaro Spallanzani decided that Needham had not boiled the broth long enough and had not removed the air from the jar before sealing it. Spallanzani improved the experiment by not only boiling the broth longer, but he drew out the air, creating a vacuum in the flask as he continued to boil the broth. He then sealed the flask. No microorganisms grew, which proved Spallanzani's theory that spontaneous generation could not occur without air. *See also* Pasteur; Redi.

Spencer-Jones' Concept for Measuring Solar Parallax: Astronomy: *Sir Harold Spencer-Jones* (1890–1960), England.

An accurate astronomic unit (AU) can be calculated by using the position of a minor planet then comparing it to the Earth's distance from the sun.

Cognizant of the fact that the minor planet Eros was to approach Earth at a distance of only 16 million miles on a specific date in 1931, Harold Spencer-Jones received the cooperation of many astronomers worldwide who all at the same time photographed Eros' position. Using these data, he established the solar **parallax** by comparing the radius of Earth from the center of the sun. His figure, 8.7904 seconds of arc, was later corrected to 8.7941 seconds of arc. This provided a more accurate figure for the astronomical unit (AU) used to measure large distances for objects in the solar system. One AU is approximately 92,956,000 miles, which is the mean distance of Earth from the sun's center. The AU is much too small a unit for measuring great distances in space. The parsec, which equals 3.258 light-years and is the distance a star would be from Earth if it had a parallax of 1 second of arc, and the distance light travels in a vacuum over one year (a light-year) are the units used for measuring distant objects in the universe. Spencer-Jones also used a very accurate quartz timepiece to determine the period of rotation of Earth on its axis. He concluded that the rotation of Earth is extremely regular but deviates very slightly each year. *See also* Cassini.

Stahl's Phlogiston Theory: Chemistry: *Georg Ernst Stahl* (1660–1734), Germany.

When substances burn, they release phlogiston. The more complete the burning, the more phlogiston the substance contains and releases.

Since ancient times, the concept of objects' burning and rusting puzzled philosopher/scientists. In the seventeenth century Johann Becher advanced the concept of phlogiston, which made sense to many people. The concept was that when an object burned, heat and light were released, while at the same time it became lighter and produced ash; thus the substance "lost" something. About a hundred years later, German chemist Georg Stahl improved the phlogiston concept to the point where it became the first rational theory of combustion. For example, when charcoal burned, almost no ash remained, meaning that charcoal contained a great deal of phlogiston, which was released during combustion. When metal was heated over charcoal, a coating or ash, which Stahl called *calx*, formed on the metal. If the metal was consumed by a very hot fire, the phlogiston was freed, leaving behind the ashlike calx. He concluded that the metal must be composed of both calx and phlogiston (metal $\rightarrow$ phlogiston + calx). Therefore, if the process was reversed and calx was heated over charcoal, it "absorbed" the liberated phlogiston from the charcoal to become metal again (phlogiston + calx $\rightarrow$ metal). He assumed charcoal was rich in phlogiston, and the release of phlogiston from burning substances seemed a rational explanation of combustion. For example, if a burning candle is placed in a closed jar, the air will soon become "saturated" with phlogiston because the candle is pure phlogiston, as evidenced by the absence of ash. This explanation is false. However, Stahl was correct when he claimed the rusting of iron was also a form of "combustion." The phlogiston theory was shattered in the late 1700s when Antoine Lavoisier developed the currently accepted theory of combustion. *See also* Cavendish; Lavoisier.

Stark's Theories: Physics: *Johannes Stark* (1874–1957), Germany. Johannes Stark was awarded the 1919 Noble Prize for physics.

Stark's Theory of the Doppler Effect on Fast-Moving Particles: *The frequencies of radiation emitted by rapidly moving particles change as the speed of the particles change.*

The Doppler effect results from the change in frequencies for both sound and light. The redshift (Hubble effect) is used to measure the distance and motion of stars. This technique is based on the Doppler phenomenon, where the frequency of the wavelengths of light from a distant fast-moving source tends to "spread" and register as longer waves as it recedes from us and thus appears red. Stark's theory applies the Doppler effect to rapidly moving particles such as electrons and photons. These fast-moving particles can be affected by both magnetic and electrical fields, which increase or decrease the frequencies of radiation they produce.

The Stark Effect: *A strong electric field can "split" spectral lines, which increases the number of lines.*

Johannes Stark was familiar with the Zeeman effect, which used strong magnetic fields to split the spectral lines of electromagnetic radiation with specific wavelengths. Since the electromagnetic spectrum consists of radiation of specific frequencies produced by the relationship between magnetism and electricity, Stark reasoned that if strong magnets can "split" a specific frequency of the spectrum, electricity could do the same. He used a strong electric field to produce a similar "splitting" or multiplying of lines of the frequencies of the electromagnetic spectrum. This was demonstrated to be a quantum effect (small changes), which Stark at first accepted but later rejected. The Stark effect was used to develop techniques to study electromagnetic radiation and subatomic particles. *See also* Doppler; Maxwell; Zeeman.

Steinberger's Two-Neutrino Theory: Physics: *Jack Steinberger* (1921–   ). United States. Jack Steinberger shared the 1988 Nobel Prize for physics with Leon Lederman and Melvin Schwartz.

A beam of neutrinos will produce two different types of neutrinos.

The law of conservation of energy states that in an isolated (closed) system, energy cannot be created or destroyed, although it may be changed from one form to another, but the sum of all forms of energy must remain constant. Physicists knew that when neutrons disintegrated and the "pieces" were measured, something appeared to be missing since the sum of the energy (or mass) of the "pieces" did not add up to the original for the neutrons. Physicists also knew that when a nucleus broke down, it emitted a beta particle (electron) plus a proton, but these two particles did not contain the total energy required by the law for the conservation of energy. In 1931 Wolfgang Pauli suggested that in addition to the electron (beta particle) and proton resulting from the disintegration of a neutron in the nucleus of an atom, another undetected particle must also be ejected. Since this undetected particle has no electrical charge and very little mass, it is difficult to detect. Pauli's theory stated that when the neutron of an atom breaks down, three, not two, particles are ejected. (See Figure F2 under Fermi.) Enrico Fermi mathematically described and named this theoretical third particle neutrino, meaning "little one" in Italian. Scientists tagged it as a "ghost particle" since it had yet to be detected except by the use of mathematics. Even so, mathematically it accounted for the missing energy when beta particles (electrons) are emitted from neutrons. Jack Steinberger developed a technique that produced and controlled a beam of neutrinos. This was somewhat of a surprise since neutrinos contain no electrical charge and practically no mass. He used this beam of neutrinos to demonstrate leptons, which are a group of particles that include electrons and neutrinos and come in pairs of opposites, thus confirming the accuracy of his theory. His theory expanded particle physics to explain the electron (e^-) and its opposite, the positron (e^+). In addition, two different types of neutrinos were discovered: the electron neutrino referred to as the e neutrino (ve) and the muon neutrino referred to as the μ neutrino ($vμ$). Steinberger's theory also helped to explain spin as a characteristic for these opposite particles. These two types of neutrinos each have a spin of one-half (opposite to each other) and a slight mass of 105.7 MeV. There are two types of muons or mu mesons: one with a positive charge ($μ^+$), the other with a negative charge ($μ^-$). Steinberger's theory advanced the concept of the duality of subatomic particles. *See also* Fermi; Pauli.

Steno's Theory for Fossil Formation: Geology: *Niles Stensen* (a.k.a. *Nicolaus Steno*) (1638–1686), Denmark.

The age of fossils can be determined by the manner and time in which solid (formerly organic) bodies are imprinted on other solid (inorganic) bodies.

Nicolaus Steno was one of the first to propose a theory for the origin as well as the nature of the formation of fossils and one of the first to propose that fossils were formed in sedimentary strata laid down in ancient seas. Two important principles are incorporated into Steno's theory of fossils. First, it is possible to identify which solidified first: the fossil material or the substance in which it was formed. Using this concept, he determined that *glossopteris*, which are seeds and remains of ancient fernlike plants, left their imprints first on the surrounding "mud" before turning into rock. Since these fossils were a hard substance found within another hard substance, he concluded it was possible to establish the date of the fossil if the date of the rock was known. Glossopteris, also known as "tongue stones," were often confused with shark's teeth because of their shapes. At that time in history, it was assumed they fell from the heavens and were the actual objects, not fossils. Steno was criticized when he claimed that "sharks' teeth" were not actual teeth but rather fossils formed by minerals replacing the original substance. His second principle stated that if both the fossil and the rock material surrounding it were similar, they then could have been formed in the same way at approximately the same time. Steno's theory, an invaluable method for interpreting fossil records, is still used today and is an example of evidence supporting Darwin's theory of organic evolution. Steno is known as the Father of Paleontology.

Stern's Theory for the Magnetic Moment of the Proton: Physics: *Otto Stern* (1888–1969), United States. Otto Stern was awarded the 1943 Nobel Prize in physics.

Protons in atoms behave like small magnets and thus assume one of two orientations within a magnetic field.

Space quantization theory states that atoms can align themselves in only a few directions when they are in a magnetic field. Otto Stern theorized that some

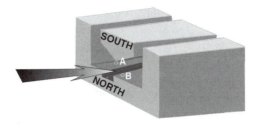

Figure S3: A much stronger magnetic field is produced at *A*, where the sharp edge of the south pole (*S*) of the magnet leaves a gap at *B* (the north pole, *N*) which generates a nonhomogeneous magnetic field that produces opposite spin orientations in atoms and is known as space quantization.

atoms, such as those in silver, could align themselves in only two directions rather than in all directions, as proposed by Newtonian physics. Stern and his colleague, Walter Gerlach (1889–1979), devised a unique experiment to test this theory. Their device was a magnet with the north pole as a flat surface and the south pole shaped as a knife edge placed close to but not touching the flat north pole surface. (See Figure S3.) This arrangement produced a nonhomogeneous magnetic field between the poles. They directed a beam of neutral silver atoms produced by heating silver metal in a vacuum, which acted like tiny atomic magnets (similar to micro compasses), through a slit between the center of the two poles of the magnet. The nonuniform magnetic field split the thin line of silver atoms being directed across the gap between the north and south poles. This caused the narrow beam of silver atoms to split into two distinct paths, each representing opposite spin orientations of the atoms (spin-up and spin-down). If the thin line of silver atoms was divided into broad bands, it would have indicated more than two orientations. The restriction of the silver atoms to just two orientations is referred to as *space quantization*. Nuclear magnetic resonance instruments used to examine atomic and molecular structures and magnetic resonance imaging medical diagnostic tools are just two of the applications that resulted from Stern's theory. *See also* Rabi; Tyndall.

Stokes' Laws of Hydrodynamics and Fluorescence: Physics: *Sir George Gabriel Stokes* (1819–1903), England.

Stokes' Law of Hydrodynamics: *A spherical body moving through a viscous fluid at a given speed produces a frictional drag.*

Sir George Stokes applied mathematics to solve many of the problems that concerned contemporary physicists. He developed a complicated equation to explain the hydrodynamics of fluids as a coefficient of viscosity: $6 \pi \eta\, rv$, where η is the coefficient of viscosity, r is the radius of the spherical body, and v is the speed of the spherical body through the fluid. This computation for the coefficient of viscosity applies only for normal conditions; the law breaks down

at extreme temperatures and pressures. Nevertheless, Stokes' law of hydrodynamics is applicable in many industries (e.g., sending oil through pipelines and
mixing of liquids).

Stokes' Law of Fluorescence: *The wavelength of fluorescent radiation is
greater than the wavelength of the radiation causing the fluorescence.*

Michael Faraday created a vacuum inside a closed glass jar and then passed
an electric current through the vacuum. He assumed that since the air had been
removed, nothing remained inside the jar to stop or slow the flow of current.
However, he noticed a greenish glow formed on the inside of the glass as the
current was turned on. Sir George Stokes named this phenomenon *fluorescence*,
which now refers to any visible light produced by fast collisions of radiation
(light, photon, electrons) with matter. It was impossible for Faraday to evacuate
all the gas molecules from the glass vessel; therefore the concept of electricity
as a fluid was not realized. Stokes proposed his law after another scientist noted
that a fluorescent beam could be diverted by exposing it to a magnetic field. He
concluded that a source of radiation (electricity) that caused the fluorescence
was always of a lesser wavelength than was the actual wavelength of the fluorescent light itself. This law is applicable only under certain conditions of atmospheric pressure within an evacuated glass container and the concentration
of gaseous molecules within that container. Stokes' law and this concept pioneered the development of the Geissler tube, computer monitors, and TV
screens. *See also* Faraday; Fraunhofer.

Stoney's Theory of the Electron: Physics: *George Johnstone Stoney* (1826–
1911), Ireland.

*Electricity is composed of small units of fundamental particles, as is matter,
and these units can carry electric charges.*

George Stoney was aware of Svante Arrhenius' work on ionic dissociation
as related to solutions of certain substances (e.g., salts that act as electrolytes
and carry electric current). Arrhenius explained that the salt dissolved to form
ions, which could then carry electrical charges. (See Figure A2 under Arrhenius.)
Thus, a better understanding of the structure of the atom resulted. George
Stoney's theory states there is an "absolute unit" of electricity with just one type
of charge, and it is carried from atom to atom when an electric current flows
through a conductor. In addition, these charges always exist in a ratio of whole
numbers, never fractions of a unit. He based his theory on the calculated mass
of the hydrogen ion given off during electrolysis. Stoney coined the term *electron* to describe this basic negative unit of electricity that could be carried by a
single atom or even a group of atoms. Several years later his term *electron* was
accepted by the science community. (See Figure T1 under Thomson.)

Using this knowledge, Stoney determined there were two different types of
molecular motion. One relates to the relative motion of gas molecules to each
other that does not result in a spectrum when exposed to radiation. The other

is the internal, random motion of molecules in a substance that will produce spectral lines related to the type of substance involved. *See also* Arrhenius; Helmholtz; Thomson.

Strasburger's Law of Cytology: Biology: *Eduard Adolf Strasburger* (1844–1912), Germany.

New nuclei of cells arise from existing cell nuclei, and through a process of mitosis, they carry factors responsible for heredity.

Eduard Strasburger based his research on plant reproduction. He was the first to describe the embryo sac in gymnosperms (conifer/pine trees) and to recognize that angiosperms (flowering plants) reproduce by double fertilization (when the two nuclei from a pollen grain fuse with two nuclei of the embryo sac). His law of cytology, in addition to cell division, described the division of nuclei to form new nuclei in plant cells. The law relates to the study of the growth, structure, reproduction, and chemical makeup of cells and basically states that new cells arise from existing cells. Strasburger's concept of mitosis occurring in the nuclei of plant cells was related to the division of these nuclei following the same principle as that of his law of cytology. From this concept of mitosis, he inferred there were factors of heredity, with which he was not completely familiar, that divided during the process of mitosis as the nuclei of plant cells divided. In mitosis, each chromosome is divided in half so that the two new (daughter) cell nuclei are exactly the same as the mother cell nucleus. Strasburger also postulated that internal physical forces (i.e., hydraulics), rather than physiological factors (cells functions), are responsible for fluid (sap) being transported in the trunks of trees and stems of plants. *See also* Schleiden; Schwann.

Struve's Theory of Interstellar Matter: Astronomy: *Otto Struve* (1897–1963), United States.

Interstellar matter appears more diffused than localized throughout the universe.

Otto Struve was the grandson of the German astronomer Frederick Struve (1793–1864), who first used **parallax** to estimate the distance from Earth of the bright star Vega. The elder Struve's major contribution was the discovery and cataloging of over 3000 binary (double) stars. Otto Struve performed spectroscopic analyses of the binary stars recorded by his grandfather and also examined the structures and atmospheres of other stars and objects. His most important theory related to just what and how much "stuff" existed in the great distances of space separating the stars and galaxies. In 1937 he discovered that the vastness of space contained great amounts of ionized hydrogen. Other elements were also found, such as helium and calcium. More recently, some astronomers have estimated that over 90 percent of all the mass (matter) in space is "dark matter." Thus, it cannot really be seen since it neither gives off nor reflects light. *See also* Arrhenius; Hoyle.

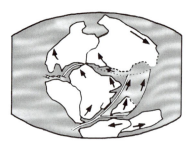

Figure S4: Eduard Suess' Gondwanaland, similar to Wegener's theory of the supercontinent Pangaea, began to separate over 200 million years ago, resulting in the formation of the modern continents. Their concept of continental drift led to the modern science of plate tectonics.

Suess' Theory of Continental Drift: Geology: *Eduard Suess* (1831–1914), Austria.

The southern continents were once combined as one large land mass.

Eduard Suess spent years studying the similarities of geological and plant fossils on the continents of Africa, Australia, and South America and the subcontinent of India. He observed similar geological structures, including mountain ranges, regions of volcanoes and earthquakes, and coastlines for these land masses, as well as examples of the ancient fossil ferns known as glossopteris that existed in the Carboniferous period of earth's development. Based on these clues, he derived his theory of a great supercontinent he named Gondwanaland, after the Gonds of ancient India. Alfred Wegener developed a theory for a similar supercontinent he named Pangaea. (See Figure S4.) The theory stated these four southern land areas were once joined and, since separation, still exhibit a pattern that indicates their common origin. More recently, computer models of the coastal margins, mountains, and other formations of these continents validate his theory. Suess' theory is an important concept for understanding continental drift in the science of plate tectonics. *See also* Wegener.

Swammerdam's Theory of Preformation: Biology: *Jan Swammerdam* (1637–1680), Netherlands.

All parts of adult animals are formed at the beginning of the egg's development.

The theory of preformation asserts that all the parts of animals are present in the female's eggs at the time of conception, or, as stated by some scientists in the seventeenth century, a tiny homunculus existed inside either the undeveloped ovum (ovists) or the sperm (spermatists) and was composed of all the parts of an adult human. Jan Swammerdam, a devoted microscopist, was the first to use the microscope in the study of zoology. He conducted excellent studies of insects and other smaller animals and was also the first to identify cells in frogs' blood. Swammerdam based his theory of preformation on the metamorphosis process of insects, which he dissected with excellent skill. He observed rudimentary parts (wings, legs, eyes, etc.) of adult insects inside the pupae and cocoons. He claimed a caterpillar did not metamorphose into a butterfly or moth, but rather continued to grow into an adult from parts existing in the caterpillar stage. He claimed the same for a tadpole changing into an adult frog (i.e., all of the body parts were contained in the egg). The preformation theory was generally accepted into the early nineteenth century until the epigenesis theory developed. Proponents of the epigenesis theory believed the egg is undifferentiated, and development occurs throughout a series of steps after fertilization by a sperm. This is basically the theory accepted today. Karl Ernst von Baer's (1792–1876) concept of the formation of a "germ layer" in the eggs of mammals, out of which all the embryonic organs develop, was the final death knell to the preformation theory. Moreover, Swammerdam's insect studies provided him with evidence that disproved the theory of spontaneous generation as an explanation for the origin of life. *See also* Haeckel.

Szilard's Theory of Neutrons Sustaining a Chain Reaction: Physics: *Leo Szilard* (1898–1964), United States.

There are certain heavy elements that when their nuclei each absorb a single neutron will split into two nuclei of different elements, and in the process emit two additional neutrons. Thus, if such an element is concentrated in a critical mass, enough neutrons will be produced to create a sustaining nuclear chain reaction, releasing a great deal of energy.

Leo Szilard followed up on the research on fissionable uranium conducted by Lise Meitner, Otto Frisch, and Otto Hahn. Once he knew it was possible to bombard uranium nuclei with neutrons to cause the nuclei to fission (split) and that this reaction would produce more neutrons than were absorbed by the nuclei, he was convinced a sustainable chain reaction was conceivable. In 1933 Szilard fled from Germany to England, where he filed a patent for the neutron chain reaction, which he later assigned to Great Britain. In 1938, he arrived in the United States, where he attempted to convince the U.S. government and scientists to develop an atomic bomb because he was convinced the German government was doing the same. He and other scientists persuaded Albert Einstein to send the famous August 2, 1939, letter to President Franklin D. Roosevelt describing the potential for developing the atomic bomb and the urgency for doing so. Despite Szilard's continued work with the atomic bomb, including the idea of a breeder reactor that produces more radioactive material than it uses as fuel, he strongly opposed the use of atomic weapons. He is known as a "scientist of conscience." *See also* Chadwick; Einstein; Fermi; Hahn; Meitner; Oppenheimer; Rutherford; Teller.

T

Tamm's Theory of the Cherenkov Effect: Physics: *Igor Yevgenyevich Tamm* (1895–1971), Russia. Igor Tamm shared the 1958 Nobel Prize for physics with Ilya Frank and Pavel Cherenkov.

When high-speed particles (electrons) pass through nonconducting transparent solids at speeds faster than light passes through the same solid, radiation is emitted.

Igor Tamm's theory is based on the quantum theory of diffused light in solid bodies. Although these high-speed particles cannot travel faster than the speed of light in a vacuum (nothing can), they do pass through certain types of solids and liquids at speeds approaching that of light (186,000 miles per second in a vacuum). At the same time, light travels through the same substances at a slower rate of speed than do high-speed particles. For instance, light traveling through water or a crystal does so at a speed less than its speed in a vacuum, while the speed of a high-energy electron can surpass the speed of light in water or a crystal. (The speed of light in a vacuum is 299,793 kilometers per second; in water, its speed is 224,900 km/s, and when traveling through a diamond, the speed of light is only is only 124,000 km/s.) Tamm's theory explained Cherenkov radiation as being similar to a shock wave produced when an object moves faster than the speed of sound through air (e.g., a bullet or jet airplane going faster than sound; *see* Mach). For both sound and Cherenkov radiation, the velocity of the object passing through the medium is greater than the shock wave created by the object's motion. This explains why water surrounding the core of nuclear reactors glows an eerie blueish color and why there are "showers" of cosmic radiation on earth. Detectors designed to count Cherenkov radiation measure the strength of high-speed particles and can also determine their velocities, which almost reach the speed of light. Tamm's explanation of the Cherenkov effect enabled physicists to understand better the operation of nuclear reactors, as well as the nature of cosmic radiation.

Tatum's Theory of Gene-Controlling Enzymes: Biology: *Edward Lawrie Tatum* (1909–1995), United States. Edward Tatum shared the 1958 Nobel Prize for physiology or medicine with George Beadle and Joshua Lederberg.

Specific genes are responsible for the production of specific enzymes that control particular biochemical reactions in living organisms.

Edward Tatum began his experiments by inducing mutations in the genes of the fruit fly. He extended this concept by exposing particular types of bread mold to x rays to induce mutant genes in the mold. He discovered that when these mutant molds were grown in different types of media, they were affected differently according to varying types of nutrients in the growing medium. He then crossbred these distinct mutant mold genes, noticing that their diet peculiarities were inherited according to the standard Mendelian percentages (*see* Mendel). Tatum and a colleague theorized that particular genes are responsible for specific enzymes in living organisms. Enzymes act as organic catalytic proteins found in living cells that control and regulate the chemical process of life. From this they concluded that all chemical processes taking place in plant and animal cells are controlled and regulated by genes.

Taylor's Theory of Gravitational Waves: Astronomy: *Joseph Hooton Taylor* (1941–   ), United States. Joseph Taylor shared the 1993 Nobel Prize in physics with Russell Hulse.

When a binary pulsar is influenced by a nearby massive object, the pulsar changes its orbital period, thereby producing gravitational waves.

Einstein used his theory of general relativity to predict that when a body accelerates, it will radiate gravitational waves. Such "waves" produced by accelerating stellar bodies are much too weak to be detected on earth, and thus are still theoretical. Taylor and his student, Russell Hulse (1951–   ), observed a binary pulsar, which is a pair of massive bodies whose orbits intersect and whose gravities affect each other's velocities and thus their orbital periods. Taylor and Hulse's pulsar was located about 1600 light-years in space, so the theoretical radiation of its gravity waves was much too weak to reach Earth. They continued to watch this observable pulsar and its companion dark neutron star, and assumed their near approach to each other would cause a slight change in the pulsar's acceleration and its orbit, which should be detectable over a period of time as a very minor alteration. After several years of analyzing his observational data, Joseph Taylor detected a very slight decrease in the pulsar's orbital period. He claimed the data supported Einstein's theory for the existence of gravitational waves. Even so, no direct radiation gravitational waves from pulsars or any other deep space objects have been measured on Earth.

Teller's Theory for the Hydrogen Bomb: Physics: *Edward Teller* (1908–   ), United States.

The production of x rays from a fission bomb will produce the pressure and temperature required for a nuclear fusion reaction.

As the atomic (fission) bomb was being developed, Edward Teller, a nuclear physicist, was already contemplating the design for a hydrogen (fusion) nuclear bomb. The distinctions between the two types of nuclear weapons are important. "Atom bomb" is really a misnomer because atomic reactions take place between and among the outer electrons of atoms during ordinary chemical reactions. It is the nuclei of atoms that are involved for both types of nuclear weapons—the so-called atomic and hydrogen bombs. The "atom" bomb involves the fission (splitting) of nuclei of heavy, unstable radioactive elements (e.g., uranium-235 or plutonium-239), which releases enormous energy and radiation, while the "hydrogen" bomb is the fusion or combining of nuclei of light elements to form nuclei of heavier elements, which also releases great quantities of energy but less radiation.

After World War II, U.S. President Harry S. Truman, concerned that the Russians also had developed and exploded their first "atomic" bomb, encouraged the development of the "hydrogen" bomb for national security. Previously Teller and other scientists had studied various designs for such weapons. Teller proposed three different designs, two of which proved impractical. The third seemed promising until a theoretical mathematician, Stanislaw Ulam, pointed out that Teller's design was not only impractical but much too expensive. Together they developed a fourth model that overcame the physical and economic problems of the other designs. One problem was that fusion, unlike fission, could not occur under normal conditions of temperature and pressure. Great force was required to slam the positively charged nuclei of hydrogen together to overcome their natural repulsion. The fusion reaction, also referred to as **thermonuclear**, requires tremendous heat and pressure to complete the reaction (e.g., the sun's conversion of hydrogen nuclei into helium nuclei). Ulam proposed construction of the fission (atomic) bomb around the H-bomb to provide the force necessary to fuse "heavy" hydrogen atoms together to form nuclei of helium. Teller improved Ulam's concept by devising a "mirror" to focus and concentrate the x rays produced by the A-bomb surrounding the H-bomb to produce the force necessary to start the fusion reaction. The following nuclear fusion reaction occurs: $_1H^2 + _1H^2 \rightarrow _2He^4 + \rightarrow$ Energy. There are two types of heavy hydrogen, deuterium $D-2$ and tritium $T-3$, both of which are used in fusion reactions because they contain extra neutrons in their nuclei. (See Figure O1 under Oliphant.) The atomic weight of two hydrogen-2 is 4.0282, while the atomic mass of helium-4 is 4.0028, representing the loss of 0.0254 mass units when hydrogen nuclei fuse to form helium nuclei. This may seem like a minute loss of mass to convert to energy ($E = mc^2$), but when trillions of nuclei are involved in a fusion reaction, this "extra" or leftover mass is converted to about ten times the energy released by a typical atomic (fission) bomb. The first successful fusion hydrogen bomb was detonated by the United States in 1951. Ulam and Teller both claimed it was their own concept to use an atomic bomb to "trigger" the H-bomb. Although many scientists gave Ulam the credit, Teller by this time had become a strong political advocate for developing thermonuclear weapons as a deterrent factor and became known as the Father of the Hydrogen Bomb. Those who opposed U.S. policy on nuclear weapons were attacked by Teller and other advocates of national nuclear policies. One of the major developers of the fission (atomic) bomb, Robert Oppenheimer, was horrified at the prospect of using the much more destructive hydrogen bomb in warfare and had his career cut short by opposing Teller's position on nuclear weapons. Edward Teller proceeded to encourage scientists to develop thermonuclear weapons and the strategic defense initiative as a means to protect the United States from long-range missile attacks. *See also* Curies; Fermi; Hahn; Meitner; Pauli; Rutherford; Szilard; Ulam.

Temin's Theory for Transcribing RNA Information into DNA: Biology: *Howard Martin Temin* (1934–1994), United States. Howard Temin shared the 1975 Nobel Prize for physiology or medicine with David Baltimore and Renalto Dulbecco.

In addition to genetic information "flowing" from DNA to RNA, a special enzyme makes it possible for DNA to receive information from RNA, thus allowing DNA to provide crucial information needed for cell growth.

While conducting research with cancer cells in chickens, Howard Temin discovered a new enzyme. He named it *reverse transcriptase* because it could reverse the flow of genetic information that at one time was believed to proceed only in one direction, from **DNA** to **RNA**. This DNA-to-RNA sequence was referred to as the "central dogma" because this sequence was required for DNA to replicate itself. This concept was generally accepted by all molecular biologists. Temin's new enzyme "transcribed" the RNA into DNA, which improved DNA's effectiveness in controlling the processes of cell metabolism. At about the same time, David Baltimore independently made the same discovery. *See also* Baltimore; Dulbecco; Gallo.

Tesla's Concept of High-Voltage Alternating Current: Physics: *Nikola Tesla* (1856–1943), United States.

High-voltage alternating current can be transported more efficiently over long distances through wires than can direct current.

In the early 1880s Thomas Edison developed the direct current generators and distribution system used by his Edison Electric Light Company in New York City. It revolutionized the use of electricity but had one major drawback: it had to be generated near the site where it was to be used. This made it useful for lighting compact cities, but since direct current (**dc**) lost much of its energy when sent over wires for some distances, it was an impractical system. Another problem with direct current was that dC dynamos (generators) and motors required a commutator with wire "brushes" to provide electrical contact with the armature and terminals. This arrangement required constant maintenance since the brushes needed frequent replacement. Nikola Tesla's concept solved this problem by using a new system of dynamos (generators) and transformers that could produce current that alternates direction many times per second and could be "boosted" to high voltages by transformers, enabling it to be sent over longer distances. An interesting result was that a Mr. Love began digging a canal to circumvent the Niagra Falls. He planned to use this diverted water to generate dc electricity for industries that could be located near his canal and thus at a greater distance than industries restricted to the Falls area due to the limits of direct current. Nikola Tesla's ac system interrupted the project, and the ditch was later filled in with waste material and became known as the infamous Love Canal. Today alternating current in the United States has a 60-cycles-per-second (Hz) rate of changing direction, at 120 volts with relatively high ampere current, while much of the rest of the world uses 50 cycle (Hz), 240 volts with low ampere current.

Nikola Tesla is best known for his insightful technical knowledge of electricity, which he applied in developing many inventions, some of which were years ahead of their time; for example, the Tesla coil/transformer used in radios and TV sets, the induction (brush-less) motor used in computer hard disk drives, telephone repeaters for long-distance phone lines and the transatlantic cable, and wireless communication devices now used in cellular phone systems. Tesla held almost 800 patents. *See also* Ampère; Edison; Faraday; Oersted; Ohm; Volta.

Theophrastus' Concepts for Plant Classification: Biology: *Theophrastus* (c.372–287B.C.), Greece.

Plants can be distinguished and classified according to their structures and physiology.

Theophrastus, a student of Aristotle, became the head of the Lyceum in Greece upon the death of Aristotle. One of Aristotle's great achievements was his classification of animals as known at that time. Theophrastus used some of Aristotle's techniques in his own pursuits. He was a keen observer who wrote excellent descriptions of plants. In his two main books, one dealing with plant structures and the second with their functions (physiology), he described over 500 plant species. His writings influenced botanists over several centuries. Theophrastus was the first to establish a relationship between the structure of flowers and the resulting fruits of plants, but his main theory distinguished between *monocotyledon* and *dicotyledon* seeds. After examining grass and wheat seeds and noticing they had only one seed "coat," he classified them as monocotyledons, while bean seeds, having two seed "coats," were classed as dicotyledons. Cotyledons are the first shoots of the new plant arising from the germinating seed; thus he considered them as "coats" or "covers." He described the differences between flowering plants (angiosperms) and cone-bearing plants (gymnosperms). He coined many new terms and names for plants and their parts and thus is considered by many biologists as the Father of Botany. *See also* Aristotle; Linnaeus.

Thomson's Electron Theory: Physics: *Sir Joseph John Thomson* (1856–1940), England.

Rays from a cathode tube can be deflected and measured by using magnetic and electric fields. Thus, cathode rays must be particles smaller than an atom and which carry an electrical charge.

J. J. Thomson experimented with radiation produced by a cathode ray tube. In 1876 Eugen Goldstein tested a Geissler tube, a vessel that enclosed a vacuum and used an internal electrode to study the "flow" of electricity. Previously, several other scientists had observed a faint fluorescence within the tube when an electric current was sent through the internal electrode. Goldstein named the tube he used a *cathode-ray tube* because the glow originated from the negative cathode inside the tube. Contrary to Benjamin Franklin's theory that electricity "flowed" from positive to negative, Goldstein noted the electricity flowed from negative to positive, while others noted this stream of glowing current could be diverted by a magnetic field. Thomson hypothesized that this beam must be composed of charged particles in order for it to interact with magnetic fields. In addition, he theorized that these charged particles should also react (be bent) in electric fields as well as magnetic fields. He installed an apparatus that measured the deflection of the cathode's rays by both electric and magnetic fields, which enabled him to determine that the ratio of the rays' electric charge to their mass was high. From earlier experiments on the unitary nature of an electric charge, he assumed that the charge he detected on the cathode rays was of the same unit. Therefore, since the ratio was high, this meant the electric charge (e) was much greater as compared to the mass (m) of these new particles, which must be much less based on the ratio e/m. He also assumed these low mass particles from the rays were parts of atoms, not the whole atom, and that they were about 1000 times less than the mass of the hydrogen atom (one proton). It was later determined that this mass equals 1/1837 of that of a proton. At first Thomson referred to these new particles as *corpuscles*. Later, he named them *electrons*, since these new particles carried an electric charge and were detected as originating from the negative electric electrode of the cathode ray tube. Electrons are the fundamental unit of electricity. It is a basic unit. No smaller electrical charge has since been discovered. Thomson designed a model of the atom using electrons embedded in the atom whose sum charge matched the positive protons' sum charge, thus making it a neutral atom. He also investigated the role of electrons in producing chemical reactions. Sometimes his model is referred to as "berries in muffins," "raisins in pudding," or "the fuzzy atom" because he presumed the electrons were more or less evenly distributed throughout the structure of the atom. (See Figure T1.) It was later discovered that electrons exist more as electrically charged particles in orbits, shells, or energy levels positioned at a relatively great distance from the comparatively small, massive, positive, centralized nucleus. This concept of an electrically neutral atom with one or more of its outer electrons joining with those of other atoms is the basis of chemistry. Among other usages, the interaction of magnetic and electric fields on a stream of electrons is used to control the signals that produce the pictures on television receivers and computer monitors. *See also* Bohr; Crookes; Faraday; Roentgen; Rutherford; Townsend.

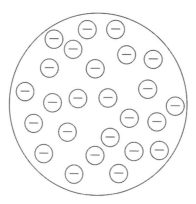

Figure T1: Depiction of J. J. Thomson's "raisins-in-a-pudding" model of the atom, which consists of a number of negative electrons randomly dispersed throughout the atom.

Tiselius' Hypothesis for Protein Analysis: Chemistry: *Arne Wilhelm Kaurin Tiselius* (1902–1971), Sweden. Arne Tiselius was awarded the 1948 Nobel Prize in chemistry.

Protein molecules carry an electrical charge; thus, it should be possible to separate them by the use of electric fields.

Arne Tiselius was familiar with electrophoresis, a procedure used to analyze chemical substances by the use of a weak electrical current. Most particles of matter contain a very small electrical charge on their surfaces. If these substances, as solutions, are applied to a special paper strip where one end of the strip is connected to a direct current source with a small negative charge and the other end of the paper is attached to the positive pole, the current will attract or repel the components of the sample substances at different rates. Depending on the sizes and individual characteristics of the component chemicals in the substance, these individual atoms and molecules will spread out on the strip of paper in very specific patterns, which can then be identified. Tiselius theorized that if this system could be improved to separate the proteins of blood, which also carry a small electrical charge, it might be possible to identify specific components of blood. He developed an improved electrophoresis system consisting of a U tube in which the proteins could be tracked as they separated. The tube could also be "disjoined" to extract specific components of the proteins for analysis. He further designed a lens system for refracting light though the different fractions, enabling a quantitative measurement of the particular protein fraction. Using this system, called the *Tiselius tube*, he identified four major components of blood serum proteins: albumins and the alpha, beta, and gamma globulin proteins. The best known is gamma globulin, whose chemical structure was first detected by Gerald Maurice Edelman (1929–   ); it stimulates antibodies in the immune system to protect against several diseases, including hepatitis

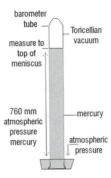

Figure T2: Torricelli's experiment demonstrated that the weight of air exerted pressure on the mercury in the dish, and it disproved the idea that the vacuum at the top of his closed tube "pulled" the mercury up the supporting column. This was the forerunner of the modern barometer.

and AIDS-related infections. Gamma globulin does not protect against the HIV virus.

Torricelli's Vacuum and Theorem: Physics: *Evangelista Torricelli* (1608– 1647), Italy.

Torricelli's Vacuum: *By filling a long glass tube, closed at one end, with mercury and inverting it into a dish of mercury, all but 760 mm of mercury will drain out of the tube, leaving a vacuum above the column of mercury.*

Evangelista Torricelli was employed by Galileo who had demonstrated that air had weight. Earlier, Jan Baptista van Helmont claimed air was not an element but rather a mixture of gases. These two concepts were incorporated into the answer as to why a pump could not raise water higher than 30 feet, which was a serious problem when draining mines that were flooded. Torricelli realized it was not the vacuum that "pulled" the water up, but rather the weight of the air (pressure) outside the pump that "pushed" the water up the pipe when low pressure was created inside the pump. In 1643 Torricelli calculated that since mercury weighs about 13.5 times that of water, air should lift quicksilver (mercury) only about 1/13.5 times as high as water. He tested this concept by filling a glass tube, closed at one end, with mercury and then inverting the tube into a dish filled with more mercury. He removed the cork from the end of the tube immersed in the dish of mercury, allowing the mercury in the vertical tube to settle as gravity pulled it down into the pool of mercury. Torricelli theorized that all the mercury in the tube would not exit the tube and end up in the dish of mercury because the weight of the mercury in the tube was the same weight as air outside the tube. The mercury in the glass tube maintained a height of about 760 millimeters. (*See* Figure T2.) The vacuum created over the mercury in the closed end of the tube was named the *Torricellian vacuum* or *torr*, in Torricelli's honor. One torr equals 1 millimeter of mercury (760 torr = 760 mm

Hg = 1 atm). After viewing the column of mercury for a few days, Torricelli noticed the height of the column varied slightly. He related this to changes in the weight of the air outside the tube of mercury caused by changes in the pressure (weight) of the air at the surface of earth. This discovery was an unintended consequence of the search for a more efficient water pump, resulting in an instrument (the barometer) that could accurately measure air pressure. *See also* Galileo.

Torricelli's Theorem: *The flow of a fluid through an opening in a standing pipe is proportional to the square root of the height of the liquid.*

Evangelista Torricelli's experience with the limitations of conventional water pumps and his concept of air pressure as the force that raises the water in "suction" pumps enabled him to develop his theorem. He observed that water escaping through holes at different heights in a standing pipe or container also escaped at different rates of flow. Being a mathematician, he believed there existed, and therefore calculated, a definite square root relationship between the height (depth) of the water and the rate of flow of water from the openings. His theorem is an important concept for industries that handle various types of fluids, since the rate of flow as related to the height of the source of the fluid can be measured. This principle is used in elevated community water storage tanks, where pipes carry water under pressure created by the height of the water in the tanks. *See also* Galileo.

Townes' Theory for Amplifying Electromagnetic Waves: Physics: *Charles Hard Townes* (1915–   ), United States. Charles Townes shared the 1964 Nobel Prize for physics with N. G. Basov and A. M. Prokhorov.

Molecules that exist in discrete energy states and absorb discrete frequencies of electromagnetic energy will emit photons of the same frequency.

Albert Einstein pointed out if an electromagnetic photon of a specific frequency struck a molecule that was in a high-energy state, the molecule would proceed to a lower-energy state while emitting a photon of the same wavelength as the one striking the molecule. Charles Townes realized this would produce two electromagnetic photons of the same frequency, which then could strike other high-energy molecules to produce more photons, resulting in a type of "chain reaction" that would produce a multitude of photons of the same wavelength and frequency. The consequence would be a flood of monochromatic (one-color) single-wavelength photons of the spectrum, all proceeding as coherent radiation (in the same direction). Townes demonstrated this theory by sending small amounts of microwave photons of a given frequency into energized ammonia molecules. The energized molecules had previously been produced by intense broadband irradiation. The microwave photons caused these molecules to drop back to their original energy level, producing new microwave photons. The result greatly amplified the original weak microwave radiation, which resulted in a flood of coherent electromagnetic radiation called the MASER (microwave amplification by stimulated emission of radiation). In 1958

Townes proposed this process could be applied to any wavelength within the electromagnetic spectrum. The concept was later improved by using just the section of the electromagnetic spectrum representing visible light, first called the "optical maser" and later named the LASER (light amplification stimulated emission of radiation). *See also* Turner.

Townsend's Theory of Collision Ionization: Physics: *Sir John Sealy Edward Townsend* (1868–1957), Ireland.

As an electric current passes through gas, some molecules become ionized; they then collide with and ionize other molecules, thus multiplying the original charge.

Sir John Townsend followed up on J. J. Thomson's discovery of the electron, at which time Thomson estimated the electron to be about 1/1000 the weight of the hydrogen nucleus, later revised to 1/1837 the mass of the proton. Thomson also determined that the electron carried the basic negative charge. Townsend calculated the amount or strength of this negative charge by forming a charged cloud of water droplets and measuring the rate of fall of the charged particles as they passed a source of electricity.

Townsend's "collision" theory answered the question of how an electric current could pass through a gas that supposedly had a weak electric field. The explanation was that as the current's electrons passed through the gas, some of the gas molecules became ionized (each gas molecule carried a charge) in the electric field. This created collisions with other gas molecules, which then became ionized, and so on, until an "avalanche" or multiplication of electrons proceeded through the gas despite the weakness of the original electric field. This theory is important in the fields of electronics and communications, where electrons cascade through multiplier tubes used to measure the radiation tracks of subatomic particles. *See also* Millikan; Thomson.

Turing's Theory for Testing Computer Intelligence: Mathematics: *Alan Mathison Turing* (1912–1954), England.

Since the brain is computable, it must be possible to program computers to acquire human intelligence and devise tests that will verify computer intelligence.

In 1937 Alan Turing designed a computing device called the Turing machine, which included a long tape divided into sections, a printer, and a correcting device. The machine had five symbols programmed to control the machine, similar to the operating systems of modern desktop computers and which could be used to make mathematical calculations. After World War II Turing developed several types of computers. He named one the ACE (automatic computing engine) and the other MADAM (Manchester automatic digital machine). By 1950 he argued that a computer could be designed to imitate human intelligence and designed a test to prove this concept, based on his idea called an "imitation game," later called the *Turing test*. The Turing test required an interrogator to ask one person and a computer (the interrogator could see neither the computer nor the person) a question that could be answered by typing out a textual answer. Turing claimed that if the interrogator was unable to judge which answer came from the human and which from the computer, then computer intelligence was proven. A similar test is used today to determine if an artificial intelligence (AI) computer program can really imitate human intelligence or "think." Later in his life, Alan Turing made distinctions between computer intelligence (AI) and thinking, emotions, and other human attributes.

Turner's Theory for Measuring Outer Energy Levels of Molecules: Chemistry: *David Warren Turner* (1927–1990), England.

The energies of outer electrons ejected from ionized gas atoms or molecules can be measured by deflecting these electrons with an electrostatic charge.

David Turner devised a technique that used a narrow beam of monochromatic ultraviolet electromagnetic radiation (MASER) to eject outer electrons from ionized atoms and molecules of gas. The energies of these ejected electrons can then be measured by their degree of deflection as they pass through an electrostatic field. This procedure is known as *molecular photoelectron spectroscopy*. Applying his theory, he assisted in developing a microscope that uses x rays to "kick" out electrons from the sample, thus measuring the sample based on the degree of deflection. *See also* Townes.

Tyndall's Theory for the Transmission of Light through Gases: Physics: *John Tyndall* (1820–1893), England.

Light passing through a clear solution of dissolved substances is not scattered, while light passing through cloudy water containing large molecules and clusters of molecules (colloids) will be scattered.

John Tyndall, experimenting with the transmission of radiant heat through different types of gases and vapors, measured the absorption and spreading out of the radiation through these gases. In 1859 he studied the effects on light when it passed through various liquids and gases and noted the degree of scattering in the path of the light. This scattering was named the *Tyndall effect*, after his theory that particles in the path of the light cause the scattering that renders the light beam visible. Nephelometry, which examines the scattering properties of small particles in air, is similar to the Tyndall effect. The scattering of the beam of light off minute particles suspended in air is more pronounced and effective when shorter-wavelength ultraviolet radiation is used. Tyndall used this effect to explain why the sky is blue overhead and why sunsets appear red. This occurs because sunlight passes through a greater number of dust particles, filtering out the ultraviolet light, allowing the longer wavelength light (red) to be seen on Earth. Tyndall is credited with first explaining the greenhouse effect as being a natural phenomenon and also determining that the dust in the atmosphere contains microorganisms. *See also* Ramsay.

U

Ulam's "Monte Carlo" System: Mathematics: *Stanislaw Marcin Ulam* (1909–1986), United States.

It is possible to obtain a probabilistic solution to complex mathematical problems by using statistical sampling techniques.

In the early 1940s, Stanislaw Ulam, a mathematician, was asked by the Los Alamos, New Mexico, nuclear development project administrators to develop a mathematical theory for nuclear reactions as applied to nuclear weapons. Before the days of analytical and digital computers, mathematically gifted people performed the tedious computing tasks and were called *computers*. Ulam's wife, Françoise, was a computer who assisted him in this task. Rather than tracking every uranium or plutonium atom in the atomic bomb models that were being devised by physicists (an impossible task), Ulam used statistical methods to simulate behaviors of individual nuclei in the reaction. He selected at random variables of the possible interactions of the nuclei, computed possible outcomes, and analyzed the results using probability statistics. Since his system was based on the *odds* as related to the probabilities of gambling odds, his method became known as *Monte Carlo statistics*, named after the famous gambling casino in Monaco. Ulam's Monte Carlo system of statistical probabilities is employed in many fields other than nuclear energy and has proven a valuable addition to our understanding of several biological processes, including life. In the late 1940s Ulam was involved with Edward Teller and others in the development of the hydrogen (thermonuclear) bomb. Ulam developed the mathematics that ensured success for the final design of the fusion H-bomb, which used a fission A-bomb as the "trigger" to provide the heat, x rays, and pressure required to accomplish the thermonuclear reaction required for the H-bomb. Teller revised Ulam's concept to "focus" the x rays produced by the A-bomb to trigger the fusion reaction. *See also* Teller.

Urey's Gaseous Diffusion and Origin of Life Theories: Chemistry: *Harold Clayton Urey* (1893–1981), United States. Harold Urey was awarded the 1934 Nobel Prize for chemistry.

Urey's Theory of Gaseous Diffusion: *Isotopes of an element in the gaseous state can be separated according to their different atomic weights.*

In 1932 Harold Urey was the first to discover and isolate deuterium (heavy hydrogen) from heavy water (D_2O). He knew liquid heavy hydrogen evaporated at a slower rate than did ordinary liquid hydrogen because regular hydrogen's nuclei are composed of just a single proton, while heavy hydrogen's nuclei contain one proton plus one neutron. Deuterium still has an atomic charge (atomic number) of $+1$, but an atomic weight of 2. (See Figure O1 under Oliphant.) Using this concept, Urey distilled several liters of liquid hydrogen to about 1 cubic centimeter of deuterium whose existence was confirmed by spectroscopic analysis. Again, using the same principle, he separated the isotopes of uranium-235 from uranium-238 by converting regular uranium-238 into a gas and "filtering" it in such a manner that the lighter, unstable uranium-235 was collected. The isotope U-235, when reaching a critical mass, can be used in a self-sustaining chain reaction. In the early 1940s U-235 was used in the first A-bomb tested in White Sands, New Mexico. The same gaseous diffusion process, based on isotopes of radioactive elements having different atomic weights, was used to separate the isotopes of plutonium, used in the second A-bomb dropped in Japan in 1945. At the end of World War II, Urey's mass production of deuterium made possible the development of the hydrogen fusion bomb. *See also* Fermi; Teller; Ulam.

Urey's Theory for the Origin of Life: *If the right mix of organic molecules existed on the primitive earth, an energy input (lightning, geothermal, or ultraviolet) may have brought life from this "soup."*

Harold Urey and others believed life began on earth some 3 to 4 billion years ago, which is approximately 1 or 2 billion years after earth was formed. These figures are today's best estimates based on fossil and cosmological research. Disagreement still exists as to the source of the organic chemicals that first self-assembled to produce organic polymers and later cells. One possibility being revived is that bacteria and microorganisms arrived on earth from comet and meteor ice and dust. Urey with his graduate student, Stanley Miller, set up an experiment to determine if several chemicals assumed to be on the **prebiotic** earth could, under laboratory conditions, be converted into organic polymers, which might, under ideal conditions, self-organize into primitive life forms. Urey and Miller formed an atmosphere of methane (CH_4), hydrogen (H_2), ammonia (NH_3), and water (H_2O) in an enclosed glass flask that could be heated. The hot evaporated gases were collected in another flask and exposed to an electric spark between two tungsten electrodes. Following this, the gases were cooled and condensed back to a liquid. They actually produced over twenty-five amino acids, some purines, and other large organic molecules, but no evidence of life itself. An important part of the experiment was the formation of amino acids that combine with some ease to form complex proteins, which are essential for life. Urey's and Miller's work led to the idea that there are at least four stages for chemical evolution of life on earth: (1) There is a nonbiological synthesis of simple organic molecules, (2) followed by the molecules forming more complex polymers (chains), (3) which form into pre- or pro-biological "clumps" or simple cells, and (4) some of the first organic substances are primitive RNA followed by DNA, which has the capability to pass on the chemical and living nature of cells from one generation to the next. *See also* Crick; Darwin; Miller; Pasteur; Redi; Watson.

V

Van Allen Radiation Belts: Physics: *James Alfred Van Allen* (1914–   ), United States.

The earth's magnetic field should react with and trap high-speed charged particles originating from space into a concentrated zone above the earth's atmosphere.

James Van Allen first used high-altitude balloons to study cosmic rays, which are high-energy particles (mostly protons) from space that constantly penetrate Earth. Three scientists—S. Fred Singer and Paul Kellogg of the United States and S. N. Vernov of the Soviet Union—first proposed the idea that radiation from space surrounded Earth. These energetic particles from the lower altitudes were confirmed by sending photographic film into space in weather-sounding rockets. In 1958 the United States sent its first satellite, *Explorer I*, which weighed 31 pounds, into orbit by using a captured German V-2 rocket. Its purpose was to detect high-energy particles in near space. It found many particles at altitudes between 200 and 300 miles but, surprisingly, none were recorded above that region. James Van Allen theorized that this low-radiation "belt" was due to Earth's magnetism, and the reason particles were not detected at higher altitudes was that the counters in the rockets were "jammed" by overwhelming masses of radiation and particles. In 1958, *Explorer IV* was launched containing a counter surrounded by a lead shield that filtered out much of this radiation, thus providing a more accurate count. Additionally, the shielded instruments recorded an increasing amount of high-energy radiation above 300 miles. After World War II the Soviet Union and the United States both exploded small atomic bombs in space to track the neutrons and energetic particles that were released. The United States also exploded three small nuclear bombs 300 miles above the South Atlantic to produce energetic particles in the upper atmosphere, which could then be detected and studied. Some radiation and particles from these space explosions persisted for several weeks to several years and were strong enough to disable a number of satellites. By 1967, this practice of detonating nuclear bombs in space was banned worldwide. Van Allen theorized that high-altitude radiation composed of charged particles was concentrated in areas or "belts" trapped by the magnetosphere, which rotates with Earth's magnetic axis. He based his theory on the concept that Earth's magnetic field extends far out into space and these particles followed the magnetic field as evidenced by their alignment with the poles of this field. This region of concentrated radiation or "belt" was first called the *Van Allen belt*. However, when it was discovered there was more than one radiation belt, the name was changed to the *magnetosphere*. The *inner radiation belt*, the one detected by the Geiger counters Van Allen used, is a rather compact area of particles in the general magnetosphere region over the equator. This belt is the result of cosmic radiation. Later, an *outer radiation belt* was discovered, which is composed of a "plasma" of energetic charged particles trapped in Earth's magnetosphere at a higher altitude and responsible for magnetic storms on Earth. This magnetosphere phenomenon was later confirmed for other planets as well.

Van de Graaff's Concept of Producing High Voltage: Physics: *Robert Jemison Van de Graaff* (1901–1967), United States.

High voltages can be produced and sustained by electrostatic generators.

Robert Van de Graaff was cognizant of the need to produce very high voltages to accelerate subatomic particles, making them useful as "bullets" to bombard the nuclei of elements to produce isotopes and nuclear changes. Regular **ac** and **dc** generators and dynamos were incapable of producing voltages in the million-volt ranges that nuclear physicists needed to create new, heavier elements as well as basic subnuclear particles. Older type "wheel-and-brush" static electricity machines were developed soon after it was learned that static electricity could be stored in a **Leyden jar** by rubbing glass or rosin with silk or wool. These machines created a spark across an inch or two in dry air. In 1931 Van de Graaff devised an improved model that used a hollow sphere with a rotating insulated rubberized belt to transfer the charge to the surface of the metal sphere. (See Figure V1.) Although his first models generated up to 100,000 volts with just a fraction of amperes (current), they were inadequate for use as particle accelerators. Van de Graaff's later models were enclosed in a tank with the insulated belt extracting negative ions from the ground (Earth), where they were stripped of their electrons and thus became positive ions. As the charges return to the grounding terminal, the negative ions are accelerated to achieve up to 10 million volts or 10 MeV of energy (but with very low amperes). Recently, even higher voltages have been achieved. This high voltage can accelerate charged particles (electrons and positrons) to very high energies, which can then be used as "bullets" to bombard target nuclei. Two of the first applications of this very high energy were the exploration of the nature of uranium nuclei fission and the production of high-energy x rays for medical and industrial use. Van de Graaff generators, when used in combination with other types of particle accel-

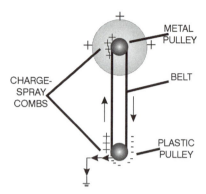

Figure V1: The Van de Graaff electrostatic generator produces an electrical charge from the ground and transfers it by the belt to the metal sphere, where it can be used as an electron source for a variety of research applications.

erators, generate the tremendous speeds (energies) of charged particles needed to knock out the many particles from nuclei. Van de Graaff generators also produce high-energy x rays to detect flaws in machinery and small cracks in airplane structures, and to inspect the interiors of explosive weapons.

Van der Meer's Theory of Particles to Confirm the "Weak Force": Physics: *Simon Van der Meer* (1925–   ), Netherlands. Simon Van der Meer shared the 1984 Nobel Prize for physics with Carlo Rubbia.

The unification of the electromagnetic and weak forces requires the existence of three heavy particles—one negative, one positive, and one neutral.

Van der Meer predicted that theoretical particles related to weak interactions were about eighty times as massive as protons, thus requiring a tremendous energy source to produce and detect them. The major obstacle to detecting these particles and confirming the theory of the weak interactions was their theoretical mass. In 1989, Van der Meer and his colleague, Carlo Rubbia (1934–   ), succeeded in generating energies sufficient to produce these massive particles. Arranging for a super-synchrotron (circular particle accelerator) to provide two beams of particles to collide head-on, they accelerated a beam of protons in one direction and a beam of antiprotons in the other. (See Figure V2.) The result was a collision of particles, each at great energies, detected in the synchrotron's target. This collider concept greatly increased the energy to over 150,000 GeV, which was adequate to produce and detect the three heavy theoretical particles. They were named W^+, W^-, and the neutral Z^0 **bosons**, which coincided with and confirmed the weak interaction now proven by Van de Meer but first suggested by Enrico Fermi. The weak interaction is similar to electromagnetic interactions of particles except the weak interaction involves neutrinos, making it a much weaker force than electromagnetic forces. At the same time weak in-

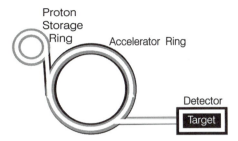

Figure V2: Particle accelerators are used to analyze and understand the nature of matter and energy. Subatomic particles are accelerated around the ring by electromagnetic forces until they approach the speed of light, which is required to smash into targets that break up into smaller particles, as well as energy that can be recorded and analyzed.

teractions are much stronger than gravitational interactions on particles. The weak interaction (and related particles) is one of the basic forces of nature, as is the "strong force," which binds neutrons and protons together in atomic nuclei. *See also* Fermi.

Van der Waals' Equation for Gas Molecules: Physics: *Johannes Diderik Van der Waals* (1837–1923), Netherlands. Johannes Van der Waals was awarded the 1910 Nobel Prize in physics.

Electrostatic forces are responsible for the attraction between gaseous molecules, thus affecting the corresponding relationships between the temperature, pressure, and volume of gases.

The ideal gas law combines the Boyles-Charles gas laws and is limited to gases at "normal" pressures and temperatures. This law is not effective for "real" gases under other than "normal" conditions—very high or low temperatures or pressures. The ideal gas law equation relates the three properties of temperature, pressure, and volume for a chemical gas, $pV = nRT$, under normal conditions (*See* Ideal Gas Laws). Johannes Van der Waals related the ideal gas law to the kinetic-molecular theory, which could account more accurately for the behavior of real gases and liquids by considering both the attractive forces between molecules as well as their actual (but limited) volumes under other-than-normal conditions. The ideal gas law might be considered the equation of the "first state," while Van der Waals' equation is an equation of the "second state," which more accurately relates the behavior of gas molecules to kinetic energy under a variety of corresponding states of temperature, pressure, and volume. Van der Waals' equation is $(p + na/V^2)(V - nb) = nRT$. In this equation the n's are amounts, a and b are constants, na/V^2 accommodates the attractive forces between molecules of gases that may be more than zero, and $V - nb$ states the volume of a real gas is never zero, which restricts the gas' molecular motions in its actual volume. The weak electrostatic attractive force between the atoms and molecules of all substances is called the *Van der Waals force*. His equation enabled other scientists to solve the problems of how to liquefy gases found in the atmosphere. *See also* Boyle; Charles; Gay-Lussac.

Van't Hoff's Theory of Three-Dimensional Organic Compounds: Chemistry: *Jacobus Henricus Van't Hoff* (1852–1911), Netherlands. Jacobus Van't Hoff was awarded the first Nobel Prize in Chemistry in 1901.

The three-dimensional symmetrical structure of organic carbon compounds accounts for their optical activity.

Until 1874 molecular structures were depicted as two-dimensional. Also in 1874, Friedrich Kekule proposed his famous structure for the carbon atom as having four electrons oriented to the corners of a square, which explained left- and right-sided isomers of some elements. While contemplating this structure for the carbon atoms, Kekule dreamed of a snake eating its tail. (See Figure K1 under Kekule.) This gave him the insight for forming a ring of carbon atoms for the benzene molecule. The benzene ring is a molecule that has six carbon atoms, each of which shares its electrons with its neighbors. The problem with his depiction was that the ring was two-dimensional, which did not explain how certain molecules (isomers) can polarize light in solution. In the same year, Jacobus Van't Hoff recast the organic carbon atom into a tetrahedral three-dimensional structure with the four bonds of the carbon atom pointed toward the vertices of the tetrahedron rather than to the corners of a two-dimensional square. (See Figure V3). His model placed the atom as suspended in the central area of the three-dimensional figure. This was not only a unique insight but

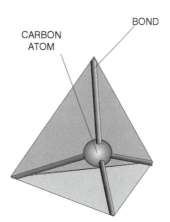

CARBON ATOM

BOND

Figure V3: The carbon atom's structure with four valence electrons may be thought of as a tetrahedron with four vertices representing the bonding of electrons. The tetrahedron has six edges and six line segments that join each pair of vertices. It is a representation of a three-dimensional triangle with the center being the carbon atom. This structure gives the carbon atom its unusual versatility and importance in forming organic and inorganic compounds.

explained how some organic isomers are structured and react in solutions. Certain isomers do polarize light in solution; others do not. The difference is in the two- or three-dimensional structures of the molecules. Van't Hoff's theory of asymmetrical three-dimensional optically active carbon (organic) compounds provided the basis for modern stereochemistry (the study of how atoms are arranged [structured] within molecules and how this affects chemical reactions). Van't Hoff's theory resulted in his law of chemical dynamics, which is important in the study of osmotic pressure in solutions. *See also* Baeyer; Fischer; Kekule; Pasteur; Van der Waals.

Van Vleck's Theory of Paramagnetism: Physics: *John Hasbrouck Van Vleck* (1899–1980), United States. John Van Vleck shared the 1977 Nobel Prize for physics with Nevill Mott and Philip Anderson.

Paramagnetic substances are independently susceptible to magnetic induction according to the temperatures involved.

John Van Vleck extended the concept of quantum mechanics associated with particles to include not only quantum aspects of waves but also magnetism. There are two basic types of paramagnetism—one involving electrons and other nuclei. Atoms of elements that have an odd number of electrons, according to quantum mechanics, cannot have a spin of zero. This results in atoms with a magnetic moment that can be affected by a magnetic field. Examples are atomic and molecular radicals (with a charge). Paramagnetic materials are magnetized parallel to the magnetic field to which they are exposed. In general, they do not become as highly magnetized as do ferromagnetic materials, and they behave differently at very high and very low temperatures. The best examples of paramagnetic materials are the atoms and compounds of the rare earths located within the transition elements of the **Periodic Table of the Chemical Elements**. Other examples are free organic radicals, nitric oxide, some low-conducting metals, and molecular oxygen. The effects of temperatures on the quantum nature of paramagnetic materials is known as *Van Vleck paramagnetism*. Paramagnetic materials are used in combination with liquid helium to remove additional heat in attempts to reach absolute zero. First they are magnetized. Then, when the magnetic field is removed, the molecules become randomly disorganized and remove more heat from the helium. A temperature of less than 0.5 degree kelvin has been achieved.

Virchow's Cell Pathology Theory: Biology: *Rudolf Carl Virchow* (1821–1902), Germany.

Diseased cells arise from other diseased cells.

Rudolf Virchow accepted the concept first stated by William Harvey that cells are derived from other cells. Virchow's contribution to the field of pathology was his belief that disease is a pathological state of cells based on observations that abnormal cells found in particular diseases arose from normal healthy cells. He believed living cells could originate only from living matter. Virchow noted this was not a rapid process. Rather, once disease infected just one or only a few cells, these cells infected other healthy cells over a period of time. Some years later, the germ theory of disease, which he rejected, made Virchow's theory less important because this new theory provided a more rational explanation for disease. Even so, Virchow is regarded as the Father of Pathology. *See also* Lister; Pasteur; Schleiden; Schwann.

Volta's Concept of an Electric Current: Physics: *Count Alessandro Giuseppe Antonio Anastasio Volta* (1745–1827), Italy.

A flowing electric current is not dependent on animal tissue and can be produced with chemicals.

Count Alessandro Volta was acquainted with both Benjamin Franklin's single-fluid theory of electricity and Charles du Fay's (1689–1739) two-fluid electricity. He also knew of Luigi Galvani's belief that moist animal tissue was required to produce a continuous flow of electricity. Volta decided these theories had weaknesses and that a combination of the correct chemicals and materials could produce an electric current. In particular, he believed that dissimilar metals, not the animal tissue of Galvani's frog experiment, generated Galvani's electricity. In 1800, Volta separated alternating sheets of zinc and silver with sheets of cardboard soaked in concentrated saltwater, which acted as the electrolyte. This was called a *voltaic pile* and resulted in a revolution in the source and use of small amounts of electricity. His "pile" enabled others to develop mechanical clappers on electric bells, the telegraph, modern dry and wet cells, and batteries of cells, which are a combination of a group of cells connected in series or parallel. The unit of electric potential (force or pressure of the current) is named after Count Volta (volts = amps × ohms). *See also* Faraday; B. Franklin; Galvani; Henry; Ohm.

Von Laue's Theory for the Diffraction of X Rays in Crystals: Physics: *Max Theodor Felix Von Laue* (1879–1960), Germany. Max Von Laue was awarded the 1914 Nobel Prize in physics.

If the wavelength of x rays is similar to the space between atoms in crystals, x rays passing through a crystal composed of atoms in a lattice arrangement should produce a diffraction pattern.

It was known for some time that x rays consist of a wave similar to light and that their wavelength is much shorter than are waves of light. Max Von Laue was aware of this fact, even though it was not yet established that x rays were one form of radiation exhibited in the electromagnetic spectrum. He also knew of the research indicating the atoms in crystals may be arranged in a very regular pattern, similar to a lattice structure, where they were lined up in rows. His theory stated that if the small crest-to-crest distance of the short wavelengths of x rays were the same as the small distances between the atoms that make up crystalline substances, then diffraction of the x ray beams should occur. He proceeded to "shoot" x rays through crystals and record the diffracted beams on photographic plates. His first attempts produced rather blurred but nonetheless expected diffraction patterns, thus proving his theory. Von Laue's theory established that x rays were part of the electromagnetic spectrum, but are of a much shorter wavelength than visible light. Just as important, his work demonstrated that atoms in inorganic crystals and some organic substances are arranged in a symmetrical and regular order. His x-ray techniques aided researchers in decoding the structure of DNA, and his work with crystals ushered in the field of solid-state physics, leading to the development of modern electronics, including semiconductor microchips and computers. *See also* Crick; R. Franklin; Maxwell; Roentgen.

Von Neumann's Theory of Automata: Mathematics: *John von Neumann* (1903–1957), United States.

Von Neumann's theory of "artificial automata" (computers) might be expressed as: (*a set of inputs*) *to* → (*a set of internal states*) *yields* → (*a set of outputs*).

After emigrating to the United States from Germany in 1930, John von Neumann attempted to solve complicated mathematical problems related to the development of the atomic bomb, hydrodynamics of submarines, missiles, weather predictions, and military strategy. Much of this work required deciphering complicated nonlinear equations, which proved to be difficult and time-consuming. Von Neumann developed a systematic mathematical theory in logic that he called "automata," which he reasoned would help develop a better understanding of both natural systems and what he called "artificial automata" (computers). The automata theory relates to three states of a system that involve sets: the input, the current internal state of the system, and the output from the system. These operations can be thought of as three sets of information and two functions. In essence, this is von Neumann's design for a computer, which is also the basic logic for current computers: the program (input), the operating system (internal state), and the data produced (output). In 1952 Von Neumann developed the MANIAC, the first modern computer using an internally stored program (operating system). It was a huge machine that filled a room and required extensive cooling to keep the vacuum tubes from overheating. It provided the basic logic (automata) and design for modern computers. *See also* Turing.

W

Waddington's Theory of Genetic Assimilation: Biology: *Conrad Hal Waddington* (1905–1975), Scotland.

By means of natural selection, acquired characteristics can be inherited genetically and through the process of evolution.

During Darwin's lifetime, the science of genetics had yet to be developed; therefore the old Lamarckian belief that characteristics acquired after birth could be inherited was still considered viable. Once research and evidence that genes are the carriers of physical characteristics became known, Lamarckism became heresy. Waddington conducted an experiment that he claimed proved his theory of "genetic assimilation" (acquired characteristics) by exposing the pupae stage of the fruit fly (*Drosophila*) to heat. He noted that a few exposed flies exhibited a different pattern of veins in their wings. Waddington separated and bred these different flies in an attempt to increase their numbers. After repeating selective breeding of these flies for several generations, he observed that a large number of offspring manifested this same pattern; thus they seemed to be breeding true. Therefore, Waddington concluded that genetic assimilation of imposed characteristics resulted through the process of natural selection. Most scientists discredited both his experiment and theory. *See also* Darwin; Lamarck; Lysenko; Wallace; Zuckerandl.

Waldeyer-Hartz Neuron Theory: Biology: *Heinrich Wilhelm Gottfried von Waldeyer-Hartz* (1836–1921), Germany.

The nervous system is composed of individual cells whose fine extensions do not join cells adjacent to them.

Heinrich Waldeyer-Hartz studied animal tissue cells and their structures, defining the "colored bodies" in cells as *chromosomes*. He also studied nerve tissue and was the first to realize that nerves are not only composed of cells, just as is other animal tissue, but that individual nerve cells do not actually touch each other; there is a gap where one nerve cell ends and the next begins. He named these individual nerve cells *neurons*.

Wallace's Theory of Evolution by Natural Selection: Biology: *Alfred Russel Wallace* (1823–1913), England.

The tendency for species to produce variations as they drift from their original types is due to a separation of their ecologies.

In the late 1800s Alfred Wallace, a contemporary of Charles Darwin, collaborated with Darwin on the development of the theory of organic evolution. Wallace proposed his concept, known as the *Wallace line*, where the separation of geographical land masses results in the development of distinct species. He based his theory on the differences of animal species he observed in Australia and Asia. Wallace claimed this "line" was created by the separation of the two continents, which, over a long period of time, also separated individual species that developed in very different directions due to natural selection created by disparate ecologies. His theory that varieties of a species tend to drift apart indefinitely from the original type is generally accepted today. The study of the geographic distribution of plants and animals is known as biogeography. *See also* Darwin.

Wallach's Theory for the Molecular Structure of Organic Compounds: Chemistry: *Otto Wallach* (1847–1931), Germany. Otto Wallach received the 1910 Nobel Prize in chemistry.

Pharmaceutical medications as well as the essences of many oils are composed of a variety of related forms of the hydrocarbon molecules.

While studying pharmacology, Otto Wallach removed the essential oils from plants using steam distillation. Many of the resulting organic compounds were used as medicines as well as in the production of perfumes, creams, and flavorings. He theorized that many of these organic substances were chemically related but was unsure of how or why their molecular structures differed. Wallach, however, identified the great variety of a particular group of compounds, all of which possessed the same general formula but with different molecular weights (isomers). One of these, terpene ($C_{10}H_{16}$), is an unsaturated hydrocarbon found in some plants and has a unit structure containing five carbon atoms (C_5). A group of organic compounds similar to terpene that has the same molecular weight but different structures is referred to as an *isoprene*. He expanded this terpene example of a hydrocarbon isoprene with the general formula $(C_5H_8)_n$ to include other hydrocarbon compounds. Wallach also discovered that these molecules can be polymerized to form other higher-molecular-weight molecules, resulting in other larger organic (hydrocarbon) molecules with formulas that are multiples of the basic terpene $C_{10}H_{16}$ formula. Some examples of these isoprenes are camphene, citrene, cinene, eucalyptine, and common terpentine. Wallach's work with the basic structure of various $(C_5H_8)_n$ isoprenes was instrumental in improving and expanding several industries, including pharmaceuticals and perfumes. *See also* Pauling.

Walton's Concept for Transmuting Atomic Particles: Physics: *Ernest Thomas Sinton Walton* (1903–1995), Ernest Walton shared the 1951 Nobel Prize for physics with John Cockcroft.

See Cockcroft for details of the theory proposing that accelerated protons split lithium nuclei into alpha particles (e.g., Lithium + proton → alpha + alpha + energy).

Watson-Crick Theory of DNA: Biology: *James Dewey Watson* (1928–), United States. James Watson shared the 1962 Nobel Prize in physiology or medicine with Francis Crick and Maurice Wilkins.

See Crick-Watson for details describing their theoretical model of the DNA double-helix.

Watson's Theory of Electricity as a Fluid: Physics: *Sir William Watson* (1715–1787), England.

Electricity is an "electrical ether" or single fluid of various densities contained in different material bodies.

William Watson improved the effectiveness of the **Leyden jar** invented by Pieter van Musschenbrock (1692–1761) and Ewald Georg van Kleist (1700–1748) by lining the interior of the glass jar with metal foil. This improved device enabled him to store a larger charge of static electricity and study the resulting larger electrical discharges, ultimately leading to his belief that electricity is a single "fluid." Watson theorized that different materials contained differing densities of this "electrical **ether**." If the density of two objects was equal, there was no sparking discharge, but if the "fluid" densities were unequal, the one with the greater density would discharge to the object with lesser density, until they were again equal. Although this theory is incorrect, it might be considered a forerunner of the concepts of equilibrium and the conservation of energy. *See also* Ampère; Faraday; Franklin.

Watson-Watt's Concept of Radar: Physics: *Sir Robert Alexander Watson-Watt* (1892–1973), England.

The interference in radio reception caused by airplanes flying over transmitting stations can be used to detect approaching aircraft.

Robert Watson-Watt knew that some radio engineers complained about radio signal interference caused by passing airplanes. In the late 1930s, he theorized that this phenomenon might be used to detect enemy aircraft. In addition to this radio "interference" by aircraft, he based his concept on the results of two other research projects: (1) the use of radio waves to determine the range in miles of different layers of the atmosphere and (2) the use of radio signals to determine the existence and distance of thunderstorms. The main problems with using this concept for a reliable aircraft detecting device at a distance of more than a few miles were the need for a very high-powered transmitter and the fact that only a very small, weak signal was "bounced" back to the receiver. Therefore, the receiver had to be capable of amplifying the signal by many factors greater than what was required for normal radio receivers. Along with scientists and engineers from the United States, he continued to develop a workable system he called *radio detection and ranging* (RADAR). By late 1938 several RADAR units were placed on the east coast of England to aid in the detection of German bombers. Since that time, RADAR units have become much smaller and more sensitive. RADAR has found many uses, including hand-held units to detect speeding vehicles on roads and highways.

Weber's Theory of Gravitational Waves: Physics: *Joseph Weber* (1919–), United States.

Gravity waves should have the same characteristics of energy and momentum as do electromagnetic waves, and thus be detectable.

Joseph Weber accepted Einstein's theory of general relativity, which included the concept that any accelerating mass generates gravitational waves as well as electromagnetic waves. Photons of light (electromagnetic radiation) exhibit both wave and momentum characteristics; thus, electromagnetic radiation must have mass. Weber reasoned that gravity waves should also exhibit momentum and thus be detectable; however, gravitational waves have not been detected because gravity is one of the weakest forces in nature. It may not seem so weak when falling to the ground, even from a low height, but compared to other forces of nature, such as the binding force of nuclear particles or even the forces that forge molecules out of chemical atoms, gravity is not very strong. In the mid-1960s, Weber designed a special barrel-like "antenna" detector 3 feet in diameter, constructed from aluminum and weighing more than 3 tons. He placed a series of piezoelectric crystals in its interior to detect gravity waves. He figured that any force, no matter how small, would alter the shape of these crystals; if there was even the slightest pressure exerted by an oscillating gravity field, the crystals would convert this distortion to an electric current that could be detected and measured. This instrument was so sensitive that the piezoelectric crystals could detect any deformity in their shape as little as 1/100th the diameter of an atom. To ensure gravity waves were being detected and not some other phenomenon, Weber erected a second detector at some distance from the first so that each antenna detector could be oriented in various direction. After several months, he claimed to have received what are called "coincident readings," meaning that when both cylindrical detectors were oriented toward the center of our galaxy, the same readings were recorded. Weber's results were never duplicated by other scientists despite the thousands of dollars spent on improved detectors. In fact, no "coincident readings" were ever recorded, even when one gravity wave detector was placed on the east coast and one on the west coast of United States. Scientists have not abandoned the theory that gravity waves exist, but rather assume they are too weak to detect. A more recent experiment "shoots" two laser beams of monochromatic light at each other from a distance of several miles. When the beams collide, they interfere with each other. If gravity waves exist, they may possibly alter the interference pattern and thus be detected. Theoretical physicists predict that sometime in the early twenty-first century, gravitational waves will be detected, a belief based on their confidence in the proven reliability of the theory of general relativity. *See also* Einstein; Curies.

Wegener's Theory of Continental Drift: Geology: *Alfred Lothar Wegener* (1880–1930), Germany.

All land on earth was once connected with the configuration of a "supercontinent," which, over time, separated and drifted apart to form the present continents.

Scientists have long speculated on the shape of the world's land masses and why this shape changed over eons. Sir Francis Bacon was the first to notice the similarity of the coast lines of eastern South America and western Africa and suggested they were once joined. In 1924 Alfred Wegener, building on Eduard Suess' theory that western and eastern land masses were once joined to form the continent Gondwanaland, called his "supercontinent" Pangaea, which means "all-land" or "earth" in Greek. (See Figure S4 under Suess.) Wegener based his theory on four important observations: (1) There is a more accurate "fit" of the edges of the continental shelves of the current continents than there is of their coast lines. (2) Current measurements indicate Greenland is moving westward from the European continent. (3) Earth's crust is composed of a lighter granite-type rock material, which floats on the heavier basalt material. Thus, the crust is composed of two layers, and the continents formed of granite "float" over the heavier basalt ocean floor. (4) Although there are significant differences in plant and animal species found on various continents, there are also great similarities of species found on the now-separated continents, indicating these continents were once connected. At first, many scientists disagreed with Wegener's theory of continental drift. Today it is accepted in an updated version to conform to the new science of plate tectonics. *See also* Ewing; Hess; Suess.

Weinberg's Grand Unification Theories: Physics: *Steven Weinberg* (1933–   ), United States. Steven Weinberg shared the 1979 Nobel Prize in physics with Sheldon Glashow and Abdus Salam.

Weinberg's Theory of the Unification of Electromagnetic and Weak Forces: *The interchange of photons and the weak force with the W and Z bosons result in the electroweak force combining with the electromagnetic force.*

Steven Weinberg knew of the dilemma of symmetry relating to photons, which are practically weightless, while bosons, which have intrinsic angular momentum, are a bit heavier than positive protons. He explained this conundrum by recounting the outset of the big bang. Weinberg used the idea of spontaneous symmetry breaking (where the symmetry of particles and energy was disturbed—chaos) to illustrate what occurred during the cooling-off period that followed the tremendous temperatures created at the outset of the big bang. This resulted in many fundamental particles assuming very different characteristics, leading to the belief that the current four primary natural forces were combined as one major force at that time. The four natural forces are:

1. *Gravity.* Although gravity is the weakest of these four forces and exhibits only an attractive force and acts over infinite distances, it is the predominant force over the entire universe.

2. The *weak nuclear force* causes the beta (electron) decay of a neutron into a neutrino and electron [neutron → proton + beta + neutrino]. It is one of the fundamental interactions of elementary particles. (See Figure F2 under Fermi.) Essentially it involves leptons and acts over extremely small distances, ranging between 10^{-9} and 10^{-10} cm.

3. The *electromagnetic force* acts on particles with electric charges and holds electrons to their orbits around the nuclei of atoms. It exhibits both attractive and repulsion forces and acts over infinite distances. The electromagnetic interactions are limited to atomic and molecular particles.

4. The *strong nuclear force* is the strongest of these four natural forces. It binds protons, neutrons (and quarks) together by gluons in the nuclei of atoms. It mostly involves hadrons and acts over very small distances, ranging from about 10^{-6} to 10^{-9} cm.

Weinberg-Salam Greater Unified Theory: *The four fundamental forces of nature interact to make up all the forces found in the universe, and thus may be integrated into a basic unified force.*

Weinberg became involved with issues related to cosmology, the origin of the universe, and the big bang. Weinberg, his colleague Abdus Salam, and other scientists continue to search for "superstrings" that may link these four basic forces. Their superstring theory states that all the known small particles of matter are not really the basic fundamental particles. Rather, an extremely small (not yet detected) vibrating string is the basic particle or energy unit, possible only 10^{-35} cm, which is smaller than anything yet to be conceived. Instead of three or four dimensions, the "strings" supposedly have six dimensions, each of which is curled up into each string. The advantage of the string theory is its ability to explain the unification of all four of the natural forces, the big bang theory, and black holes. Weinberg currently believes we are on the verge of uncovering the final theory, which Albert Einstein referred to as the unified field theory. It has also been referred to as the grand unification theory (GUT), the theory of everything (TOE), and the "answer." *See also* Einstein; Hawking; Salam; Witten.

Weizsacker's Theories of Star and Planet Formation: Physics: *Baron Carl Friedrich von Weizsacker* (1912–), Germany.

Weizsacker's Theory of Star Formation: *A nuclear chain reaction involving a "carbon-cycle" occurs inside a condensed mass of gas, resulting in the for-mation of a star that produces heat and light.*

In 1929 George Gamow was the first to propose that the source of a star's "core" energy is a nuclear reactor where hydrogen nuclei are converted into helium nuclei by the process of nuclear fusion, resulting in the release of vast amounts of energy. Several years later, when more was known about nuclear reactions, Hans Bethe provided the details of how hydrogen fusion could occur in the sun's core without exploding the star. Baron Carl Weizsacker advanced a similar theory with the addition of what is known as the "carbon cycle" or the "carbon-nitrogen cycle" (not to be confused with the biosphere carbon cy-cle). Weizsacker believed that in massive stars, a carbon molecule attracts four protons (hydrogen nuclei), and through a series of theoretical nuclear reactions, it produces one carbon nucleus and one helium nucleus while emitting two positrons and tremendous heat and light energy [$C + 4\,^+H \rightarrow C + ^{++}He + 2\,p^+$]. Since stars are composed mainly of hydrogen, this process, which takes place at their centers, can continue until all the hydrogen is converted to helium. However, a time span of billions of years must pass before "death" occurs for most stars. In addition to great amounts of heat and light produced by stars, the triple-alpha process, a nuclear reaction in stars, fuses three helium atoms to form carbon, which makes carbon-based life on Earth possible.

Weizsacker's Nebula/Planetary Hypothesis: *As a nebula of swirling gases and small particles condenses, turbulence is created that will form the planets in their orbits.*

There is a long history of ideas, concepts, hypotheses, and theories to explain the origin of the solar system and its planets. One of the more popular ideas was the "passing star" theory, which explained how matter was pulled off two stars as they passed close to each other to form planets orbiting around one or both stars. In 1944 Weizsacker applied mathematics related to the science of magnetohydrodynamics to explain how a mass of thin gas moving in a giant magnetic field in space could, through angular momentum, "push" the energy of the moving gas outward, thus providing angular momentum for the planets to remain in their orbits. One problem was that planets exhibit more angular momentum than Weizsaker predicted, and angular momentum is always con-served; it cannot be created nor destroyed, just transferred. Weizsacker's nebula theory, with modifications that use the sun's magnetic field to increase the an-gular momentum of the planets (provided by Fred Hoyle), is currently the best explanation for the formation of our solar system. *See also* Bethe; Gamow; Hoyle; Laplace.

Wheeler's "Geon" Theory: Physics: *John Archibald Wheeler* (1911–), United States.

Geometrodynamics (Geon) is an electromagnetic field maintained by its own gravitational attraction.

John Wheeler searched for a theory to unify two seemingly unrelated fields:

gravity and electromagnetism. This involved a method to demonstrate the concept of "action at a distance." Since the days of Aristotle, it was believed that something had to push or pull an object continually in order to make it move or cease moving. Neither could an object be moved by a force not in direct contact with that object. This was implicit in Newton's third law of motion (for every action, there is an equal and opposite reaction). Wheeler and his colleague, Richard Feynman, offered a solution that proposed a retarded effect on an object rather than an instantaneous effect. Their solution, somewhat like one of Einstein's "thought experiments," does not require any laboratory or equipment. Wheeler and Feynman suggested that two objects (1 and 2) be set up exactly one light-minute apart (1/525,600 of a light-year). Then any light (or any electromagnetic wave) sent from object 1 will take exactly 1 light-minute to reach object 2. Thus, it could be said that there was a delay from the signal to the reception of 1 light-minute, or since the action was received after it was sent, it was "retarded." In addition, there was no direct or instantaneous contact between the force exerted by object 1 with the retarded action by object 2. Both gravity and electromagnetism exhibit some properties of "action at a distance," which is one reason Wheeler attempted to unify them into a single theory. One problem was Newton's third law (if there is a "forward" effect from object 1 to object 2, there should also be an effect acting "backward," from object 2 to object 1). This problem could be eliminated only if "retarded" effects were considered. Geon unification theory has never been proved. *See also* Feynman.

Whipple's "Dirty Snowball" Theory of Comets: Astronomy: *Fred Lawrence Whipple* (1906–), United States.

Comets are composed of ice, dust, gravel, some gases, and possibly a small, rocky core. They are similar to a dirty snowball.

In 1949, astronomer Fred Whipple based his comet theory on the spectroanalysis of their light and their evolution as they made return trips on elliptical paths through the solar system. He theorized that comets are basically formed of ice and contain a mixture of sandlike dust, gravel, and possibly some gases, such as carbon dioxide, methane, and ammonia, and some comets may have a rocky core. Whipple explained that when a comet approached the sun, even millions of miles distant, the comet's ice vaporized, expelling the dust and gas to form a hazy tail, which always pointed away from the sun as it continues on its orbit. This is a major feature of Whipple's theory. Comets have three basic parts: the head, which is the brightest, varies in size from 0.5 to about 5 or 7 miles wide; a halo, which may be 50,000 to 75,000 miles wide, that glows around the head; and the tail, which is a much fainter glow and may extend 50 to 75 million miles in front of the head. Sunlight and solar wind create radiation pressure on the comet, which forces the gaseous ice/dust of a comet's tail always to point away from the sun. In 1986 a U.S. spacecraft investigating and gathering data on Halley's comet confirmed Whipple's theory of a comet's structure, with one exception: rather than being a "dirty snowball" of dust, it is now believed to be more like an "icy dustball," since the ice is condensed on the outside of the dust particles, and after each pass around the sun, more and more of the ice is lost, meaning that the comet becomes less and less brilliant as it ages. *See also* Halley; Oort.

Wien's Displacement Law: Physics: *Wilhelm Carl Werner Otto Fritz Franz Wien* (1864–1928), Germany. Wilhelm Wien received the 1915 Nobel Prize in physics.

As the temperature rises for electromagnetic radiation, the total amount of radiation increases, while the wavelength of the radiation decreases.

Wilhelm Wien knew that the amount of electromagnetic radiation increases as temperatures rise (a glowing red-hot stovetop element feels hotter than one that is not glowing and appears black). He also knew that very long and very short wavelengths are less abundant in nature than those near the center of the electromagnetic scale. After measuring various wavelengths, he determined these central "peak" wavelengths vary inversely with the **absolute temperature**. This is known as *Wien's displacement law*: wavelength and amount of thermal radiation are determined by the temperature of the radiation. This law is demonstrated by heating a hollow metal ball with a hole in it, called a "black body," and then measuring the wavelength and amount of radiation emitted. As the temperature of the black body increases to the red-hot stage, longer wavelength radiation is emitted. When the temperature becomes even greater, white-hot shorter wavelength radiation is detected. The "amount" of radiation peaks at about the same wavelength range as that of visible light on the electromagnetic radiation scale. The law can be expressed as: $\lambda T =$ constant, where λ is the wavelength, T is the temperature, and the constant is equal to 0.29 cm k. Wien used this law to indicate the distribution of energy in the spectrum as being a function of temperature. The law is applicable for shorter wavelengths but breaks down for longer wavelengths. The black body radiation distribution law (the beginning of quantum theory), developed by Max Planck, is equivalent to Wien's displacement law when the frequency is very large. Planck's law is correct at any frequency, whereas Wien's is correct only for high frequencies. *See also* Bohr; Boltzmann; Einstein; Helmholtz; Planck; Schrödinger.

Wigner's Concept of Parity/Symmetry in Nuclear Reactions: Physics: *Eugene Paul Wigner* (1902–1995), United States. Eugene Wigner shared the 1963 Nobel Prize with Maria Goeppert-Meyer and J. Hans Jensen.

Parity is conserved in nuclear reactions because nature cannot differentiate between left and right orientations or between time periods.

Eugene Wigner, a theoretical physicist, contributed to the understanding of nuclear physics by applying quantum theory to fundamental symmetry principles. He stated that parity is conserved in nuclear reactions. (Any two integers have parity if they are both even or both odd, and fundamental physical interactions do not distinguish between right or left nor clockwise or counterclockwise, thus ensuring symmetry, which is a major physical concept.) Wigner's theory stated that for all matter, energy, and time in the universe, nature makes no distinction between the physical orientation in space of particles, or of more or less time. This relates to nuclei and subnuclear particles' having mirror images, as they are involved in all types of chemical and nuclear reactions. In other words, it does not matter if the molecules or nuclei of matter are oriented as mirror images of each other. The results will be identical in the same time period. Or if a particle is ejected from a nucleus, no distinction is made as to whether it leaves from the right or left. This theory was accepted until 1958, when weak nuclear reactions were discovered. An example of a weak nuclear interaction is when a neutron decays into a proton plus a beta particle (electron) and a neutrino; parity is not conserved. Even so, this exception does not eliminate the concepts of parity or symmetry. Wigner's concepts of parity and symmetry are related to the premise that the greater the "cross section" of a nucleus, the more likely it is that the nucleus can absorb a neutron. This idea contributed to the successful production of a sustained chain reaction in the first nuclear pile which took place in Chicago in 1942. *See also* Boltzman; Fermi; Schrödinger; Weinberg; Wu; Yang.

Wilkinson's Concept of "Sandwich Compounds": Chemistry: *Sir Geoffrey Wilkinson* (1921–1996) England. Geoffrey Wilkinson shared the 1973 Nobel Prize with Ernst Fischer.

Homogeneous catalysts can be formed by adding hydrogen to the double bonds of alkenes.

Geoffrey Wilkinson, primarily an inorganic chemist, explored the attachment of hydrogen to metals to form complex compounds (hydrides) composed of molecules sandwiched together with hydrogen bonds that could be used as catalysts, later known as *Wilkinson's catalyst.* Using these catalysts, he developed systems that could alter the nature of organic compounds by adding hydrogen to the double bonds of some hydrocarbon type molecules. By bonding hydrogen to compounds known as *alkenes*, which have unsaturated molecules, he converted them into branched-chained, hydrogen-saturated, paraffin-type compounds. This process, known as addition hydrogenation, converts unsaturated vegetable liquid oils (e.g., corn oil) to solid fats (e.g., margarine) by adding hydrogen to the double bonds of the oil molecules. This hydrogenization process may also rupture these organic bonds, resulting in hydrocracking, hydroforming, or catforming, which splits off sections of hydrocarbon molecules by using low heat and a platinum catalyst. This process converts crude petroleum into more useable branched-chained fractions (e.g., gasoline, ethene, propene). It is also known by a more generic name, *hydrogenolysis*, which converts bituminous coal into a variety of useful hydrocarbon products, including coal tar dyes, medicines, cosmetics, lubricants, and other petroleum-like products (e.g., "coal-oil" or kerosene).

Williamson's Theory of Reversible Chemical Reactions: Chemistry: *Alexander William Williamson* (1824–1904), England.

A chemical reaction will reach dynamic equilibrium when, under correct conditions of concentration, temperature, and pressure, it becomes reversible.

Alexander Williamson demonstrated it was possible to produce a number of different organic compounds by replacing one or more hydrogen atoms in inorganic compounds, thus forming organic radicals. From this, he developed chemical formulas for a number of compounds, such as alcohols and ether. While experimenting with these new substances, he discovered that some chemical reactions are reversible. A mixture of two compounds will react to form two very different compounds. However, using the correct amounts of the initial substances along with the correct temperature, concentration, and pressure, the reaction will reverse itself and the new compounds will revert to the original substances. In other words, once the first two compounds form the second two, the second ones will revert to the original two compounds ($A + B \leftrightarrow C + D$). Under these conditions the entire system is considered to be in dynamic equilibrium. This concept is vital to the chemical industry, concerned with the conditions necessary to ensure a chemical reaction is *not* in equilibrium so that it will proceed in the desired direction, resulting in the preferred product.

Wilson's Hypothesis of Cloud Condensation: Physics: *Charles Thomson Rees Wilson* (1869–1959), England. Charles Wilson received the 1927 Nobel Prize in physics.

If dust-free, supersaturated moist air is rapidly expanded, the moisture condenses on both fine nuclei and ions (particles).

While experimenting with supersaturated water vapor in a laboratory vessel, Charles Wilson rapidly expanded the volume of this moist air, which formed a cloudlike formation in the chamber. Since the air was dust free, he assumed some type of "nuclei" were present, which provided a base for the moisture to condense into water droplets. He theorized that the recently discovered radiation called x rays might also cause condensation tracks to form in the moist air in his chamber. Subsequently he discovered that supersaturated air became conductive when x rays passed through this moist air, and much more condensation was produced than could be caused by just expanding the air's volume. Wilson then developed his famous Wilson cloud chamber, based on the work of J. J. Thomson and Ernest Rutherford, which detects radioactive radiation of all kinds as well as very small, almost weightless subatomic particles as they form ionized curved paths through supersaturated air in the chamber. These ionized paths form water droplets and can be photographed and studied to determine the characteristics of the radiation or nature of the subatomic particle. *See also* Compton; Millikan; Rutherford; Thomson.

Wilson's "Out-of-Africa" Theory: Biology: *Allan Charles Wilson* (1934–1991), New Zealand.

The ratio of mitochondrial DNA differences between humans and great apes indicates a divergence of lineages 5 million years ago, which achieved a complete separation between species 200,000 years ago.

Allan Wilson studied the DNA found in the mitochondria of cells which, unlike regular DNA, is found outside the cell nucleus. Mitochondria exist in the **organelles**, structures located in the cytoplasm that produce the energy required for cell growth and life. This extranuclear DNA is referred to as $_{mt}$DNA and is carried only by the mother's cells. It is also believed that genetic variations arise from mutations of the $_{mt}$DNA and accumulate through the maternal side at a rather steady rate, which provides a means to calculate statistically, through maternal $_{mt}$DNA, the age of ancestors. In other words, $_{mt}$DNA becomes a *molecular clock*. Wilson therefore theorized that all human mitochondrial $_{mt}$DNA must have originated with a very old, common, female ancestor. He collected a sample of mitochondria cells from individuals of all races from all parts of the world and discovered there are just two basic genetic branches, both of which originated in Africa. His theory that the maternal ancestor for all humans lived on the African continent became known as the "Out-of-Africa" theory and was later dubbed the "Eve hypothesis" by journalists. Wilson's next research dealt with the age of this "common" female ancestor. He found the ratio of $_{mt}$DNA between chimpanzees and humans was 1:25, and since the beginning of the separation of the *Homo* species from the great apes was about 5 million years, he calculated that 1/25 of this 5 million years was equal to 200,000 years. (More recently it has been established that chimpanzees and humans share over 98 percent of the same DNA.) He theorized this was the time a complete separation from our nonhuman ancestors resulted in a human species. Some scientists claim humans diverged and became a separate species in more than one geographical region. Some scientists claim this divergence occurred more than 200,000 years ago, while others believe it occurred less than 200,000 years ago. A competing theory, called "multiregionalism," proposes that ancient humans originated in several different regions of the world and over time migrated, interbred, and produced hybrids that became some of the now-extinct species of the *Homo* group (e.g., Neanderthal man). Most scientists now accept that the extinct species of *Homo* who walked erect on two legs developed about 100,000 to 200,000 years ago in Africa, 60,000 years ago in Australia, 40,000 years ago in Europe, and 35,000 years ago in Northeast Asia and appeared in the northwestern part of America about 15,000 to 30,000 years ago. The more recent species of man, *Homo sapiens-sapiens* (intelligent man), appeared in Europe or Eurasia about 10,000 to 15,000 years ago. More fossil evidence will need to be found and analyzed before the argument concerning the origin of humans as a separate species can be settled.

Wilson's Theory of Dynamic Equilibrium of Island Populations: Biology: *Edward Osborne Wilson* (1929–   ), United States.

Geographically isolated species will establish a dynamic equilibrium of their populations.

E. O. Wilson is an entomologist and sociobiologist who studies ants and other social insects. Wilson and his colleague, Robert MacArthur (1930–1972), theorized that, in time, related species would develop distinct differences in order to resist interbreeding, and a "dynamic equilibrium" of their populations would naturally be established. They based their concept on what they called "character displacement," which takes place when isolated species are once again brought back into close geographic proximity to each other. To prove their theory, they eliminated all insects on a small island off the south Florida coast and waited to see how it would be repopulated as compared to the original number of species. After several months, they returned to find that the island had been repopulated by the same number of species in the same ratios as before, thus proving their theory that for isolated geographic areas, a dynamic equilibrium among species populations will develop. (They assumed the insects' eggs were not completely destroyed or adults arrived from other nearby land areas.) Wilson also contends that individual animals and groups (both insects and humans) use their genetically driven cultural attributes, which are the result of natural selection, to control their populations and make sacrifices for the group. *See also* Darwin; Wallace.

Witten's Superstring Theory: Physics: *Edward Witten* (1951–), United States.

Events at the nuclear level unify general relativity by combining gravity, quantum mechanics, and space in ten dimensions.

There are two major theories of physics: (1) the very small (quantum theory and the uncertainty principle as related to atoms, molecules, subatomic particles, and radiation) and (2) the very large (Einstein's theory of general relativity, gravity, the cosmos, black holes, and so forth). Both are related to the Standard Model of quantum mechanics (*see* Schrödinger). Edward Witten was convinced that the string theory could resolve the problems encountered when combining these two great theories that deal with the small and large. Usually a minute elementary particle is defined as a "point." Witten redefined fundamental particles as a vibrating string or looped string that has different states of oscillation with harmonics similar to a vibrating violin string, making them somewhat "fuzzy." Therefore, a single string can have several harmonics and can consist of a large grouping of different types of elementary particles. This results in a spectrum of particles that then can be "quantized" and related to the "graviton," referred to as the *quantization of gravitational waves*, which in itself makes gravity a priori of the string theory. All types of minute particles and subatomic particles (e.g., electrons, protons, muons, neutrinos, and quarks) fit into the string theory. Thus, Witten claims to have combined the quantum mechanic aspects of the electromagnetic/small with the quantum of relativity/gravity of the large. Based on pure mathematics, Witten proposed how a space consisting of two, four, six, or ten dimensions can explain superstrings and how particles can interact within such a geometric formation, and that six of these dimensions are "folded" into the four known dimensions (height, width, depth, and time). The original string theory was based on the notion that just after the big bang, as the universe cooled, there were cracks and fissures in space that contained great masses, energy sources, and gravitational fields. Recently Steven Hawking claimed that no evidence exists to support the existence of strings; thus the great unification of Einstein's general relativity and gravity with electromagnetism remains elusive. *See also* Einstein; Hawking; Schrödinger; Weinberg.

Wohler's Theory for Nonliving Substances Transforming into Living Substances: Chemistry: *Fredrich Wohler* (1800–1882), Germany.

By applying heat, it is possible to convert nonliving (inorganic) molecules to living (organic) molecules.

Fredrich Wohler's experiments challenged "vitalism," the prevailing theory dealing with organic chemistry, which stated it was not necessary to explain the compounds that make up living organisms since there was a "spirit" connected to life. This spirit was the God-given "vital essence" in all living things, including organic molecular compounds. Vitalism was accepted as the reason humans cannot and should not transform nonliving chemical into living substances. Wohler proved otherwise, which not only resulted in the beginning of the end of vitalism, but provided understanding of new concepts for inorganic and organic chemistry. In 1928 he used heat to decompose ammonium isocyanate, an inorganic chemical, into urea, an organic chemical found in urine (NH_4NCO + heat $\rightarrow NH_2CONH_2$). (These are two very different compounds but with the same molecular formula, thus they are isomers.) Vitalism is a persistent theory and still has adherents. In the field of chemistry, Wohler's work pioneered modern organic chemistry (carbon chemistry), particularly as related to human physiology of respiration, digestion, and metabolism.

Wolfram's Theory of Complex Systems: Physics: *Stephen Wolfram* (1959–   ), England.

Complex systems are driven by one-dimensional cellular automata that follow specific rules.

Stephen Wolfram was interested in a theoretical model for parallel computing that would increase computational power. This idea is based on the ability to understand entities consisting of a group of "cells" that are controlled by a series of rules leading to complexity and chaos. Complex systems are based on the concept of *cell automata* first proposed by John von Neumann. There are several rules for one-dimensional automata cells:

- All cells in a "set" may or may not be filled in.

- The pattern may alternate from "filled in" to "not filled in" and continue to change, but each set must be one way or the other.

- These patterns may "grow" and continue to form the same patterns in ever increasing complexity, as in self-replicating fractal patterns.

- These patterns will continue to become increasingly complex and chaotic. (See Figure W1.)

Figure W1: Fractals are "self-similarities" or "self-replicating" patterns that become increasingly complex and chaotic as they progress. They are similar to Penrose tiles connected in repeated diminishing patterns of geometric shapes as related to chaotic behavior.

This theory of complexity and self-organization explains many natural systems, including how simple organic molecules combined to form increasingly complex patterns until they could become self-replicating and thus living. The theory also describes formal language theory related to the evolution of grammar and original languages into the modern languages of the world. *See also* Penrose; Von Neumann.

Wolf's Theory of the Dark Regions of the Milky Way: Astronomy: *Maximilian Franz Joseph Cornelius Wolf* (1863–1932), Germany.
The dark areas of the Milky Way galaxy are regions where dense "clouds" obscure some of the stars.
Maximilian Wolf designed the *Wolf diagram* used to measure not only the absorption of light but also the distance from Earth to what was called "dark nebula." He attached a camera to the eyepiece of a telescope, enabling him to expose photographic plates and thus record observations over long periods of time. By using time exposure, more light from distant and dim objects was gathered, allowing him to view images on the photographic plates that could not be seen otherwise. Using these methods, Wolf theorized that the dark areas in the Milky Way are gas clouds. More recently it was proposed that over 90 percent of all matter in the universe is composed of **dark matter**, gas, or even **neutrinos**, none of which can be seen by visible light and which outweighs all the trillions of stars.

Woodward's Theory of Organic Molecular Synthesis: Chemistry: *Robert Burns Woodward* (1917–1979), United States. Robert Woodward received the 1965 Nobel Prize in chemistry.

Molecules of organic substances maintain an orbital symmetry enabling them, by geometric orientation, to rotate 180 degrees on their axes and thus become a "negative" mirror image of themselves.

Robert Woodward used the principle of symmetry related to molecular orbits to develop his theory of how some molecules, through a series of addition-reactions, can be synthesized into many useful chemical products. He expanded organic synthesization to the formation of complicated molecules, some involving as many as fifty sequences or series of chemical reactions. Some examples of the products resulting from his addition-reactions are quinine, cholesterol, cortisone, lysergic acid (LSD), reserpine, strychnine, chlorophyl, and vitamin B_{12}.

Wu's Theory of Beta Decay: Physics: *Chien-shiung Wu* (1912–1997), United States.

The direction of the emitted beta particle is related to the direction of spin of the nucleus from which it originates.

In 1934 Enrico Fermi verified Wolfgang Pauli's concept of beta decay, where a neutron disintegrates into an electron and neutrino, leaving behind a proton: neutron $\rightarrow$ electron (β) + neutrino + proton. (See Figure F2 under Fermi.) This process is also known as the *nuclear weak force*, which is stronger than gravity but much weaker than the *strong nuclear force* that holds nuclei together. But there were problems with the Pauli/Fermi theory. In 1957, Chien-shiung Wu theorized that the problem was the direction of the beta decay. She demonstrated that the direction of emission of the beta particle was related to the spin orientation of the nucleus that was decaying. Thus, the emission process of the system is not identical to the mirror image of the system, and therefore parity (right-left symmetry) is not conserved during beta emission. Parity means that two systems that are mirror images of each other are the same in all respects except for this left-right or mirror image phenomenon, and therefore should retain identical symmetry just as humans have a left-and-right side (mirror image) but also have bilateral symmetry. *See also* Fermi; Feynman; Pauli; Wigner; Yang.

Y

Yang's Theory of the Nonconservation of Parity in Weak Interactions: Physics: *Chen Ning Yang* (1922–), United States. Chen Ning Yang shared the 1957 Nobel Prize in physics with Tsung Dao Lee.

The physical law of conservation of parity (symmetry) will break down during weak interactions of subnuclear elementary particles such as beta decay.

Chen Ning Yang, a theoretical physicist, predicted in 1956 that the basic physical law of conservation of parity, first proposed by Eugene Wigner in 1927, would break down when the weak interactions (forces) related to the decay of elementary subnuclear particles were involved. *Parity* refers to the symmetrical quantum-mechanical nature of physical systems. *Parity conservation* refers to the basic physical concept of symmetry, which states that fundamental physical interactions cannot distinguish between right- or left-handedness, clockwise or counterclockwise, or mirror images of physical systems. In addition, for the conservation of parity, no distinction for the particle's orientation in space or the direction of time exists. Yang's theory predicted this concept was violated during the weak interactions of basic atomic elementary particles. These weak interactions are the fundamental forces that take place among elementary particles, including beta decay of nuclei, which produces neutrinos and electrons. Therefore, it is also known as beta interactions. These weak interactions are weaker than electromagnetic forces but stronger than gravitational interactions. (The strong interaction involves the force that holds the nucleus together.) Unlike electromagnetic and gravitational interactions, whose forces fall off as with the square of the distance and are thus become less strong over long distances, the weak interactions fall off very rapidly, and thus are not effective beyond the size of the atom from which they originate. Yang and his collaborator, Tsung-duo Lee (1926–) theorized that a subnuclear particle called **kaon** would break down into two *pions*, which would maintain and conserve parity. But at times some of the kaons broke down into three pions, and thus, in this example of weak interactions, parity was not conserved (e.g., two odds and one even, or it could be described as two pions that spin clockwise while the other spins counterclockwise). This means the symmetry of left and right was not equal and electrons would exit the reactions in one direction more than in the other (nonsymmetrically). This resulted in the conclusion that parity would be conserved for electromagnetic and strong interactions but not for weak interactions. Their theory was confirmed by Chien-shiung Wu, who demonstrated that parity is not conserved in beta disintegrations. Physicists now believe there may be other antiparticles or energies that could account for this uneven symmetry. The concept of parity conservation for weak interaction of elementary subatomic particles is important for understanding the basic nature of matter. *See also* Fermi; Feynman; Gell-Mann; Pauli; Wigner; Wu; Yukawa.

Young's Wave Theory of Light: Physics: *Thomas Young* (1773–1829) England.
Light is transmitted through the aether as a wave front of beams that are both identical and singular.

Thomas Young studied the functioning of the human eye. His theory that the lens of the eye changed shape to adjust to light and distance led him to explore the nature of light and how it traveled from one object to another. In the early nineteenth century, there were two conflicting concepts of the nature of light. One claimed light was a stream (emission) of particles sent out by objects, which were received by the eye; the other stated that light consisted of minute standing waves and was transmitted by the **aether**. But the prevailing concept was the corpuscular theory for the emission of light, which claimed the polarization of light was possible only if light was a collection of single tiny particles originating from an object. This concept did not conform to Young's experimental evidence or to the mathematics of that time. Young proposed the transmission theory of light as a wave front of identical "beams" (not corpuscles), passing through a medium that he and others referred to as aether. Young experimented with a beam of light that he focused through two pinholes in a barrier to the path of a light beam. This produced two separate beams of light emanating from the pinholes on the other side of the barrier. These new standing wavelets exhibited two curved wave fronts whose matching crests showed up as alternate areas of light on a back screen. (See Figure Y1.) Later, two narrow slits were used instead of pinholes. This phenomenon is known as diffraction, where the waves spread and bend as they pass thought the small openings in the barrier. At the point where the crests of the light wavelets were "matched," they intensified each other to form bright strips of light. Conversely, where the crests of the waves were not matched (interfere), they counteracted or blocked each other to form dark images. Diffraction (the splitting of the light beam into wavelets) and interference (the matching, or not matching, of the wavelet's crests) are basically the same phenomenon and now apply to all forms of electromagnetic radiation. Young's interference experiment was a classic demonstration proving the wave nature of light. It took some time for other physicists to understand

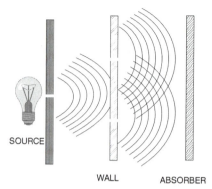

SOURCE

WALL   ABSORBER

Figure Y1: Young's experiment demonstrated the wave nature of light. The light source passes through the hole in the first barrier, proceeds as a wave front to the second wall, where it passes through two holes, and emerges as two wave fronts that are recorded on the absorber wall. The light areas on the third wall occur when the crest of the light waves are "in phase" and add to their brightness, while the dark areas are where the waves are "out of phase" and interfere with each other.

the importance of Young's wave theory, but once accepted, it was used to exhibit why the different colors of the spectrum have different wavelengths. Using the wave front theory of light, Young and other physicists explained transverse wave propagation, the mechanical quality of the light medium, polarization, reflection and refraction, and other optical phenomena. His theory later assisted in determining the speed of light in air and water. His wave theory was augmented by the "quantum/photon" theory of light proposed by Schrödinger and Einstein, which resulted in the wave/particle duality of light. *See also* Einstein; Fresnel; Hertz, Huygens; Maxwell; Schrödinger.

Yukawa's Meson Theory for the "Strong Interaction": Physics: *Hideki Yukawa* (1907–1981), Japan. Hideki Yukawa was the first Japanese to be awarded the Nobel Prize for physics (1949).

Nuclei containing more than one positive proton must be held together by a force stronger than that of the protons' opposing positive charges.

Hideki Yukawa knew the "electroweak" force or the "weak interaction" involved in beta decay was much weaker than the force that binds nucleons of the nucleus together. Beta decay is the simplest type of radioactivity: neutron decays into a proton, an electron, and what was later discovered to be a neutrino, which is considered massless. (See Figure F2 under Fermi.) Yukawa believed there must be a heavier particle that could fuse protons and neutrons within the nucleus of atoms. Using electromagnetic forces as an analogy, he applied quantum theory to predict that a stronger force was responsible for "binding" protons and neutrons in nuclei. The difference was that electromagnetic photons (visible light), which are considered massless, interact over infinite distances, while Yukawa's predicted nuclear binding "strong interaction" particle would be many times heavier than an electron and could react only over a distance less than the diameter of an atom (about 10^{-12}). In 1935 Yukawa predicted a new particle would bind **nucleons** in a nucleus. A few years later the discovery of this new elementary particle was confirmed by Carl Anderson. It was named the **meson**, and later *mu-meson*, which is now called a *muon*. Muons did not interact frequently enough with the nucleons (quarks, neutrons and protons) to "glue" them together adequately. It was later discovered the muon was a decay product of another particle with 265 times the mass of an electron. This heavier particle discovered by Cecil Powell (1903–1969) in 1947 was first called the **pi-meson** and was later named the *pion*. The decay of the pion confirmed Yukawa's prediction for the "strong interaction" (force) that binds particles in nuclei. *See also* Anderson; Fermi; Feynman; Gell-Mann; Pauli; Wigner; Wu; Yang.

Z

Zeeman's Theory of the Magnetic Effect on Light: Physics: *Pieter Zeeman* (1865–1943), Netherlands. Pieter Zeeman shared the 1902 Nobel Prize in physics with Hendrik Lorentz.

The spectral lines of light emitted from atoms are split into either two or three lines when the atoms emitting the light are subjected to a magnetic field.

It was earlier demonstrated that the light given off from burning elements, when viewed through a spectroscope, would form distinct patterns of colors and dark lines. Sodium was the commonly used element for spectroscopic viewing. Its spectral lines are referred to as the "D-lines" due to their position in the electromagnetic spectrum. Pieter Zeeman, who undertook to verify Hendrik Lorentz's theory on atomic structure, set up a spectroscope to view these D-lines, placing an electromagnet between the scope and the sodium light source. He noticed that when the magnetic field was oriented perpendicular to the path of the light, the spectral lines were split into three distinct lines. When the magnetic field was oriented parallel to the light path, the lines were split into two images. This phenomenon, which is the splitting of spectral lines of a light source when passing through a magnetic field, became known as the *Zeeman effect*. Zeeman calculated the ratio of the electrical charge to the mass of the vibrating sodium ions, which proved it had a negative charge. *See also* Bohr; Lorentz; Maxwell.

Zeno's Paradoxes: Physics: *Zeno of Elea* (c.490–430 B.C.), Greece.

If space can be continually divided into an infinite number of units, it will take an infinite amount of time to pass through all these units of space. Therefore, motion is an illusion.

Zeno of Elea, a pre-Socratic philosopher, devised paradoxes as arguments to contradict his philosophical opponents. The "theory" that motion cannot exist is only one of several of Zeno's paradoxes, based on the "dichotomy" that motion cannot exist because before it can reach where it is going, it must first reach a midpoint (half of its destination). He continued by stating that before this midpoint could be reached, it must reach one-fourth of its course, and before this, its one-eighth point, its one-sixteenth point, and so on. If one continues with this concept of motion, it can never proceed from where it starts. A similar but flip-side paradox is best explained by Zeno's story of the race between Achilles and the tortoise. Both Achilles and the tortoise start from the same point, but the tortoise is allowed to start first and reaches point A (half the distance of the race). Before Achilles can pass the tortoise, he must also reach point A, but by this time the tortoise has proceeded to point B. Now Achilles must run to point B, but the tortoise has proceeded to point C and so on. In a race so designed, Achilles will never catch the tortoise because as hard as he tries, he can cut the remaining distance only in half each time; thus the tortoise is always ahead and Achilles cannot win. This is an example of dividing the race into an infinite number of tasks, just as Zeno also stated that a line or space can be divided into an infinite number of units. This argument was used by Democritus to determine the atomic nature of matter by continually dividing a handful of soil into halves, over and over again, into an almost infinite number of times, until one tiny piece of matter so small it cannot be further divided remains—thus the atom. Zeno's paradox remained unsolved for two thousand years until it was explained by the use of calculus as the *convergence series*, an infinite series with a finite sum. *See also* Atomism Theories.

Zinn's Concept of a "Breeder" Reactor: Physics: *Walter Henry Zinn* (1906–   ), United States.

When irradiated by neutrons, uranium-238 can be converted into fissionable plutonium-239.

It was known for some time that when neutrons were slowed down, they could penetrate the nuclei of uranium, thus causing the uranium nuclei to split into nuclei of lighter elements while collectively giving off great amounts of energy. Walter Zinn and his mentor, Leo Szilard, demonstrated that a small mass of the split uranium nucleus was converted into energy, as predicted by Einstein's formula, $E = mc^2$. At the beginning of World War II, Zinn worked with Enrico Fermi on the Manhattan Project to build the first **nuclear pile**. Zinn was the person who slowly pulled the control rod from the pile to allow more neutrons to interact with purified uranium. At the time, no one knew if it would work or blow up. The pile was successful and demonstrated a sustainable fission reaction was possible, which led to the first "atomic" bombs. Zinn was also in charge of dismantling the reactor. In 1951 Zinn developed the first breeder reactor that used neutrons emitted from the core of a reactor to change a blanket of U-238 surrounding the core into plutonium-239. Pu-239, first identified in 1940 by Glenn T. Seaborg, who used a cyclotron, is fissionable with a long half-life. About 0.66 pound is needed to reach a critical mass and become a nuclear bomb, which is about one-third as much required of the less plentiful

U-238. Pu-239 is highly radioactive but relatively easy to produce and can be used in lightweight reactors to produce heat and electricity. *See also* Fermi; Seaborg; Szilard.

Zuckerandl's Theory for Measuring the Rate of Evolution: Biology: *Emile Zuckerandl* (1922–   ), Austria.

The differences in the hemoglobin chains in mammals can be used as a "clock" to measure the evolution of species.

While comparing the amino acids in the hemoglobin of the blood of different animals, Emile Zuckerandl discovered that out of the 146 amino acids in one of the human hemoglobin chains, only one was different from the gorilla. But there were more differences in the amino acids for other mammals. He used this concept to formulate a "mean" difference of twenty-two hemoglobin chains for all mammals. He then considered the time during which these animals "split" off from a common ancestor to be about 80 or 90 million years, surmising it takes approximately 7 or 8 million years for evolution to change one pair of amino acids. This theory was improved and became valuable for future biologists to estimate the rate of organic evolution. *See also* Darwin; Waddington; A. Wilson.

Zwicky's Theory for Supernovas and Neutron Stars: Astronomy: *Fritz Zwicky* (1898–1974), United States.

Supernovas are brilliant stellar explosions, distinct from novas, that collapse into neutron stars under their own gravitational force.

Fritz Zwicky was the first to theorize that supernovas are different from other bright objects in the sky. (*Nova* means "new" in Latin.) His theory stated that supernovas were stellar explosions that produce great brightness. Zwicky claimed that only about two or three supernovas are ever discovered during each thousand-year period. He determined that supernovas have a brightness about fourteen to fifteen times that of the sun, making them visible at the great distances where galaxies are found. He also calculated that when a supernova burned out and there was no longer radiation to maintain its size and brilliance, it would collapse. According to the law of gravity, it would attract all of its mass into a dense core and end its existence as a neutron star. A neutron star is as massive as a regular star but is only 7 to 9 miles in diameter. In other words, its density is so great that a teaspoonful would weigh many tons. Thirty years later in 1932, Zwicky's prediction was borne out when the existence of superdense neutron stars was discovered. His work with supernovas resulted in theories of galaxy evolution, including galaxy clusters and superclusters of galaxy clusters, which indicate that distant matter in the universe may not be evenly distributed. *See also* Baade; Hubble; Rossi.

Glossary

absolute temperature. In theoretical physics and chemistry, refers to the kelvin scale, specifically for absolute zero, which is $-273.16°$Celsius or $-459.69°$F. This is the temperature at which all matter possesses no thermal energy and at which all molecular motion ceases. It has never been reached.

ac. Abbreviation for *alternating current*. Electric current in a circuit that reverses its direction at repeated intervals.

acid. A substance that releases hydrogen ions when added to water. Strong acids are sour tasting, turn litmus paper red, and react with some metals to release hydrogen gas.

adiabatic. Refers to thermodynamic process in which there is no transfer of heat into or out of a closed (isolated) system.

adsorption. Adherence or collection of atoms, ions, or molecules of a gas or liquid to the surface of another substance, called the *adsorbent*; for example, hydrogen gas collects or adsorbs to the surface of several other elements, particularly metals. An important process in the dyeing of fabric.

aether. Early scientists assumed a "medium" called aether, which in Greek mythology typifies the upper air, occupied all space, and thus was believed to be required for the transmission of electromagnetic waves. Also referred to as *ether*.

agar. A gelatinous substance extracted from a specific species of red marine algae. Used mainly as a gelling agent in bacterial culture media. Also an ingredient in creams, ointments and commercial laxatives.

AIDS. Acquired immune deficiency syndrome; disease that compromises the immune system through persistent opportunistic infections and malignancies and is believed to be caused by the human immunodeficiency virus (HIV).

alchemist (alchemy). A forerunner of modern chemistry (chemists) practiced from approximately 500 B.C. through the sixteenth century. It had a twofold philosophy: the search for and use of the philosophers' stone to transmute base metals into gold and to prepare and perfect medicine for people, called the *elixir vitae*.

alkaloid. In organic chemistry, a basic nitrogenous compound obtained from plants, soluble in alcohol and insoluble in water (e.g., morphine, nicotine, caffeine, cocaine).

alpha particle. A nucleus of a helium atom (H^{++})—that is, two positive protons and two neutrons, without any electrons. Alpha particles, along with beta and gamma particles, constitute the three basic forms of radiation resulting from nuclear decay.

amino acid. An organic compound comprising both an amino group (NH_2) and a carboxylic acid group (COOH). They are polymerized to form proteins and peptides. Amino acids occur naturally and also have been synthesized in laboratories. It is believed that products of a naturally occurring synthesis of amino acids may be the building blocks of life.

amplifier. A device capable of increasing the level of power or the magnitude, for example, an electric current that uses a transistor or an electron tube with an electric signal, that varies with time and does not distort the shape of the electrical waves.

anode. The positively charged electrode in an electrolytic cell, electron tube, or storage battery; also, the collector of electrons.

antibody. A blood serum protein, sometimes occuring normally or generated in response to an invading antigen, that specifically reacts with a complementary antigen to produce immunity from a number of microorganisms and their toxins.

antimatter. *See* antiparticle.

antineutrino. The antiparticle to the neutrino. *See also* antiparticle.

antiparticle. A subatomic particle—a positron, antiproton, or antineutron—with the identical mass of the ordinary particle to which it corresponds but opposite in electrical charge or magnetic moment. Antiparticles make up antimatter, the mirror image of the particles of matter that make up ordinary matter as we know it on Earth. This is a theoretical concept devised to relate relativistic mechanics to the quantum theory.

antiproton. The antiparticle to the proton. *See also* antiparticle.

aperture. An opening (i.e., hole or slit) through which light waves, radio waves, electrons, or radiation can pass. This may be the adjustable opening on optical instruments (e.g., cameras and telescopes).

aplanatic lens. A lens whose surfaces are not segments of a sphere. It is used to correct imperfect focusing called *spherical aberration*.

asteroid. Derived from the Greek word *asteroids*, which means "starlike." They are small bodies that revolve around the sun. They are sometimes called small "planetoids," and when they become fragmented and land on Earth, they are considered meteorites. Most asteroids are found in a planetary orbit called the *asteroid belt*, located between the orbits of Mars and Jupiter.

atomic number (proton number). The number of positively charged protons found in the nucleus of an atom, on which its structure and properties depend. This number determines the location of an element in the Periodic Table of the Chemical Elements. For a neutral atom, the number of electrons equals the number of protons.

atomic weight (atomic mass). The total number of protons plus neutrons in an atom.

autocatalytic. The theoretical process whereby primordial organic molecules may have replicated themselves in early prebiotic environments.

autonomic. Independent, spontaneous, or involuntary. Relates to the autonomic nervous system and autonomic reflex system in vertebrates and other animals, as well as the autonomic movement in plants.

autopoiesis. The self-maintenance of an organism.

axiom. An assumption on which a mathematical theory is based.

base. An alkali substance that reacts with (neutralizes) an acid to form a salt; for example, $4HCl + 2Na_2O \rightarrow 4NaCl + 2H_2O$ [hydrochloric acid + sodium hydroxide yields sodium chloride (table salt) + water].

biosynthesis. The natural synthesis (fusion) of an organic chemical compound by living organisms.

black holes. Theoretically, they are thought to be vortex areas in space where massive stars have collapsed, creating such great gravity that not even light can escape into space.

bonding (chemical). Electrostatic force that holds together the elements that form molecules of compounds. This attractive force between atoms is strong enough to hold the compound together until a chemical reaction causes the substance to form new bonds or break the bonds that form the molecule. *See also* covalent bond; ionic bonding.

bosons. One of the two main classifications of subatomic particles, they are weak "force" particles (photons, pi mesons, gluons, positive W and negative W particles, neutral Z particles, graviton). Fermions is the other main classification of subatomic particles.

capacitor. A storage device (condenser) for static electricity consisting of two or more metal surfaces (conductors) separated from each other by a dielectric. It stores the electrical energy and impedes the flow of direct current. *See also* Leyden jar.

catalyst. Any substance that affects the rate of a chemical reaction without itself being consumed or undergoing a chemical change. Platinum/palladium pellets in automobile catalytic converters are chemical catalysts. A biological catalyst (e.g., an enzyme) affects chemical reactions in living organisms.

cathode. A negatively charged electrode or plate, as in an electrolytic cell, storage battery, or electron tube similar to a TV. Also, the primary source of electrons in a cathode ray tube such as the Crookes tube.

chromatography. Any of a group of techniques used to separate complex mixtures (i.e., vapors, liquids or solutions) by a process of selective adsorption (not to be confused with absorption), the result being that the distinct layers of the mixture can be identified. The most popular techniques are liquid, gas, column, and paper chromatography.

chromosomes. The complex, DNA-containing, threadlike material inside the nuclei of the cells of living organisms that determines hereditary characteristics of that organism.

cloud chamber. A device for detecting the paths of high-speed particles as they move through a chamber filled with air or gas saturated with water vapor. The device is fitted with a piston that, when moved outwardly, affects the expansion of the gas and the cooling of the vapor. A fog or cloud of minute droplets then forms on any nuclei or ions present in the chamber. Also known as the *Wilson cloud chamber*.

coke. The residue produced after bituminous coal or other carbonaceous materials, such as petroleum or pitch, are heated to extremely high temperatures in the absence of air. Primarily consisting of carbon, it is used as a fuel in blast furnaces.

comet. A nebulous celestial formation consisting of rocks, ice, and gases. Comets are composed of three main parts: the *nucleus*, which is the center made of rock and ice; the *coma*, which is composed of the gases and dust that form around the nucleus as it evaporates; and the *tail*, which is made up of the gases and spreads out from the coma.

compound. A substance in which two or more elements are joined by a chemical bond to form a substance different from the combining elements. The combining atoms do not vary their ratio in their new compound and can be separated only by a chemical reaction, not a physical force. *See also bonding*.

conductor. Substances that allow heat or electricity to flow through them.

cosmology (**cosmos, cosmological**). The study of the universe on both the smallest and largest of scales in terms of time, space, and the makeup of the universe. It includes theories about the origin of the universe and everything in it, the evolution of the universe from past to present to future, and the structure of the universe and its celestial bodies at various stages of their evolution.

covalent bond. Sharing of electrons by two or more atoms to form a pair of electrons. This type of bonding always produces a molecule. Also known as *electron pair bond*. *See also* bonding.

critical mass. The minimum mass of fissionable material (U-235 or PU-239) that will initiate an uncontrolled fission chain reaction, as in a nuclear (atomic) bomb.

critical temperature. The temperature above which a substance cannot be converted from the liquid to the gaseous state or vice versa, regardless of the pressure applied. Also, the temperature at which a magnetic material will lose its magnetism.

cryogenics. Study of the behavior of matter at very low temperatures, below $-200°C$. The use of the liquefied gases (oxygen, nitrogen, hydrogen) at approximately $-260°C$ is standard industrial practice.

cyclotron. A particle accelerator made up of two hollow cylinders (similar to two opposing D structures) that are connected to a high-frequency alternating voltage source in a constant magnetic field. The charged particles, which are injected near the midpoint of the gap between these two hollow cylinders, are then accelerated in a spiral path of increasing expanse so that the path traveled by these accelerated particles increases with their speed where a deflecting magnetic field deflects them to a target. *See also* particle accelerator.

dark matter. Nonluminous matter that is assumed to be present in the Milky Way and other galaxies that explains the motions of the stars and clouds of gases in those galaxies. Cosmological theory states that such dark matter makes up over 90 percent of all matter in the universe and must exist in order to achieve the critical density necessary to close the universe.

dc. Abbreviation for direct current. Electric current that flows in only one direction.

deterministic/determinism. The doctrine that all phenomena are causally determined by prior events. It has also been stated as the relationship between a cause and its effect, particularly natural phenomena, or as the hypothesis stating a set of precisely determined conditions will always repeat the same effect, or that an event cannot precede its cause. Also known as *causality*.

diastolic. Refers to the rhythmic relaxation and dilation of the heart's chambers, particularly the ventricles. The diastolic reading on a blood pressure monitor that records the lowest arterial blood pressure during the time the ventricles fill with blood.

DNA. Abbreviation for deoxyribonucleic acid. The complex ladder-like, double-stranded nucleic acid molecule present in chromosomes that forms a double helix of repetitive building blocks and shapes the inherited genetic characteristics of all living organisms, with the exception of a small number of viruses.

Doppler effect. The apparent change or shift in the observed frequency of a sound or electromagnetic wave due to the relative movement between the source and the observer. The same principle applies when determining the distance of stars in the galaxy. The Doppler frequency or shift is based on the color shift (frequency of light) related to the star's velocity. The light frequency for a star receding from Earth is redder (longer wavelengths) than a star approaching Earth, which emits a blue light (shorter wavelength).

ecology. The scientific study of the interrelationships of organisms to each other and their physical, chemical, and biological environments.

electrolysis. A process in which an electric current is passed through a liquid, known as an electrolyte, producing chemical changes at each electrode. The electrolyte decomposes, thus enabling elements to be extracted from their compounds. Examples are the production of chlorine gas by the electrolysis of sodium chloride and the electrolysis of water to produce oxygen and hydrogen.

electrolyte. A compound that, when molten or in solution, will conduct an electric current. The electric current decomposes the electrolyte.

electromagnet. A strong magnet composed of a wire coil wrapped around a soft-iron core through which a current of electricity is passed and becomes demagnetized when the flow of electric current is suspended.

electron. An extremely small, negatively charged particle that moves around the nucleus of an atom. The interaction of the electrons of atoms is the chemistry of Earth's elements.

electroscope. An apparatus that detects the presence and signs of minute electrical charges using a process of electrostatic attraction and repulsion.

empirical. Relates to actual observation, practical experience, and experimentation rather than scientific theory.

entropy. Disorganization, randomness. In thermodynamics, it is the function of the system where the amount of heat transfer introduced in a reversible process is equal to the heat absorbed by the system from its surroundings, divided by the absolute temperature of the thermodynamic system.

enzyme. Any of a number of proteins or conjugated proteins produced by living organisms that act as biochemical catalysts in those organisms.

equinox. One of two points or moments on the celestial sphere when the center of the sun intersects the celestial equator, in either a north or southbound direction.

ether. *See* aether.

eugenics. The genetic principles of heredity to improve a species, most often associated with breeding or engineering of a "superior" race of humans while discouraging the breeding of those considered "inferior." Animal and plant breeding, as well as genetic counseling might be considered less extreme applications of eugenics.

eukaryotic. Describes the state of a cell (eukaryote which comprises all living things except bacteria and cyanobacteria) containing a definitive nucleus, in which nuclear material is surrounded by a membrane and cytoplasm-containing organelles. Along with prokaryotes, they are the two major groups into which organisms are divided.

fermion. A subatomic particle (electron, proton or neutron) having odd half-life integral angular momentum, which obeys the Pauli exclusion principle: no more than one in a set of identical particles may occupy a particular quantum state.

fission. The splitting of an atom's nucleus with the resultant release of enormous amounts of energy and the production of smaller atoms of different elements. Fission occurs spontaneously in the nuclei of unstable radioactive elements, such as U-235 and PU-239, and is used in the generation of nuclear power, as well as in nuclear bombs.

fluorescent. Consisting of a gas-filled tube with an electrode at each end. Passing an electric current through the gas produces ultraviolet radiation, which is converted into visible light by a phosphor coating on the inside of the tube. This emission of light by the phosphor coating is called fluorescence.

forensics. Relates to public discussion, particularly in legal proceedings, concerning engineering practices, medical evidence, chemical studies, and so on, where the findings are presented as legal evidence in a court of law.

fractal. An irregular or fragmented geometrical shape whose intricate structure is such that, when magnified, the original structure is reproduced (self-similarity). Fractals are important in the study of certain branches of physics, as well as in chaos theory and computer-generated graphics.

fusion. An endothermic nuclear reaction yielding large amounts of energy in which the nuclei of light atoms (e.g., forms of heavy hydrogen, such as deuterium or tritium) join or fuse to form helium (e.g., energy of the sun or the hydrogen bomb). The opposite of fission.

galaxy. A huge grouping of millions, or even billions, of stars held in one of several shapes by their mutual gravity. There are elliptical, irregular, and spiral galaxies.

galvanometer. An instrument that measures a small electrical current using mechanical motion derived from the electrodynamic or electromagnetic forces produced by the current.

gene. The basic unit of hereditary material that is composed of a sequence of nucleotides of a section of DNA or RNA molecules. The sequence of nucleotides determines the structure of amino acids in proteins, which is fundamental to all other biological processes. Genes, individually or in groups, determine inherited characteristics.

genetics. The science of biological heredity and the mechanisms by which characteristics are passed along to succeeding generations.

geomagnetism. Refers to Earth's magnetism and, in a broader sense, the magnetic phenomena of interplanetary space.

glaciation. The alteration of the surface of Earth by passage of glaciers, mainly by erosion or deposition.

global warming. The increase in global temperatures reportedly caused by the emission of industrial gases, along with other natural air pollutants, which traps heat from the sun. A natural cloud cover acts as an "insulating blanket," which keeps the heat of Earth and the lower atmosphere from radiating into the outer atmosphere and on into space. Speculation, unproven to date, holds that the addition of pollutant clouds into the atmosphere has increased Earth's temperature, which causes climatic, often catastrophic, changes in the environment. Also referred to as the *greenhouse effect*.

gluon. A hypothetical, massless, neutral elementary particle that carries the strong force (interactions) that binds quarks, neutrons, and protons together. Gluons can also interact among themselves to form particles that consist only of gluons without quarks and are called *glueballs*.

graviton. A hypothetical (not yet discovered) carrier particle presumed to be the quantum of gravitational interaction, having a mass and charge of zero and a spin of 2.

hadron. An elementary particle, part of the largest family of elementary particles, that has strong interactions, usually producing additional hadrons during high-energy collisions.

half-life. The time required for one-half of the atoms of heavy radioactive elements to decay or disintegrate by fission into lighter elements.

halogens. Electronegative monovalent nonmetallic elements of Group 17 (VIIA) of the Periodic Table of the Chemical Elements (fluorine, chlorine, iodine, bromine, astatine). In pure form, they exist as diatomic molecules (e.g., Cl_2).

heliocentric. Refers to the belief that the sun is the center of the solar system or universe.

hominid. Member of the mammal family of which *Homo sapiens* is the only surviving species.

hominoid. Manlike; an animal that resembles a human. (Humans and anthropoid apes are usually included in the superfamily commonly referred to as hominoids.)

humoral theory. Pertains to the practice of medicine, primarily in the Middle Ages, whereby the body was governed by four principal humours or fluids (blood, phlegm, choler, and black bile). These were present in varying proportions in each person, the balance of which was essential for continued good health. If any of these four "humours" was out of balance, a procedure (e.g. blood letting) was performed by the physician in an effort to restore "balance."

hydrostatic. The study of liquids at rest (e.g., liquids contained in dams, storage containers, and hydraulic machinery).

hypotenuse. In a right triangle, the side opposite the right angle.

in vitro. Meaning "in glass" in Latin. Refers to an observable biological reaction that occurs under artificial conditions outside a living organism, usually in a test tube or a petri dish.

ion. An atom or a group of atoms that have gained or lost electron(s) and thus have acquired an electrical charge. The loss of electrons gives positively charged ions. The gain of electrons results in negatively charged ions. If the ion has a net positive charge in a solution, it is a *cation*. If it has a net negative charge in solution, it is an *anion*. An ion often has different chemical properties from the atoms from which it originated.

ionic bonding. Donating of electrons from one element to another element, forming positively and negatively charged ions, respectively. The electrostatic attraction between the oppositely charged ions constitutes the bond. Also known as *electrovalent bond*.

ionization. The chemical process for producing ions in which a neutral atom or molecule either gains or loses electrons, giving it a net charge, thus becoming an ion.

irrational numbers. Any real number that is not the quotient of two integers. They are usually algebraic (roots of algebraic equations) or transcendental numbers.

isomer. In chemistry, chemical compounds with the same molecular composition but different chemical structures. For example, butane has two isomers, C_4H_{10} and $C_2H_4(CH_3)_2$. In nuclear physics, isomers refer to the existence of atomic nuclei with the same atomic number and the same mass number but different energy states.

isostasy. The theoretical gravitational equilibrium existing in Earth's crust. If there is a disturbance on the surface of Earth (e.g., erosion or glacier movement, which is also referred to as deposition), there are counterbalancing movements in Earth's crust. The areas of deposition will sink, while the areas of erosion will rise. The same counterbalancing effect also occurs in the oceans as the lack of density in ocean water is compensated by an excess density in the material under the ocean's floor.

isotopes. Atoms of the same element with different numbers of neutrons. All atoms of an element always contain the same number of protons. Thus, their proton (atomic) number remains the same. However, an atom's nucleon number, which denotes the total number of protons and neutrons, can be different. These atoms of the same element with different atomic weights (mass) are called *isotopes*. Isotopes of a given element all have the same chemical characteristics (electrons and protons), but they may have slightly different physical properties.

kaon. An elementary particle that is a subclass of the hadrons. Mesons consist of quark-antiquark pairs. They have zero spin, a nonzero strangeness (quantum) number, and a mass of approximately 495 MeV. It is the lightest hadron to contain a strange quark. Also known as a *K meson* in particle physics.

kinetic energy. Energy associated with motion.

lepton. In particle physics, any light particle. Leptons have a mass smaller than the proton mass and do not experience the strong nuclear force. They interact with electromagnetic and gravitational fields and essentially interact only through weak interactions.

Leyden jar. An early and improved form of capacitor (condenser). Metal foil was placed on both the inside and outside of the glass jar, allowing the glass to act as a dielectric or nonconducting substance to separate the electrical charges. A charge of stored static electricity occurred as the wire touched the inside foil, which was fed through the cork on the top of the jar. A circuit was completed when the wire conducted the electricity to the foil on the outside of jar, or a spark jumped to a finger brought near the wire exiting the jar. *See also* capacitor.

magnet. A body or an object that has the ability to attract certain substances (e.g., iron). This is due to a force field caused by the movement of electrons and the alignment of the magnet's atoms.

magnetic moment. In physics, the ratio between maximum torque that is exerted on a magnetized body, electric current-carrying coil, or magnetic domain and the strength of that magnetic domain or field. Also called *magnetic dipole moment*.

magnetohydrodynamics. In physics, the study of motion or dynamics of electrically conducting fluids (plasmas, ionized gases, liquid metals) and their interactions with magnetic fields. Also known as *hydromagnetics* or *magnetofluid dynamics*.

magnetosphere. The comet-shaped regions surrounding Earth and the other planets where the charged particles are controlled by the planet's own magnetic field rather than the sun's. Earth's geomagnetic field is believed to begin at an altitude of about 100 kilometers and extends to the far-away borders of interplanetary space.

maser (microwave amplification by stimulated emission of radiation). A device that converts incident electromagnetic radiation from a wide range of frequencies to one or more discrete frequencies of highly amplified microwave radiation.

mass. The quantity (amount) of matter contained in a substance. Mass is constant regardless of its location in the universe. Mass should not be confused with weight.

mean. Determined by adding all the values or a set of numbers and dividing the sum by the total numbers or values. Usually associated with the term *average*.

meiosis. A type of division of the nuclei of cells during which the number of chromosomes is reduced by half.

meson. An elementary particle with strong nuclear interactions, having a baryon number zero. Mesons are unstable and decay to the lowest accessible mass states.

metabolism. A chemical transformation that occurs in organisms when nutrients are ingested, utilized, and finally eliminated (e.g., digestion, absorption, followed by a complicated series of degradations, syntheses, hydrolysis, and oxidations utilizing enzymes, bile acids, and hydrochloric acid). Energy is an important by-product of the metabolizing of food.

meteorite. A small portion of a larger meteor, meteoroid, or a disintegrated chunk of an asteroid that has not completely vaporized as it entered and passed through Earth's atmosphere and that eventually lands on Earth's surface.

molecule. The smallest particle of a substance containing more than one atom (e.g., O_2) or a compound that can exist independently. It is usually made up of a group of atoms joined by covalent bonds.

muon. The semistable second-generation lepton, with a mass 207 times that of an electron. It has a spin of one-half and a mass of approximately 105 MeV, and a mean lifetime of approximately 2.2×10^{-6} second. Also known as *mu-meson*.

nebula. An immense and diffuse cloudlike mass of gas and interstellar dust particles, visible due to the illumination of nearby stars. Examples are the Horsehead nebula in Orion and the Trifid Nebula in Sagittarius.

neutrino. An electrically neutral, stable fundamental particle in the lepton family of subatomic particles. It has a spin of one-half and a small or possibly a zero at-rest mass, with a weak interaction with matter. Neutrinos are believed to account for the continuous energy distribution of beta particles and are believed to protect the angular momentum of the beta decay process.

neutron. A fundamental particle of matter with a mass of 1.009 (of a proton) and having no electrical charge. It is a part of the nucleus of all elements except hydrogen.

Nuclear pile. A nuclear fission reactor.

nucleon. A general term for either the neutron or proton, in particular as a constituent of the nucleus.

nucleotide. The structural unit of nucleic acid found in RNA and DNA.

nucleus. The core of an atom, which provides almost all of the atom's mass. It contains protons, neutrons, and quarks held together by gluons (except hydrogen's nucleus, which is a single proton) and has a positive charge equal to the number of protons. This charge is balanced by the negative charges of the orbital electrons.

organelle. A distinct subcellular structure, with a specific function and defined shape and size, found in the cytoplasm of the cell (e.g., mitochondrion).

oxide. A compound formed when oxygen combines with one other element—a metal or nonmetal (e.g., magnesium oxide).

ozone layer. The layer found in the upper atmosphere, between 10 and 30 miles in altitude. This thin layer of gases contains a high concentration of ozone gas (O_3), which partially absorbs solar ultraviolet (UV) radiation and prevents it from reaching Earth. It is mostly formed over the equator and drifts toward the North and South Poles. It seems to have a cyclic nature. Also called the ozonosphere.

parallax. The apparent change in direction and/or position of an object viewed through an optical instrument (e.g., telescope), which occurs by the shifting position of the observer's line of sight.

particle. A very small piece of a substance that maintains the characteristics of that substance. Also known as fundamental particles found in atoms.

particle accelerator. A machine designed to speed up the movement of electrically charged subatomic particles that are directed at a target. These subatomic particles, also called *elementary particles*, cannot be further divided. They are used in high-energy physics to study the basic nature of matter, as well as the origin of life, nature, and the universe. Particle accelerators are also used to synthesize elements by "smashing" subatomic particles into nuclei to create new, heavy, unstable elements, such as the superactinides. *See also* cyclotron.

Periodic Table of the Chemical Elements. An arrangement of the chemical elements in sequence in the order of increasing atomic numbers. It is arranged in horizontal rows for periods and in vertical columns for groups and illustrates the similarities in properties of the chemical elements.

phage. A parasitic virus in a bacterium that has been isolated from a *prokaryote*. Also called *bacteriophage*.

phlogiston. The hypothetical substance believed to be the volatile component of combustible material. It was used to explain the principle of fire before oxidation and reduction were known and prior to the discovery of the principle of combusion.

photon. The quantum unit of electromagnetic radiation or light that can be thought of as a particle. Photons are emitted when electrons are excited in an atom and move from one energy level (orbit) to another.

photosynthesis. Process by which chlorophyll-containing cells in plants and bacteria convert carbon dioxide and water into carbohydrates, resulting in the simultaneous release of energy and oxygen.

phyla (plural of **phylum**). A primary taxonomic ranking of organisms into groups of related classes. Phyla are grouped into kingdoms, except in most plants where kingdom is replaced by division.

pi. The transcendental number $3.141592\ldots$ for the ratio of the circumference of any circle to its diameter, using the symbol π.

pi-meson. A short-lived elementary particle classified as a meson, which is primarily responsible for the strong nuclear force. It exists in three forms: neutral, positively charged, and negatively charged. The charged pions decay into muons and neutrinos, and the neutral pion decays into two gamma ray photons. Also called *pion*.

polygon. A simple closed curve in the plane that is bounded by three or more line segments.

positron. The positively charged antiparticle of an electron; e^+ or p^+.

prebiotic. Refers to the period on Earth before the existence of organic life.

primordial. The original or first in a sequence, usually referring to the earliest stage of the development of an organism or its parts.

prism. A homogeneous, transparent solid, usually with a triangular base and rectangular sides, used to produce or analyze a continuous spectrum of light.

prokaryote. Any organism of the Procaryote kingdom in which the genetic material is not enclosed within the cell nucleus and possesses a single double-stranded DNA molecule. Only bacteria and cyanobacteria are prokaryotes. All other organisms are eukaryotes.

proton. A positively charged particle found in the nucleus of an atom.

quantum. The basic unit of electromagnetic energy that is not continuous but occurs in discrete bundles called "quanta." For example, the photon is a small packet (quantum) of light with both particle and wave-like characteristics. A quantum unit for radiation is the frequency v to the product $\hbar v$, where $\hbar$ is Planck's constant. The quantum number is the basic unit used to measure electromagnetic energy. To simplify, it is a very small bit or unit of something.

quark. A hypothetical subnuclear particle having an electric charge one-third to two-thirds that of the electron. Also known as the *fundamental subatomic particle*, which is one of the smallest units of matter.

radical. Also known as "free radical." A group of atoms having one unpaired electron. Also, in mathematics, a given root of a quantity.

radioisotope. The isotopic form of a natural or synthetic element that exhibits radioactivity. The same as a radioactive isotope of an element.

rectifier. A device (diode) that converts alternating current (ac) to direct current (dc).

reduction. The acceptance of one or more electrons by an atom or ion, the removal of oxygen from a compound, or the addition of hydrogen to a compound.

retrovirus. An animal virus containing *RNA* in which the genome replicates through reverse transcription and which has two proteinaceous structures, enabling it to combine with the host's *DNA*. Retroviruses contain oncogenes, which are cancer-causing genes that become activated once the virus enters the host's cell and begins to reproduce itself.

RNA. Abbreviation for *ribonucleic acid*. The linear, single-stranded polymer of ribonucleotides, each of which contains sugar (ribose) and one of four nitrogen bases (adenine, guanine, cytosine, uracil). It is present in all living cells (prokaryotic and eukaryotic) and carries the genetic code, which is transcribed from the DNA to the ribosomes within the cell where this genetic information is reproduced.

semiconductor. Usually a "metalloid" (e.g., silicon) or a compound (e.g., gallium arsenide), which has conductive properties greater than those of an insulator but less than those of a conductor (metal). It is possible to adjust their level of conductivity by changing the temperature or adding impurities.

solenoid. An electromagnetic coil of insulated wire that produces a magnetic field within the coil. Most often it is shaped like a spool or hollow cylinder with a movable iron core that is pulled into the coil when electric current is sent through the wire. It then is able to move other instruments (e.g., relay switches, circuit breakers, automobile ignitions).

species. The lowest ranking in the classification of organisms. It is the distinguishable group with a common ancestry, able to reproduce fertile offspring, and that are geographically distinct. (Related species are grouped into a genus.)

spectroscopy. The analysis of chemical elements that separates the unique light waves either given off or absorbed by the elements when heated.

Standard Model. In particle physics, a collection of established experimental knowledge and theories that summarize the field. It includes the three generations of quarks and leptons, the electroweak theory of weak and electromagnetic forces, and quantum chromodynamic theory of strong forces.

steroid. A class of lipid proteins, such as sterols, bile acids, sex hormones, or adrenocortical hormones, that are derived from cyclopentanoperhydrophenanthrene. A shorter term for *anabolic steroid*.

subatomic particle. A component of an atom whose reactions are characteristic of the atom (e.g., electrons, protons, and neutrons).

superconductivity. A property of a metal, alloy, or compound that at temperatures near absolute zero loses both electrical resistance and magnetic permeability (is strongly repelled by magnets), thereby having infinite electrical conductivity.

supernova. A great explosion of a large star that collapses because of its gravitational force, sending great bursts of electromagnetic radiation (light) into space.

systolic. Refers to *systole*, which is the rhythmic contraction of the heart, particularly the ventricles, by which blood is driven through the aorta and pulmonary artery after each dilation or **diastole**.

thermionic emission. The emission of electrons or ions, usually into a vacuum, from a heated object, such as the cathode of a thermionic tube.

thermodynamics. The study of energy and laws governing the transfer of energy from one form to another, particularly relating to the behavior of systems where temperature is a factor (i.e., direction of the flow of heat and availability of energy to perform work).

thermonuclear. Release of heat energy when the nuclei of atoms split (fission, atom bomb, or nuclear power plant) or when nuclei combine (fusion, hydrogen bomb).

transistor. A device that overcomes the resistance when a current of electricity passes through it, used widely in the electronics industry.

ultraviolet (UV). The radiation wavelength in the electromagnetic spectrum from 100 to 3900 angstroms (Å), between the x-ray region and visible violet light.

universe. All the space, matter and energy that exists, including that which existed in the past and is postulated to exist in the future.

valence. The whole number that represents the combining power of one element with another element. Valence electrons are usually, but not always, the electrons in the outermost shell.

vector. A quantity specified by magnitude *and* direction whose components convert from one coordinate system to another in the same manner as the components of a displacement. Vector quantities may be added and subtracted.

velocity. The time rate at which an object is displaced. Velocity is a vector quantity whose quantity is measured in units of distance over a period of time.

Bibliography

Adair, Eleanor, R., ed. *Microwaves and Thermoregulation*. New York: Academic Press, 1983.

Adams, Fred, and Greg Laughlin. *The Five Ages of the Universe*. New York: Free Press, 1999.

Asimov, Isaac. *Asimov's Biographical Encyclopedia of Science and Technology*. New York: Doubleday, 1964.

———. *Beginnings: The Story of Origins—of Mankind, Life, the Earth, the Universe*. New York: Berkeley Books, 1987.

———. *Asimov's Chronology of Science and Discovery*. New York: Harper & Row, 1989.

———. *Isaac Asimov's Guide to Earth and Space*. New York: Fawcett Crest, 1991.

Bacon, Francis. *Novum Organum*. 1620. Reprinted as *Physical and Metaphysical Works of Lord Bacon, Including the Advancement of Learning and Novum Organum*. St. Clair Shores, MI: Scholarly Press, 1976.

Baeyer, Hans Christian von. *Maxwell's Demon: Why Warmth Disperses and Time Passes*. New York: Random House, 1998.

Barnes-Svarney, Patricia, ed. *The New York Public Library Science Desk Reference*. New York: Macmillan, 1995.

Barrow, John D. *Theories of Everything: The Quest for Ultimate Explanation*. Oxford: Oxford University Press, 1991.

Beckmann, Petr. *A History of PI (Π)*. New York: St. Martin's Press, 1971.

Bolles, Edmund Blair, ed. *Galileo's Commandment: An Anthology of Great Science Writing*. New York: Freeman, 1997.

Boorstin, Daniel. *The Discoveries: A History of Man's Search to Know His World and Himself*. New York: Vintage Books, 1985.

Bruno, Leonard C. *Landmarks of Science: From the Collection of the Library of Congress*. 1987. Reprint. New York: Facts on File, 1990.

Bunch, Bryan. *Handbook of Current Science and Technology*. Detroit: Gale, 1996.

Campbell, Norman. *What Is Science?* New York: Dover; 1953.

Campbell, Steve. *Statistics You Can't Trust: A Friendly Guide to Clear Thinking about Statistics in Everyday Life*. Parker, CO: Think Twice Publishing, 1999.

Carnap, Rudolph. *An Introduction to the Philosophy of Science*. New York: Dover, 1995.

Close, Frank E., and Philip R. Page. "Glueballs." *Scientific American* (November 1998).

Concise Science Dictionary. 3rd ed. New York: Oxford, 1996.

Coutts, Timothy J., and Mark C. Fitzgerald. "Thermophotovoltaics." *Scientific American* (September 1998).

Crombie, A. C. *The History of Science, from Augustine to Galileo*. New York: Dover, 1995.

Cromer, Alan. *Uncommon Sense: The Heretical Nature of Science*. New York: Oxford University Press, 1993.

Crystal, David, ed. *The Cambridge Paperback Encyclopedia*. Avon, UK: Cambridge University Press, 1993

Daintith, John, Sarah Mitchell, Elizabeth Tootill, and Derek Gjertsen. *Biographical Encyclopedia of Scientists*. 2d ed. 2 vols. Bristol, England and Philadelphia: Institute of Physics Publishing, 1981–1994.

Darwin, Charles. *On the Origin of Species*. New York: Mentor, 1958.

Davis, Paul. *About Time*. New York: Touchstone/Simon & Schuster, 1995.

Derry, T. K., and Trevor I. Williams. *A Short History of Technology: From Earliest Times to A.D. 1900*. New York: Dover, 1960.

Disney, Michael. "A New Look at Quasars." *Scientific American* (June 1998).

Dobzhansky, Theodosius. *Genetics and the Origin of Species*, 1937. Reprint. New York: Columbia University Press, 1982.

Ehrlich, Paul. *The Population Bomb*. New York: Sierra Club/Ballantine, 1968.

Einstein Revealed. A NOVA Production by Green Umbrella Ltd. for WGBH/Boston in association with BBC-TV and Sveriges Television. 1996. Videocassette. *Encyclopedia Brittannica CD*, Chicago, 1999.

Euclid. *Elements of Geometry: The Thirteen Books of Euclid's Elements*. Ed. Gail Kay Haines. New York: Dover, 1989.

Ferris, Timothy. *Coming of Age in the Milky Way*. New York: Doubleday, 1988.

Feynman, Richard P. *Six Easy Pieces: Essentials of Physics*. New York: Helix/Addison-Wesley, 1995.

Forrester, Jay. *World Dynamics*. 2d ed. Cambridge, MA: Wright-Allen, 1973.

Freeman, Ira M. *Physics Made Simple* (revised by William J. Durden). New York: A Made Simple Book/Doubleday, 1990.

Gamow, George. *Mr. Tompkins in Paperback*. Reprint. Cambridge, UK: Cambridge University Press, 1994.

Gerstein, Mark, and Michael Levitt. "Simulating Water and the Molecules of Life." *Scientific American* (November 1998).

Gleick, James. *Chaos: Making a New Science*. New York: Penguin, 1987.

Gonzalez, Frank I, "Tsunami!" *Scientific American* (May 1999).

Gribbin, John. *The Search for Superstrings, Symmetry, and the Theory of Everything*. Boston: Little, Brown, 1998.

The Handy Science Answer Book. Compiled by the Science and Technology Department of the Carnegie Library of Pittsburgh. Detroit: Visible Ink Press, 1994.

Hawking, Stephen. *A Brief History of Time*. New York: Bantam, 1988.

Henry, Patrick, Ulrich G. Briel, and Hans Bohringer. "The Evolution of Galaxy Clusters." *Scientific American* (December 1998).

Hogan, Craig J., Robert P. Kirshner, and Nicholas B. Suntzeff. "Survey Space-Time with Supernovae." *Scientific American* (January 1999).

Holmyard, E. J. *Alchemy*. New York: Dover, 1990.

Horgan, John. *The End of Science*. Reading, MA: Addison-Wesley, 1996.

Howard, Philip K. *The Death of Common Sense*. New York: Random House, 1994.

Isaacs, Alan, ed. *A Dictionary of Physics*. New York: Oxford University Press, 1996.

Jaffe, Bernard. *Crucibles: The Story of Chemistry*. New York: Dover, 1976.

Jones, Judy, and William Wilson. *An Incomplete Education*. New York: Ballantine, 1995.

Kaku, Michio. Visions: *How Science Will Revolutionize the 21st Century*. New York: Anchor Books, 1997.

Kohn, Alexandre. *From the Closed World to the Infinite Universe*. Baltimore: Johns Hopkins University Press, 1957.

Krebs, Robert E. *The History and Use of Our Earth's Chemical Elements*. Westport, CT: Greenwood, 1998.

———. *Scientific Development and Misconceptions through the Ages*. Westport, CT: Greenwood, 1999.

Krupp, E. C. *Beyond the Blue Horizon*. New York: Oxford University Press, 1991.

Landy, Stephen D. "Mapping the Universe." *Scientific American* (June 1999).

Lasota, Jean-Pierre. "Unmasking Black Holes." *Scientific American* (May 1999).

Lerner, Rita G., and George L. Trigg. *Encyclopedia of Physics*. 2d ed. New York: VCH Publishers, 1991.

Levi, Primo. *The Periodic Table*. New York: Schocken/Random House, 1984.

Livingston, James D. "100 Years of Magnetic Memories." *Scientific American* (November 1998).

Luminet, Jean-Pierre, Glenn D. Starkman, and Jeffrey R. Weeks. "Is Space Finite?" *Scientific American* (April 1999).

MacDonald, Ian R. "Natural Oil Spills." *Scientific American* (November 1998).

Macrone, Michael. *Eureka! What Archimedes Really Meant*. New York: Cader, 1994.

Malthus, Thomas R. *An Essay on the Principle of Population*. Ed. Anthony G. Flew. New York: Penguin, 1985.

Margotta, Roberto. *The History of Medicine*. London: Reed International Books, 1996.

Margulis, Lynn, and Dorion Sagan. *Slanted Truths: Essays on Gaia, Symbiosis, and Evolution*. New York: Copernicus, 1997.

McGraw-Hill Concise Encyclopedia of Science and Technology. 3d ed. New York: McGraw-Hill, 1994.

McGraw-Hill Dictionary of Scientific and Technical Terms. 5th ed. New York: McGraw-Hill, 1994.

Motz, Lloyd, *The Story of Physics*. New York: Avon, 1989.

———, and Jefferson Hane Weaver. *Conquering Mathematics: From Arithmetic to Calculus*. New York: Plenum, 1991.

Moyer, Albert E. *A Scientist's Voice in American Culture: Simon Newcomb and the Rhetoric of Scientific Method*. Berkeley and Los Angeles: University of California Press, 1992.

National Academy of Sciences. *Science and Creationism: A View from the National Academy of Sciences*. 2nd ed. Washington, DC: National Academy Press, 1999.

Nesse, Randolph M., and George C. Williams. "Evolution and the Origins of Disease." *Scientific American* (November 1998).

Newton, Isaac. *Philosopiae Naturalis Principia Mathematics*. 1650. *The Mathematical*

Papers of Isaac Newton. Ed. D. T. Whiteside. Cambridge, UK: Cambridge University Press, 1967–1981.

Newton, Roger G. *The Truth of Science: Physical Theories and Reality.* Cambridge, MA: Harvard University Press, 1997.

New York Academy of Sciences. *The Flight from Science and Reason.* Ed. Paul R. Gross, Norman Levitt, and Martin W. Lewis. New York: New York Academy of Sciences, 1996.

Norris, Christopher. *Against Relativism: Philosophy of Science, Deconstruction and Critical Theory.* Malden, MA: Blackwell Publishers, 1997.

North, John. *Astronomy and Cosmology.* New York: Norton, 1995.

Oxlade, Chris, Corinne Stockeley, and Jane Wertheim. *The Usborne Illustrated Dictionary of Physics.* London: Usborne Publishing, 1986.

Pannekoek. A. *A History of Astronomy.* New York: Dover, 1961.

Read, John. *From Alchemy to Chemistry.* Toronto: General Publishing, 1995.

Ronan, Collin. D. *Lost Discoveries: The Forgotten Science of the Ancient World.* New York: McGraw-Hill, n.d.

Rudin, Norah. *Dictionary of Modern Biology.* New York: Carron's Educational Series, 1997.

Scerri, Eric R. "The Evolution of the Periodic System." *Scientific American* (September 1998).

Schneider, Herman, and Leo Schneider. *The Harper Dictionary of Science in Commonplace Language.* New York: Harper & Row, 1988.

Science and Technology Desk Reference. Compiled by the Science and Technology Department of the Carnegie Library of Pittsburgh. Detroit: Gale, 1996.

Silver, Brian L. *The Ascent of Science.* New York: Oxford University Press, 1998.

Silvers, Robert B., ed. *Hidden Histories of Science.* New York: New York Review of Books, 1995.

Singer, Charles. *A History of Scientific Ideas.* New York: Barnes & Noble, 1966.

Sokal, Alan, and Jean Bricmont. *Fashionable Nonsense: Postmodern Intellectuals' Abuse of Science.* New York: Picador, 1998.

Spangenburg, Ray, and Diane K. Moser. *The History of Science: From Ancient Greeks to the Scientific Revolution.* New York: Facts on File, 1993.

———. *The History of Science: In the Eighteenth Century.* New York: Facts on File, 1993.

———. *The History of Science: In the Nineteenth Century.* New York: Facts on File, 1994.

———. *The History of Science: From 1946 to the 1990s.* New York: Facts on File, 1994.

Stachel, John, ed., with the assistance of Trevor Lipscombe and others. *Einstein's Miraculous Year: Five Papers That Changed the Face of Physics.* Princeton, NJ: Princeton University Press, 1998.

Strahler, Arthur N. *Understanding Science: An Introduction to Concepts and Issues.* Buffalo, NY: Prometheus, 1992.

Thain, M., and M. Hickman. *The Penguin Dictionary of Biology.* Reprint, New York: Penguin, 1995.

Trefil, James. *1001 Things Everyone Should Know about Science.* New York: Doubleday, 1992.

Waldrop, Mitchell, M. *Complexity.* New York: Touchstone, 1992.

Webster, Charles. *From Paracelsus to Newton: Magic and the Making of Modern Science*. New York: Barnes & Noble, 1982.

Webster's II New Riverside University Dictionary. New York: Riverside Publishing, 1994.

Weinberg, Steven. *The First Three Minutes: A Modern View of the Origin of the Universe*. 1977. Updated ed. New York: BasicBooks, 1993.

Weissman, Paul R. "The Oort Cloud." *Scientific American* (September 1998).

Whitfield, Philip. *From So Simple a Beginning: The Book of Evolution*. New York: Macmillan, 1993.

Wilson, Edward O. *Naturalist*. Washington D.C.: Island Press, 1994.

———. *Consilience: The Unity of Knowledge*. New York: Knopf, 1998.

Yonas, Gerold. "Fusion and the Z Pinch." *Scientific American* (August 1998).

INDEX

Page numbers for entries in text are **bold.**

About the Author

ROBERT E. KREBS, former science teacher, science specialist for the U.S. Government, and university research administrator, retired as Associate Dean for Research at the University of Illinois Health Sciences Center, Chicago. Dr. Krebs now lives on South Padre Island, Texas, where he writes science books while overlooking the Laguana Madre Bay.